Protection of Electrical Power Distribution Systems

Protection of Electrical Power Distribution Systems

Smart Grid, Microgrid, AI, and Cyber Security

Tariq Masood
College of Business, Engineering, Computing & Analytics (BECA)
Frostburg State University, MD USA

Jamil Abdo
College of Business, Engineering, Computing & Analytics (BECA)
Frostburg State University, MD USA

Atif Iqbal
College of Engineering
Qatar University, Doha Qatar

IEEE Press
WILEY

Published by John Wiley & Sons, Inc., Hoboken, New Jersey.

Library of Congress Cataloging-in-Publication Data Applied for

Hardback ISBN: 9781394327645

Cover Image: © Kenneth Bagge Jorgensen/Shutterstock
Cover Design: Wiley

Set in 9.5/12.5pt STIXTwoText by Straive, Chennai, India

This book is dedicated to our lovely parents and family members (sons, daughters, brothers and sisters), who have selflessly sacrificed their time, making this project a grand success.

This book is also dedicated to all the engineers who, through their pioneering efforts, have significantly contributed to the development of smart grid, microgrid, and FACTS technology.

The first author dedicates this textbook to the late Professor Dr. Abdel-Aty Edris, a true legend in FACTS technology. He was not only the author's PhD mentor and role model but also an invaluable external supervisor and former Director at EPRI. Also, extended dedication honors to the late Professor Dr. Ned Mohan, a distinguished figure in the field of Power Electronics.

Textbook Package Components

This Textbook package is an integrated learning package comprising the following elements.

- Workbook-Exercises: *Smart Protection of Power Distribution Systems*
- Marked assignment
- Self-Assessment Questions Answers

You will find guidelines for the use of all these materials in your book package containing the handbook with exercises, Marked Assignments and Self-assessment Questions and their answers. Self-assessment questions (SAQs) are used throughout the workbook to make sure that you understand all the concepts. Worked solutions are provided in a further booklet but do not turn to them too early. Test yourself!

In addition, there are numerous exercises in each section which help the flow of the argument, and for which answers usually appear in the text that follows. Occasionally an exercise might ask you to consider an aspect from your own experience or working environment. In these cases, the "answer" you give will be specific and personalized and for this reason a "solution" as such cannot be provided. Like the SAQs it is a good idea to do the exercise as you work through the workbook. Try not to skip any. Some of them may seem elementary and it can be tempting to hurry on to the next paragraph, but they all have a purpose.

It is a good idea to make your own notes as you go along; these will be useful when you are revisiting for examination for certification and to assess and modify the existing relay operations and control to maintain seamless power distribution systems operations at Power Utilities. To help reader identify essential concepts, you will find that each section starts with a list of learning objectives. Ensure you understand them before moving on to the next section.

Contents

Author Biographies

Tariq Masood, MPhil, PhD, PE, is an Assistant Professor of Electrical Engineering, Senator, and Member of the Academic Affairs Committee, Academic Standard Committee, and Institutional Review Board at Frostburg State University. He earned his PhD in Electrical Power System Smart/Microgrid and Renewable Energies from the University of Bath, UK. He has over 25 years of work experience as an **Operations Technical Coordinator** at Qatar Energy; **Vice President of Technical Operations** at Energy and Utilities Service in New Jersey, USA; and also served as a full-time professor, visiting professor, adjunct professor, and invited speaker in many international universities. Dr. Masood received outstanding performance awards, including being honored by the Deputy Prime Minister/Minister of Energy, Qatar, and the President of IEEE. He has authored or co-authored more than 70 research articles and four books and is also an associate editor of *IEEE Access Journal*.

Jamil Abdo, PhD, is a Professor of Mechanical Engineering and the Chair of the Department of Engineering at Frostburg State University. He earned his PhD in Mechanical Design from Southern Illinois University, Carbondale, USA. Dr. Abdo has extensive research experience in machine element design, failure analysis, product development, and renewable energy.

Atif Iqbal, DSc, PhD, is a Full Professor of Department of Electrical Engineering at Qatar University. Dr. Iqbal is also a Fellow of IEEE, IET, IE, and AAIA. Dr. Iqbal has been listed in top 2% highly cited scientists of the world (Stanford University, USA) since 2019. He has supervised 21 PhD. He was also a recipient of the Outstanding Faculty Merit Award for 2014–2015 and the Research Excellence Awards at Qatar University, in 2015, and 2022. He has authored or coauthored more than

580 research articles, 8 patents, and 4 books and several chapters in edited books. He has supervised several large research and development projects worth several million USD. He is serving as the Vice-Chair of the IEEE Qatar Section. He is an Associate Editor of the *IEEE Transaction on Industrial Electronics* and Senior Editor of *IEEE Access*.

Foreword

In an era marked by the rapid transformation of the electrical power industry, the protection and control of power distribution systems have become more complex and critical than ever before. With the growing integration of renewable energy sources, the rise of smart grid technologies, and the increasing focus on cybersecurity, modern power distribution demands innovative solutions and a deep understanding of advanced protection mechanisms. *Protection of Electrical Power Distribution Systems: Smart Grid, Microgrid, AI, and Cybersecurity* aims to address these challenges, providing a comprehensive exploration of both fundamental and cutting-edge aspects of power system protection. This book is meticulously crafted to serve as an indispensable guide for professionals, engineers, researchers, and students navigating the complexities of today's power distribution networks. It opens with a thorough discussion of non-unit and unit protection of distribution feeders, which are the cornerstone of safe and reliable power delivery. By delving into the principles and practical applications of protection transducers, readers are introduced to the latest technologies that ensure precise and effective system responses under various fault conditions, providing a sense of support and guidance.

Key topics such as transformer protection, busbar protection, and motor protection are presented in detail, underscoring their importance in maintaining the stability and security of power systems. The book further delves into embedded generation protection systems, which are crucial for incorporating distributed energy resources into the grid without compromising system integrity. Autoreclose relaying and coordinated protection and control are also explored, offering insights into optimizing system performance and minimizing downtime in the event of disturbances. Going beyond foundational protection strategies, this publication takes a forward-looking approach by examining smart grid operation and control. It discusses the dynamic nature of microgrid operations, emphasizing the need for adaptable and robust control mechanisms that cater to the evolving landscape of decentralized energy production and consumption.

The book's comprehensive coverage of artificial intelligence (AI) and machine learning applications in power systems illuminates how these technologies can enhance decision-making, predictive maintenance, and system resilience, thereby improving operational efficiency and making the audience feel prepared for the future. Given the increasing prevalence of digital threats in the power sector, the book dedicates a critical section to cybersecurity and information protection programs. This segment equips readers with strategies and best practices for safeguarding power systems against cyber-attacks, ensuring the integrity and confidentiality of sensitive information. As the electric grid becomes more interconnected and reliant on data, the implementation of robust cybersecurity measures is paramount.

This work is not merely a technical manual but a comprehensive roadmap that combines theoretical concepts with practical applications. By covering a wide array of topics, from traditional protection systems to the latest innovations in AI and cybersecurity, it addresses the pressing needs of the modern power distribution landscape. Readers will find this book invaluable, whether they are looking to deepen their understanding of current technologies or to stay ahead in this rapidly evolving field, making them feel understood and catered to. *Protection of Electrical Power Distribution Systems: Smart Grid, Microgrid, AI, and Cybersecurity* is an authoritative resource poised to contribute significantly to the development and implementation of more secure, efficient, and reliable power distribution systems. It embodies a holistic approach to electrical power protection, guiding engineers, researchers, and policymakers in their efforts to build a safer and more resilient grid for the future.

Saifur Rahman, PhD
2023 IEEE President
Joseph Loring Professor & Director
VT Advanced Research Institute
Arlington, VA 22203, USA
www.srahman.org

Preface

The transition toward the protection of power systems, smart grids, and microgrid systems is a technical evolution and a key enabler for achieving and strengthening global sustainability and asset reliability in the energy sector. This book, a practical guide and a key contributor to the global effort to advance the **UN Sustainable Development Goals (SDGs)**, particularly in promoting sustainable energy (SDG 7), industry innovation **(SDG 9)**, and climate action **(SDG 13)**, provides actionable frameworks that can be implemented in power distribution systems to safeguard operations and equipment. The content is not just theoretical insights, but also practical solutions that can be applied, with the ultimate goal of ensuring safe and reliable energy production, while advancing sustainability in a way that aligns with the broader global movement toward achieving the SDGs.

Practical Solutions for Sustainable Energy Operations
Power distribution systems, particularly those embedded within Smart Grid and Microgrid architectures, must prioritize health, safety, and security to ensure operational efficiency and the well-being of personnel and the broader public. As the leader and principal author of *Operational Excellence 20 Initiatives in the Energy Sector*, I emphasize the critical need for a robust maintenance strategy that ensures both preventive and corrective maintenance. This strategy directly supports SDG 9, which calls for innovation in industry infrastructure. By maintaining asset reliability and system integrity, energy utilities can reduce downtime, enhance performance, and minimize environmental impacts, contributing to sustainable production – a core aim of SDG 12 (Responsible Consumption and Production). This book provides a step-by-step guide for implementing these maintenance strategies, focusing on optimizing OPEX (Operating Expenses) and CAPEX (Capital Expenditures). This approach is essential for securing the continuity of energy delivery – safeguarding sustainable energy production processes and optimizing operational costs, aligning with SDG 7 for affordable and clean energy.

Supporting Health, Safety, Security, and Environmental (HSSE) Protection
Safe and sustainable energy production is at the core of achieving SDG 7: Affordable and Clean Energy. Smart Grids enable the seamless incorporation of renewable energy sources such as solar, wind, and hydropower, reducing the carbon footprint of energy production.
In alignment with SDG 3 (Good Health and Well-Being) and SDG 8 (Decent Work and Economic Growth), the techniques outlined in this book also focus on Health, Safety, Security, and Environmental (HSSE) Protection. Through safe production practices, reliable maintenance strategies, and minimizing operational risks, energy producers can foster a working environment that prioritizes the health and safety of employees while reducing environmental impact. The HSSE framework presented in this book is not just about protecting human and ecological resources; it's about ensuring that energy production can continue safely and sustainably, contributing to climate action (SDG 13).

Inspiring Innovation in Achieving Sustainable Goals in the Energy Sector
The reliability and integrity of grid assets are fundamental to operational efficiency and long-term environmental sustainability. Unreliable systems result in energy waste, increased emissions, and higher maintenance costs. By focusing on predictive maintenance and real-time asset management, smart distribution systems ensure that infrastructure remains robust against natural disasters, climate change, and technical failures. Ultimately, this book seeks to inspire action by providing actionable and proven methods that power utilities can use to ensure safe and reliable energy delivery. By promoting innovation in maintenance strategies, the content aligns with the themes of equality, sustainability, and environmental protection set out by the UN SDGs. This effort, alongside the UN's mission, underscores the importance of maintaining asset reliability and integrity and fostering an environment, where safe and sustainable production is possible. Through ongoing training, improvement initiatives, and risk management, this book equips energy professionals with the tools they need to contribute to the global SDG effort.

A comprehensive maintenance strategy is crucial to prolonging the life of grid assets and ensuring they operate safely and sustainably. Through AI-enhanced predictive maintenance, grid operators can foresee potential issues before they escalate, thereby preventing hazardous and costly outages. **This book fully aligns with the UN SDGs, emphasizing sustainability, asset reliability, and safe production**. It serves as a valuable resource for energy sector stakeholders committed to fostering a sustainable, secure, and resilient energy future, and its strong alignment with the UN SDGs underscores its potential impact on the global sustainability effort.

This book does not discuss the details of distribution system design and system analysis, which have already been the subject of several good, published books.

Section 1: This book provides the basics and advances of the art and science of protection for distribution systems. It will empower those interested in this area, enhance their knowledge, and enable them to delve deeper into the specifics of the subject. Although several leading textbooks cover the topic, the most excellent understanding of the subject can be best achieved by working with the schemes and exploring the history behind their development.

Section 2: By the end of this section, you will have a solid understanding of the general principles of the protection overlay. This knowledge will enable you to differentiate between non-unit or nonrestricted protection and unit or restricted protection. You will also be equipped to divide a power system network into manageable protection zones and determine their ideal boundaries. Moreover, you will be able to assess the capabilities of the protection system to safeguard the power system as a whole and identify fundamental weaknesses in a protection scheme, thereby applying your learning in practical scenarios.

Section 3: In this section, you will be introduced to the practical application of non-unit protection to safeguard distribution feeders through a series of exercises. These exercises will guide you on how to operate a fuse to protect distribution feeders, choose suitable values to preserve a radial distribution feeder, and appreciate fuses with circuit breakers for high-voltage feeders. You will also learn to select the appropriate setting for inverse definite minimum time-delayed overcurrent relays to protect a radial distribution feeder system and understand the limitations of these relays when protecting interconnected feeder circuits. Ultimately, you will gain a comprehensive understanding of the application of directional overcurrent relays to protect interconnected distribution feeders.

Section 4: In this section, you will learn the operation and application of current and voltage transformers (also known as protection transducers) used in the protection of distribution systems. You will be able to distinguish clearly between the primary differences in design criteria for the two types of transducers for the protection system. The section will also define the accuracy requirements of the two transducers, emphasizing the need for precision in protecting each piece of equipment in the power distribution system. This focus on precision will make you feel more attentive and detail-oriented in your approach.

Section 5: In this section, you will be able to learn to define the basic principle of unit protection, identify the need for bias, and derive both bias and operating quantities. You will also assess the advantages and disadvantages of different differential protection schemes (including pilot wire and digital) used in practice.

Finally, you will have an appreciation of directional earth fault and rough balance protection.

Section 6: In this section, you will be able to learn and list the main types of faults affecting transformers, describe the different types of transformer protection, and demonstrate some of the typical constraints and application considerations for various types of transformer protection.

Section 7: In this section, you will be able to learn and list the main types of faults affecting the busbar, describe the different kinds of busbar protection available, and determine some of the typical constraints and application considerations for various types of busbar protection.

Section 8: In this section, you will be able to learn and define the essential motor characteristics involved in protection, determine the motor protection requirements, and describe the basic principles of motor protection against winding faults, overload, and phase unbalance.

Section 9: In this section, you will be introduced to the implications of introducing embedded generation into a utility supply network. Understanding these implications is crucial, as it prepares you for the future challenges in the field. You will also learn about the techniques used to protect the embedded generator and the utility intertie, equipping you with the knowledge to navigate the evolving landscape of electrical engineering.

Section 10: In this section, you will be able to learn and list the leading causes of faults in distribution systems, appreciate the likelihood of faults being permanent, and describe the operation of single-shot auto-reclose relays.

Section 11: In this section, you will be able to learn and list the primary protection, control, and monitoring functions found in the substation. You will explore why are conventional protection and control schemes are challenging to test, maintain, and expand? You will also understand how the coordination of protection and control gear can be achieved in the substation using computer-controlled equipment. You will describe the relevant digital technology used in coordinated substations, mainly digital communications, and the structure of a modern coordinated substation.

Section 12: At the end of this section, you will learn what a smart grid's features are. How do distribution automation field devices like those discussed in 10 case studies function? How do we detect and locate a fault in a distribution system? Which areas are significantly impacted by essential volt-var devices, and how do these devices play a crucial role in the smart grid's operation? What is the precise definition of distributed automation? What are the specific critical functions of the SCADA system?

Section 13: At the end of this section, you will learn about the Microgrid, a topic directly relevant to the field of study in electrical engineering. This includes independent active and reactive power control, correction of voltage sag and system imbalances, and meeting the grid's load dynamics requirements. This section also covers the classification of Microgrids, the merits and demerits of AC and DC microgrid functionality, and the functionality of hybrid Microgrids. Finally, this section will discuss the performance comparison of DC and AC microgrids, providing insights directly applicable to the work and studies in electrical engineering.

Section 14: At the end of this section, you will have a solid introduction to the general principles of the AI components used for industrial applications. This knowledge is crucial for future professionals in the field. You will start by discussing the thinking probabilistically and the potential for AI machine power and disruption. You will learn how to design reliable systems by integrating AI, followed by a detailed exploration of AI Machine Prediction rules in Microgrid and Smart grid Operations. The section concludes with a Tutorial on simulating AI models and a discussion on the practical challenges and future directions of AI.

Section 15: At the end of this section, you will be able to know the practical principles of control and mitigation plan for the cybersecurity components used for industrial applications in the energy sector by large. You will learn the basic cybersecurity principles, operational challenges in Electrical Power Systems, and their impact and solutions. More importantly, you will validate the cybersecurity process mapping, data collection, and mapping, gaining confidence in your understanding. Finally, you will learn how to design two levels of project scope, first for IT penetration and Assessment project scope, and second for OT penetration and Assessment project scope in Electrical Power Systems, mainly distributed electrical Power Generation.

Acknowledgments

As a principal author, I express my deep gratitude to the **Academic Senior Leadership** for their unprecedented support and encouragement in making this project a grand success, as listed **in their short profile**. Their pivotal role in arranging my workshops and lectures on **Electrical Power Systems (smart grid, microgrid, and renewable energy resources)** from time to time since 2004 has been a cornerstone of my professional journey. The sheer number of these events, totaling over 200 workshops/lectures, underscores the breadth of their support and the depth of my dedication. As an **adjunct professor, a visiting professor, a full-time professor, and an invited speaker at OIT, UTT, Texas A&M Qatar, UTA, TTU, UND, UDC, UET, MUET, QU, DHA-UET, NEDUET, NUST, NFC IET, MNSUET, NTDC, BZU,** University of Granada, University of Salamanca, University of La Laguna Internationality.

I have delivered these workshops and lectures and taught different courses part-time and full-time with great enthusiasm and commitment, and I am deeply thankful for their support.

I also sincerely appreciate the **Academic Senior Leadership's** encouragement and supportive role in making this project a grand success:

Mark H. Weichold, former Dean, Executive Vice President, and CEO at Texas A&M University; **Mazen O. Hasan**, Chairman of the National Committee for Qualifications and Academic Accreditation in the State of Qatar, Former **Professor** of the Dean college of Engineering, and the Vice President and CEO of the **Qatar University**, Doha, Qatar; **Ken L. Fincher**, Vice President and Executive Director Board Secretary at Oregon Tech; **Manuel Pérez-Donsión**, President of the European Association for the Development of Renewable Energies, Environment and Power Quality (EA4EPQ) at the University of Vigo, Spain; **Abdul Jabbar**, Managing Director of the National Transmission Dispatch Company, Lahore, Pakistan; **Akhtar Kalam**, Professor and Head of the **Smart Energy Research Unit** at the Institute for Sustainable Industries and Liveable

Cities at Victoria University, Australia; **Afzal Haque**, Vice-Chancellor of Sir Syed University and former VC of NED UET, Karachi, Pakistan; Adel **El Banna**, Former Head of Asset Integrity and Optimization, Qatar Energy**; Hassan El-Kishky**, Professor and Chair of the Department of Electrical Engineering at the University of Texas, Tyler; **Muhammad Kamran**, Vice Chancellor of Muhammad Nawaz Sharif University of Engineering and Technology, Multan, Pakistan; **Saeed Minhas**, Vice Chancellor of **DHA/SUFF** University in Karachi, Pakistan; **Suhail A Qureshi**, Former Professor and Dean of UET, Lahore, Pakistan. **Tahir Imran Qureshi, P**rofessor and Head of the Environmental Sciences and Engineering Department at NFC IET, Multan, Pakistan. **D.P. Kothari**, Former Vice Chancellor of IIT, India; **Arshad Ali**, Former Professor and Dean at NUST, Pakistan**;** **Bhawani Shankar Chowdhry**, Former Professor and Dean at MUET, Pakistan; **Ahad Ali**, Professor and Director of Industrial Engineering at LTU; **Amir Shahirinia**, Professor and Director of the Center of Excellence for Renewable Energy (CERE) at the University of the District of Columbia; **Amjad Ali**, Professor at Hamad Bin Khalifa University, Qatar; **Esther T. Ososanya**, Professor and Chair, Department of Electrical and Computer Engineering, University of the District of Columbia; Florin **POPENTIU**, Professor of the Department of Control Engineering and Industrial Informatics at University **"Politehnica"** of Bucharest; **Michael Giesselmann**, Professor and Chair of the Department of Electrical and Electronics Engineering at the Texas Tech University; **Navarun Gupta**, Professor and Chair, Department of Electrical Engineering at the University of Bridgeport; **Aziz Rahman**, Professor at Texas A&M University, Qatar (TAMUQ); **Raziq Yaqub**, Professor at Alabama A&M University; **Robert Melendy**, Professor of Electrical Engineering at George Fox University, Newberg, Oregon; **Farhan Malik**, Professor at ADPoly, Abu Dhabi; **Abdul Sattar Malik**, Professor and Chair of the Department of Electrical and Electronics Engineering at Bahauddin Zakariya University, Pakistan; **Hardeep Singh**, Professor at Amritsar Group of Colleges, India; **Adnan Kavak**, Professor of the Department of Electrical and Electronics Engineering at Kocaeli University, Turkey; **Murtaza Hashami,** Equipment Specialist at ARAMCO, Saudi Arabia**; Muhammad Zubair Ikram** at Bose Corporation, Meta-Reality Labs Research, and Texas Instruments.

Their time, support, and willingness to provide their short profile with full conviction have been invaluable, as listed below. Their endorsement of this project as a reference book for their institutions' **undergraduate and graduate students and a copy in the central library** is not just a testament to its value but also a significant influence on its success. I am genuinely grateful for their important and impactful role in this project.

I also express my deep gratitude to Professor **Dr. Saifur Rahman**, the **IEEE President and CEO in 2023** and the President of the **IEEE Power** and Energy Society (PES) for 2018 and 2019, for being part of my book's acknowledgment page. As an **Executive Secretary/Treasurer, I contributed to the IEEE Qatar Section from 2004 to 2014** and received numerous outstanding performance awards honored by the **IEEE President. These** serve as a testament to my value and integral role in dynamically growing the **IEEE Qatar section**. I also led and organized the **IEEE Regional** Meeting in Qatar, a source of pride and honor for all of us and a testament to our value and integral role in the IEEE community.

I am also profoundly grateful to the **Senior Leadership at FSU**, including the **Jamil Abdo**, Professor and Chair, Department of Engineering; **Sudhir Singh, Dean**, College of Business, Engineering, Computational & Mathematical Sciences; **Tarki L. Tyler**, Senior Advisor to the President, **Former Provost and Vice President** for Academic Affairs; and **Al Delia**, Acting President at **Frostburg State University**. Their recognition of my contribution as a faculty member at FSU and their inclusion in my acknowledgment page is not just a token of appreciation, but a reflection of their significant role in our research. Their influence and guidance have been invaluable.

Last, but not least, I am very much indebted to and appreciate the Honorable **Nasser Jeham Abdel-Aziz Al Kuwari**. I have had the privilege of working under his leadership at **Qatar Energy**[1] for over 12 years. We led high-level technical committees and accomplished many projects, particularly the GCC Production and Maintenance Committee. We also shared and implemented many Production Best Practices initiatives across the GCC (Qatar – QE, Oman – PDO, Bahrain – BP, Kuwait – KOC, United Arab Emirates – ADNOC, Saudi – ARAMCO) oil and gas industries in the energy sector. Over this extended period, his guidance and mentorship played a pivotal role in my career progression.

Mr. **Al Delia** is serving as the Acting President of Frostburg State University (FSU) in Frostburg, Maryland, during an extended leave of absence by the University President. He was appointed by USM Chancellor **Jay Perman and** FSU President **Ron Nowaczyk** in November 2023. In this role, Mr. Delia is responsible for all aspects of the University's functions. Mr. Delia joined FSU in July 2018 as the inaugural Vice President for the Division of Regional Development and Engagement. In that capacity, he oversees the University's efforts to help advance the region it serves through engagement with business, industry, community organizations, and government. Before returning to Maryland in 2018, Al served in various positions in academia and government in North Carolina.

Traki L. Taylor, PhD, is the Senior Advisor to the President. Dr. Taylor formally served as Provost and Vice President for Academic Affairs at Frostburg State

1 Formerly known as **Qatar Petroleum**, the upstream production facilities, at Dukhan Operations, represent the main field of QP, encompassing almost 1500 oil and gas production wells and gas and water injection wells located 70 miles from Doha Metropolitan City.

University. Before that leadership opportunity, she was the Chief Diversity, Equity, and Inclusion Officer and Assistant Vice Chancellor for Academic Affairs at the State University System of Florida. Dr. Taylor provided direction for DEI initiatives at the 12 SUS institutions and led the work of the academic affairs unit. Dr. Traki L. Taylor was named one of Maryland's Top 100 Women (2023) by the Daily Record. She earned her PhD and Master of Arts degrees from the University of Illinois, Urbana-Champaign, in Educational Policy Studies, and her Bachelor of Science degree in Special Education at Coppin State College in Baltimore, Maryland.

Sudhir Singh, PhD, is the Dean of the College of Business, Engineering, and Computational and Mathematical Sciences and Professor of Finance at Frostburg State University, Frostburg, Maryland. He earned a Master of Management Studies in Finance and Marketing from the Birla Institute of Technology & Science in Pilani, Rajasthan, India; a Master of Business Administration in Finance and Accounting from the Georgia Institute of Technology in Atlanta, Georgia; and a PhD in Finance from the Virginia Polytechnic Institute and State University in Blacksburg, Virginia. Over an extensive academic and professional career of more than 32 years at Frostburg, Dr. Singh has provided executive leadership experience to the college, advancing the themes of innovation, interdisciplinarity, and collaboration within its academic ethos.

Nasser Jeham Abdel-Aziz Al Kuwari is a distinguished leader in the oil, gas, and petrochemical industries, with over 30 years of expertise. He has served as CEO of both Qatar Fuel Additives Company (QAFAC) and Qatar Chemical Company (Q-Chem), leading these organizations to new heights. In addition, he has held numerous board positions across major oil and gas companies in Qatar, where his strategic vision and leadership have had a lasting impact on the industry.

Saifur Rahman is the founding director of the Advanced Research Institute at Virginia Tech, USA, where he is the Joseph R. Loring Professor of Electrical and Computer Engineering. He is a Life Fellow of the IEEE and an IEEE Millennium Medal winner. He was the IEEE President and CEO in 2023 and was the president of the IEEE Power and Energy Society (PES) for 2018 and 2019.

Mark H. Weichold has worked for General Dynamics Ft. Worth Division, Motorola in Austin, TX, and the U.S. Army Electronic Technology and Devices Laboratory in Ft. Monmouth, NJ. He joined the Electrical Engineering faculty at Texas A&M University in 1982 and now holds the rank of Professor. He is the recipient of three US patents and has over 100 journal and conference publications. Dr. Weichold served as the Associate Provost for Undergraduate Programs at Texas A&M, and in January 2007, he became CEO and Dean of Texas A&M University's branch campus in Doha, Qatar. After his return to Texas A&M

from Qatar, he served in administrative positions as the Senior Associate Dean for Academic Programs in the College of Engineering, as well as the Interim Provost and Executive Vice President. In addition, he served as the Interim Provost and Vice President at Texas A&M University-San Antonio and as the Associate Vice Chancellor for Academic Affairs at the Texas A&M University System. Dr. Weichold has returned to the classroom and is currently an ABET Commissioner. He is also engaged as a member of the Board of Governors at Habib University in Karachi, Pakistan, for more than 10 years.

Mazen Hasna holds a PhD in Electrical Engineering from the University of Minnesota, Twin Cities. He has served in several senior positions in Qatar University, including Dean of Engineering, and Vice President and Chief Academic Officer. Since 2022, he has been the founding chairman of the national committee for qualifications and academic accreditation in the State of Qatar. Dr. Hasna is an established researcher with more 250 journal and conference publications, and close to 10,000 citations in the general field of wireless communications.

Akhtar Kalam, Emeritus Professor, is an Academic Director at Texila College of Australia. He has been at Victoria University (VU) since 1984 and is recently associated with VU's Institute for Sustainable Industries and Liveable Cities. He is the editor-in-chief of the *Australian Journal of Electrical & Electronics Engineering* and has held Distinguished Professorship positions in many national and international institutions. He has been recognized internationally and nationally for his research.

Afzal Haque, an alumnus of NED University of Engineering and Technology and the University of Manchester, United Kingdom, has over 47 years of varied and versatile experience and has served in responsible positions at national and international levels in the fields of Engineering and Technology and Education. Presently, he is serving as the Vice-Chancellor of the **DHA Suffa** University, Karachi. Previously, Dr. Haque served as the Vice-Chancellor of Sir Syed University of Engineering and Technology and NED University of Engineering and Technology.

Ken L. Fincher is the Vice President and Executive Director Board Secretary. It was wonderful working with Dr. Masood during his time at Oregon Tech. His work with our students was remarkable and was on full display during the Oregon Tech **IdeaFest** event. As the **Vice President** of University Advancement and Executive Director of the Foundation at Oregon Tech, it is always rewarding to see Professors' impact on students, research and the academy.

Arshad Ali is the former executive of the National University of Science and Technology (NUST), Islamabad, Pakistan. He is a highly accomplished individual with extensive experience in management, strategy planning, and industry engagement, spanning over 40 years. Dr. Ali has actively collaborated on numerous research and technology development projects with renowned institutions such as **CERN, Stanford, and Caltech Universities**.

Bhawani Shankar Chowdhry is a Distinguished National Professor, Professor Emeritus, and former Dean of the Faculty of Electrical, Electronics, and Computer Engineering at the Mehran University of Engineering and Technology Jamashoro, Sindh, Pakistan.

Ahad Ali is an Associate Professor and Director of Industrial Engineering Programs (BSIE & MSIE), Director of Graduate Certificate in Lean Six Sigma, and Director of Smart Manufacturing and Lean Systems Research Group, A. Leon Linton Department of Mechanical, Robotics, and Industrial Engineering at the Lawrence Technological University, Southfield, Michigan, USA. He established Siemens Electro-Matic Industrial Engineering Lab at LTU in 2019. Dr. Ali was the primary author of the ABET self-study report of the BSIE Program at LTU in 2022. He is serving as an Executive Director of the IEOM Society International.

Amir Shahirinia has a robust background in Power Systems, Power Electronics, AI, Renewable Energy, and Smart Grids. His research and projects span innovative solutions in renewable energy integration and smart grid technologies. Dr. Shahirinia is the Director of the Center of Excellence for Renewable Energy (CERE) and the Director of Smart Grids and Artificial Intelligence (SGAI) at the University of the District of Columbia.

Amjad Ali received his PhD in Electronics and Radio Engineering from Kyung Hee University, South Korea, in 2015. From 2015 to 2022, he was an Assistant Professor at the COMSATS University Islamabad, Pakistan. He was a Research Professor at Korea University from 2018 to 2019, from where he earned the Best Researcher Award. Currently, he is a Postdoctoral Researcher at the Hamad Bin Khalifa University, Qatar. He has published over 80 papers in top-tier journals and conferences, with 2600 citations. His research focuses on future Internet technologies, cybersecurity, and applied AI/ML.

Arabella Bhutto is a Professor of Innovation and Entrepreneurship. Her master's and PhD degrees are from the United Kingdom, and her postdoc is from Utah, USA. She is the author of many research papers and a book on her new academic tool developed for the capacity building of STEM graduates to prepare them for entrepreneurship. She is funded by the National Science Foundation for developing a new global curriculum supporting entrepreneurial ecosystem for entrepreneurial Universities.

Esther T. Ososanya is the Chair and Professor of Electrical and Computer Engineering, University of the District of Columbia, Washington, DC. His areas of expertise are microcomputer architecture, VLSI and ASIC designs, embedded systems, nanotechnology, and renewable energy.

Florin Popentiu Vlădicescu is at present a PhD Supervisor at the Military Technical Academy "Ferdinand I," and an Associate Professor in the Department of Control Engineering and Industrial Informatics at the University "Politehnica" of Bucharest. Following the approved Joint Council of Europe/UNESCO "Convention on the Recognition of Qualifications Concerning Higher Education in the European Region" (Lisbon, 1997), Dr. Florin Popentiu Vlădicescu has founded the first UNESCO Chair in 1998 at City, University of London, in Information and Communication Engineering. He has been a Visiting Professor at many universities.

Hassan El-Kishky, Professor and Chair of the Department of Electrical Engineering at the University of Texas, Tyler. He received a B.Sc. degree in Electric Power Systems and Rotating Machines from Ain Shams University, Cairo, Egypt, in 1984 and his MSc and PhD degrees in Electrical Engineering from Assiut University, Assiut, Egypt, and Arizona State University, Tempe, AZ, in 1990 and 1995, respectively.

Manuel Pérez-Donsión received an MSc in Industrial Engineering from the Polytechnic University of Catalonia in 1980, and his PhD from the University of Vigo in 1986. In 1987, he was promoted to full Professor at the University of Vigo. He is the President of the European Association for the Development of Renewable Energies, Environment and Power Quality (EA4EPQ) and the Chairman of the Steering Committee of the International Conference on Renewable Energies and Power Quality (ICREPQ).

Michael Giesselmann, P.E., received his doctoral degree in Electrical Engineering in 1986 from the Technical University in Darmstadt, Germany. At present, he holds the rank of professor and was the Chair of the Department of Electrical and Computer Engineering at Texas Tech University from 2010 until 2021. Dr. Giesselmann is a Life Senior Member of the IEEE and a licensed Professional Engineer in Texas. Dr. Michael Giesselmann has been a principal faculty member and key researcher in the Pulsed Power Laboratory since 1986.

Muhammad Kamran is the Vice Chancellor, of Muhammad Nawaz Sharif University of Engineering & Technology, Multan. He has received the Engineering Excellence Award from the Institution of Engineers Pakistan in 2024. He served as Dean of the Faculty of Electrical Engineering at UET Lahore, from April 2021 to September 2022. His areas of expertise include power system quality, power system protection, EHV engineering, and power system analysis.

Navarun Gupta is the Chair and Professor of Electrical Engineering at the University of Bridgeport. He received his PhD in Electrical Engineering from Florida International University in 2003. Dr. Gupta's research interests lie in applications of signal processing to audio and bio signals. He is also active in the engineering education research.

Rahman is an Associate Professor in the Petroleum Engineering Program at Texas A&M University at Qatar (TAMUQ). Before his appointment at TAMUQ, he was an Assistant Professor at Memorial University of Newfoundland and an Instructor at the University of Alberta, Canada. Dr. Rahman received his PhD from the University of Alberta, Canada, in 2010. Since then, he has been involved in several research collaborations with companies, including Total Energies, Qatar Energy LNG, Schlumberger, North Oil Company, NEL, Syncrude Canada, GRi Simulations, and Petroleumsoft.

Raziq Yaqub has over 25 years of experience in industry, government, and academia. He holds 85 issued patents and has been an inductee of the New Jersey Inventors Hall of Fame since 2014. Currently an Associate Professor at AAMU, he has previously served as a Director at Toshiba, guiding 4G wireless communication, and at NIKSUN, leading cybersecurity technology training. At Tecvox, he spearheaded wireless charger development. As a Senior Consultant to the State of New Jersey, he secured an $87 M grant for 4G/LTE deployment for first responders. Additionally, he was a Faculty Fellow at NASA Glenn Research Center, leading AI-based fault detection.

Robert Melendy is an Associate Professor of electrical engineering at George Fox University in Newberg, OR. Previously, he served as an Associate Professor of Electrical Engineering and Renewable Energy at the Oregon Institute of Technology, where he worked with Prof. Dr. Tariq Masood, forming a collaborative relationship in electrical energy and power systems.

Saeed Minhas is the Vice Chancellor of the University of DHA Suffa University, Karachi Pakistan. He is an arms control and disarmament expert with an experience of over 8 years. He has served as Director of Research at Pakistan's National Defense University (NDU), Islamabad.

Suhail A Qureshi is a renowned Academician and Researcher. He has served as teacher for more than 40+ years as a Professor and Dean of the Faculty of Electrical Engineering at UET, Lahore. He has presented over 150 research papers and written more than six books. He received the Best Teacher Award from HEC Pakistan and was a three-time recipient of the ORS from England.

Tahir Imran Qureshi is working as an ***E-waste Management Consultant*** for the Crisis Resilient Social Protection (CRISP) Project of the World Bank,

Islamabad, Pakistan. He had been working as ***Director*** of Post Graduate Studies Center, and ***Head*** of the Environmental Sciences and Engineering Department at NFC Institute of Engineering and Technology, Multan. **Dr. Qureshi** has received an ***MSc*** in Chemistry from Bahauddin Zakariya University, Multan and an ***MS*** in Biochemistry and ***PhD*** in Environmental Engineering from Kyungpook National University, South Korea.

Muhammad Zubair Ikram has worked at Bose Corporation, Meta Reality Labs Research, and Texas Instruments. His area of interest is signal processing for audio, communication, and radars. He has published over 40 research papers and holds over 20 patents. Muhammad earned his PhD in Electrical and Computer Engineering from the Georgia Institute of Technology and his MBA from the University of Texas at Dallas.

Abdul Jabbar currently serving as Managing Director, National Transmission Despatch Company, Pakistan. He has more than 30 years of multidisciplinary experience as a professional electrical engineer and energy executive in utilities, consulting, IPPs, and power development sector, particularly in high distribution, transmission, operation and maintenance, renewable energy, and managing large-scale distribution companies. He has provided leadership directions and project management support to execute strategic business plans. He successfully led the top profit-making DISCOs/SoEs, such as FESCO and IESCO, as Chief Executive Officer, and managed one of the largest power sector organizations of the country.

Farhan Malik, an esteemed academic and researcher with 14 years of experience in teaching, research, and industry, has been a valued member of ADPoly since August 2018, bringing his expertise in electrical engineering. He previously served as an Assistant Professor at the American University of the Middle East in Kuwait and contributed significantly as a researcher at Aalto University in Finland, where he earned his Doctor of Science degree in 2016 with a thesis on "Flexible Loads in Smart Grids."

Abdul Sattar Malik received the BS and MS degrees in Electrical Engineering from the University of Engineering & Technology, Lahore, Pakistan, in 2000 and 2005, respectively. He received his PhD degree from Beijing Institute of Technology, Beijing, P.R. China, in December 2009. Currently, he is working as Professor/Chairman of the Department of Electrical Engineering, Bahauddin Zakariya University ZU Multan, Pakistan. He is also an elected member of the Governing Body of Pakistan Engineering Council for the tenure 2024–2027. His research interests include low-power ad-hoc wireless network protocol architecture and distributed clustering protocols for wireless sensor networks.

Adel El Banna holds a BSc in Chemical Engineering and a Post-graduate Diploma in Plant Design, and he has acquired 35 years of experience extended from plants operations, commissioning, start-up, test run, troubleshooting, optimization up to projects design, and safety review for big oil, gas, and petrochemical companies.

Hardeep Singh, having 20 years' experience, is working as a Professor at Amritsar Group of Colleges, Amritsar. He is a life member of various bodies and has been honored as a "Patron Member" by World Academic-Industry Research Collaboration Organization (WAIRCO) for the period of life time. His profile was published in "World Book of Researchers 2018" at Oxford, United Kingdom. The University Grants Commission (UGC) has certified him as Resource Person for Indian Knowledge Systems (IKS). He has been honored with "Letter of Appreciation" from the "World Book of Records," London, for commendable achievements, and was honored as a Senior Researcher by the Indian Academy of Emeritus Professors in 2022.

Dwarkadas Prahladadas Kothari an educationist and Professor who has held leadership positions at engineering institutions in India, including IIT Delhi, Visvesvaraya National Institute of Technology, Nagpur, and VIT University, Vellore. Currently, he serves as an Honorary Adjunct Professor in the Electrical Engineering Department . As a recognition of his contributions to engineering education, he was honoured as an IEEE Fellow.[2] Previously, he served as the Vice-Chancellor at VIT University.

Adnan Kavak is a Professor at the Department of Computer Engineering, the Director of Wireless Communications and Information Systems Research Center, and a member of Artificial Intelligence Research Lab at Kocaeli University in Turkey. He received his MSE and PhD degrees in Electrical and Computer Engineering from the University of Texas, Austin, USA. His current research interests include future wireless network technologies and e-health systems, with a focus on applying AI/Machine learning techniques on these systems.

Murtaza Hashami is an experienced Condition Monitoring Specialist with a demonstrated history of working more than 25 years in academia, Research and Development (R&D), and industrial environments. He possesses strong expertise in reviewing and evaluating new technologies and innovation ideas for implementation in power systems for enhanced reliability and extended lifetime of high-voltage (HV) equipment. He has enriched experience to develop roadmap, deploy, and implement advanced intelligent technologies for remote, online, and continuous condition monitoring of substation equipment to drive digital transformation in the future smart grid environment.

Definition/Abbreviations

Abbreviations	**Definition**
AGC	Automatic generation control
CPM	Controller Performance Monitoring
CT	Current transformer
CVR	Conservation voltage Reduction
DCS	Distribute control system
DER	Distributed energy resources
DMS	Distribution Management System
DOE	Department of Energy
DSGU	Distributed storage and generation units
EHV	Extra high voltage
EMS	Energy Management System
FDIR	Fault Detection, Isolation and Restoration
FM	Frequency modulation
HRC	High rupturing capacity
IDMT	Inverse definite minimum time delay overcurrent relay
IPU	Instantaneous pickup
LAN	Local/ area network
LSI	Long time, short time, and Instantaneous
LTD	Long Time Delay (LTD)
LTPU	Long Time Pick Up
LVPCB	Low voltage power circuit breaker
LVRT	Low-voltage ride-through
MCCB	Moulded Case circuit breaker
MEPS	Modern Electric Power System
ML	Machine Learning
NUG	Nonutility generators
OLTC	On-load tap changers
OMS	Outage Management System

PCC	Concept of Standard Coupling
PSM	Plug scale (or setting) multiplier
QE	Qatar Energy, former name Qatar Petroleum
RPA	Relay Protection and Automation
SCADA	Supervisory Control and Data Acquisition
SCR	Short circuit ratio
STD	Short Time Delay
STPU	Short time Pick Up
STS	Static Transfer Switch
TCC	Time current curve
TMA	Time multiplier setting
UPS	Uninterruptible power supply
VSC	Voltage Source Converter
VT	Voltage transformer
VVO	Volt-Var Optimization
WAN	Wide area network (WAN)

Web Catalog

Comprehensive reference on safeguarding electrical power systems from faults and failures, exploring emerging technologies such as AI and ML.

Supported by case studies of real-world power system faults, **Protection of Electrical Power Distribution Systems** offers an in-depth examination of the methods and technologies used to safeguard electrical grids from faults and failures. The book begins with a comprehensive introduction to power system fundamentals, emphasizing the critical role of protection mechanisms in maintaining system stability and ensuring operational safety. It thoroughly explores various types of faults, including short circuits, ground faults, and overloads, detailing their potential impacts on system performance, reliability, and safety.

The book introduces foundational concepts such as protection overlay, unit protection, and non-unit protection, before diving into the core principles of power system protection. These include the key attributes of **selectivity**, **sensitivity**, **speed**, and **reliability**, which are essential for designing effective protection systems. A detailed discussion follows on the wide range of protective devices used in modern power systems, such as fuses, circuit breakers, and relays, along with strategies for their coordination to ensure optimal performance.

To enhance reader engagement and understanding, the book includes practical exercises and self-assessment questions that focus on protection system design and implementation. These tools encourage deeper learning and provide valuable insights into the real-world application of power system protection technologies.

Protection of Electrical Power Distribution Systems includes information on:

- Differences between restricted and unrestricted protection and criteria to determine ideal boundaries of protection zones.
- Modern technologies applied to protection such as digital relays and intelligent electronic devices (IEDs).
- Different kinds of busbar protection and the main types of faults affecting the busbar.

- Types of transformer protection and their constraints and the operation of current and voltage transformers.
- Future trends in the field such as the application of AI and ML in predictive maintenance and fault diagnosis.
- To ensure the resilience of IT and OT systems in critical infrastructure operations, a comprehensive cybersecurity assessment framework must be implemented. This framework should identify vulnerabilities, assess risks, and deploy adaptive security measures to protect critical infrastructure against evolving cyber-threats.

Protection of Electrical Power Distribution Systems serves as an essential reference for engineers, researchers, and students, offering a deep understanding of power system protection principles and practices to enhance the reliability and safety of electrical power systems.

Exercises Discussed in this Book

Figures Discussed in this Book

Equations Discussed in this Book

Tables Discussed in this Book

1

Introduction

1.1 Objectives

When you have completed this section, you will know the aims and content of this book.

1.2 Introduction

Electricity has become the principal source of energy for modern society, and for most of the population, it is now almost a necessity of life. The distribution system is a part of the electricity supply industry, carrying electricity to most of the customers and therefore having the greatest impact on the quality of service provided to those customers. Although it is the least charismatic of the three main areas of the electricity supply system, namely generation, transmission, and distribution, it attracts a major part of the capital investment in the industry.

The protection system is a fundamental part of any electricity supply system and has an essential role in ensuring that distribution companies achieve an acceptable level of quality at an economic price. It is particularly important for the distribution system, since this is that part of the supply system which suffers from the greatest number of short-circuit faults.

The role of the protection system is principally to ensure that short-circuit faults or other abnormal operating conditions do not detract from the quality of supply to most customers connected to the distribution network. Operations of the protective system either isolate or initiate the isolation of faulted sections of the power supply network to maintain the continuity of supply elsewhere on the system. Power accuracy and coordination of the protection system are essential to ensure that the minimum amount of plant or ideally the faulted Plant only is disconnected from the network due to a short-circuit fault. The protection system

Protection of Electrical Power Distribution Systems: Smart grid, Microgrid, AI, and Cyber Security, First Edition. Tariq Masood, Jamil Abdo, and Atif Iqbal.

must operate as quickly as possible to minimize the amount of damage and disturbance caused by the fault and to minimize safety hazards to utility personnel and the general public. However, although speed is important, it is always secondary to accuracy.

The traditional form of protection was all based on electromechanical technology, and many of these schemes are still in use today. The first electronic relays were introduced in the 1970s, and microprocessor-based relays arrived in the 1980s. The introduction of microprocessor relays enabled the relay to monitor the power system's performance, to communicate with remote supervisory and control computers, and in some cases to have settings that can be adjusted remotely.

This book provides a basic introduction to the art and science of the techniques and technologies used in protecting distribution systems; it follows on from the basics in protection provided in book one. It does not attempt to cover all aspects of protection nor the subtleties of the subject, since these can only be acquired from a close involvement with a particular electricity supply system and an understanding of the history behind the development of that system. It is a companion to the next book, which covers the protection of the transmission system.

The book starts with the concepts of protection overlay, unit protection, and nonunit protection. These are explained in the next section, which includes the role and importance of providing backup protection to ensure that the power system is still protected even when the protection system fails. The most widely used protective system is the fuse, and which the subject of the opening part, Section 3. This covers the nonunit protection of distribution feeders and overcurrent protection in particular. Following fuses, the use of time-delayed protection relays is described together with their application to protect both standard and more complex supply networks.

An essential element in the operation of protection relays is the operation of the primary transducers, which are addressed in Section 4.

Section 5 considers unit protection schemes applied to distribution feeders, and the advantages and disadvantages that these provide over the nonunit schemes described in Section 3.

Techniques used to protect specific units of power system plant are considered in Sections 6–8, which cover the protection of transformers, busbars, and motors, respectively. The growth in the use of embedded generation and the difficulties in protecting these systems from both short-circuit faults and abnormal operating conditions are considered in Section 9. Autoreclose schemes are considered in Section 10, and recent advances in the coordination of protection and control schemes are examined in Section 11. This latter area is of growing interest due to the almost revolutionary advances made possible by the use of microprocessor-based relaying systems.

This book provides the basics of the art and science of protection for distribution systems and will enable those interested in this area to delve deeper into the specifics of the subject. Although several leading textbooks cover the topic, the greatest understanding of the subject can be best achieved by working with the schemes and exploring the history behind their development.

2

The Protection Overlay

2.1 Objectives

At the end of this section, you will have had an introduction to the general principles of the protection overlay, and you will be able to:

- Explain the difference between nonunit or nonrestricted protection and unit or restricted protection
- Divide a power system network into manageable zones of protection
- Determine the ideal boundaries of the protection zones
- Assess the capabilities of the protection system to protect the power system as a whole
- Identify fundamental weakness in a protection scheme

2.2 Introduction

Since the earliest days of providing bulk supplies of electrical energy, the general approach to protecting an electrical power system has been to divide it into manageable components and to protect each of these with dedicated protective devices. This distributed approach has enabled the protection system to develop alongside the power system plant and to accommodate changes in the power system and its configuration. Although over the years the concept of centralized protection systems has been proposed many times, distributed protection systems has been irrespective many times, distributed protection systems has been proposed many times, distributed protection is still the norm irrespective of the protection technology used.

The electrical power system represents the largest electrical circuits devised by human beings; therefore, its protection presents a major challenge. Their role is to carry bulk supplies of electricity from generation, transmit it from the areas of

Protection of Electrical Power Distribution Systems: Smart grid, Microgrid, AI, and Cyber Security, First Edition. Tariq Masood, Jamil Abdo, and Atif Iqbal.

generation to those of application, and finally distribute it to consumers. Through the operation of switchgear, a power system's configuration is inherently variable, and it is continually changed to suit the particular conditions prevailing at any one time.

The power system is designed to operate both efficiently and effectively, in that it is required to provide an economic and a continuous service for both suppliers and consumers of electricity. Economy of operations is expressed in terms of the utilization of resources, comprising the cost of the power system plant, the cost of operating it, and the losses incurred in transporting the power. Continuity of service is expressed in terms of the availability of the system to transport the power away from the generators and to deliver a continuous supply to consumers. Naturally, power system operators and customers want the greatest continuity of service at the lowest cost. There is a natural conflict between these requirements, and a compromise is continually being sought.

An important aspect of the management of a modern power system is the need to ensure that the economy and continuity of service provided by the system are not unduly compromised by a fault on the system or by an abnormal operating condition. This is associated with the protectability of the system and involves the provision of protective gear to detect faults and abnormal operating conditions and to implement appropriate switching actions when they occur. In an ideal world, power system faults and abnormal operating conditions would not occur, and hence there would be no need for protective gear. Unfortunately, although in the real world, faults in power system plants and abnormal operating conditions rarely arise, if they do occur, the consequences can be dramatic. Therefore, there is a continuous need for protection systems to identify when the system-affected plant can be disconnected from the network. The decision-making process used in protective gears must be accurate in that they must correctly identify the power system plant that is faulted or operating abnormally, such that it, and it alone, is disconnected from the network. This protective gear must also operate with the minimum delay, such that the affected plant can be disconnected from the rest of the power system as quickly as possible, both to minimize potential damage and to maintain system stability.

2.3 Nonunit Protection and Unit Protection

Before introducing the concept of the zone of protection, it is necessary to introduce the concept of the nonunit or nonrestricted type of protection and the unit or restricted type protection. Although there are several interesting schemes that exhibit the characteristics of both, all protective devices and schemes can be classified as either nonunit or unit protection.

2.3.1 Nonunit or Nonrestricted Protection

Nonunit protection represents the largest classification of protective devices and includes fuses and all types of overcurrent relays and distance relays. Nonunit protection operates in a stand-alone mode. It uses information, which is obtained locally, that is, from its relaying point. This point on the power system is defined by the position of the transducers used to obtain information for the protection system. In most cases, it is defined by the location of the current transducers. It is therefore commonly referred to as single-point protection.

Within the classification of nonunit protection, there are two subclassifications: those protections that use time to discriminate between faults or abnormal operating conditions affecting the protected plant and those affecting adjacent plant, and those protections that use a second source of information obtained at the relaying point.

2.3.1.1 Nonunit Protection Using Time for Discrimination

For those protections that use time for discrimination, the operation of nonunit protection is not restricted to faults on a particular unit of the plant. Therefore, this type of nonunit protection provides inherent backup protection for a fault on the adjacent plant should the protection associated with that plant fail to disconnect that plant from the network. Time is frequently used to provide discrimination between the operations of protective gear protecting adjacent parts of the power system network. The protection systems are chosen in such a way that the protection directly associated with the unit of plant operates faster for faults in that plant than the protection associated with the adjacent plant. Arguably, the simplest type of nonunit protection and the most widely used is the electric fuse. Considering the radial system shown in Figure 2.1, fuse B_n would carry the fault current

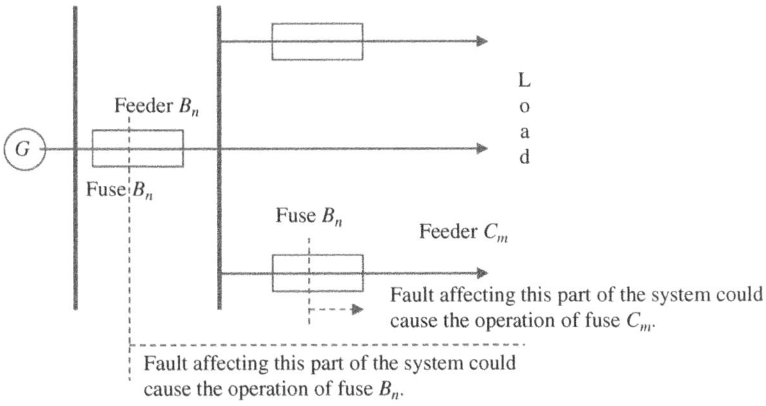

Figure 2.1 Nonunit protection applied to a radial feeder network.

caused by short circuit faults on feeder B_n and that the part of the network fed by that feeder and therefore could operate for these faults. For short circuit faults on feeder C_m, however, fuse C_m is chosen such that it operates more quickly than fuse B_n carrying the same fault current, and hence it will disconnect feeder C_m from the source of supply before fuse B_n can operate. If, however, fuse C_m fails to operate or clear the fault for whatever reason, fuse B_n *will eventually* operate and clear the fault. For a short circuit fault of feeder B_n and B_n will operate alone, and hence, disconnect feeder B_n and all following feeders from the source of supply. In this scenario, fuse C_m will not be subjected to fault current and hence will not operate.

The operation and application of fuses and overcurrent relays are both considered in more detail in the following subsection.

2.3.1.2 Nonunit Protection Not Using Time for Discrimination

Protection using a second source of information obtained at the relaying point uses that second input to constrain the reach of the protection such that it will only respond to faults or abnormal operating conditions on a section of the protected plant. Therefore, they will not operate for faults or abnormal operating conditions beyond that reach or affecting adjacent plant. The best example of this type of protection is the distance relay. This uses voltage to constrain the reach of the protection and hence that part of the power system which is protected. The setting is chosen so that it will not operate for faults or abnormal operating conditions affecting the plant adjacent to that being protected by the protective device. When the protection is intended to provide enhanced protection and backup protection for part of an adjacent system, time is used to provide discrimination. The time of delay is chosen such that it enables the protective gear, which should clear the fault, to operate first. The use of stepped distance relays to protect feeder circuits is illustrated in Figure 2.2, which shows that the relay provides instantaneous protection for part of the protected feeder, time-delayed protection for the remainder of the protected feeder, and backup protection for the first part of the following feeder. The operation and application of distance relays are not considered in this book; however, Protection of Electrical Power Transmission Systems: by Tariq Masood, which covers transmission system protection.

2.3.2 Unit or Restricted Protection

Unit or restricted protection represents those types of protective devices that will only respond to faults or abnormal operating conditions on the unit of plant they are protecting and will never operate for faults or abnormal operating conditions on adjacent plant. The best-known type of unit protection is the current differential protection illustrated in Figure 2.3.

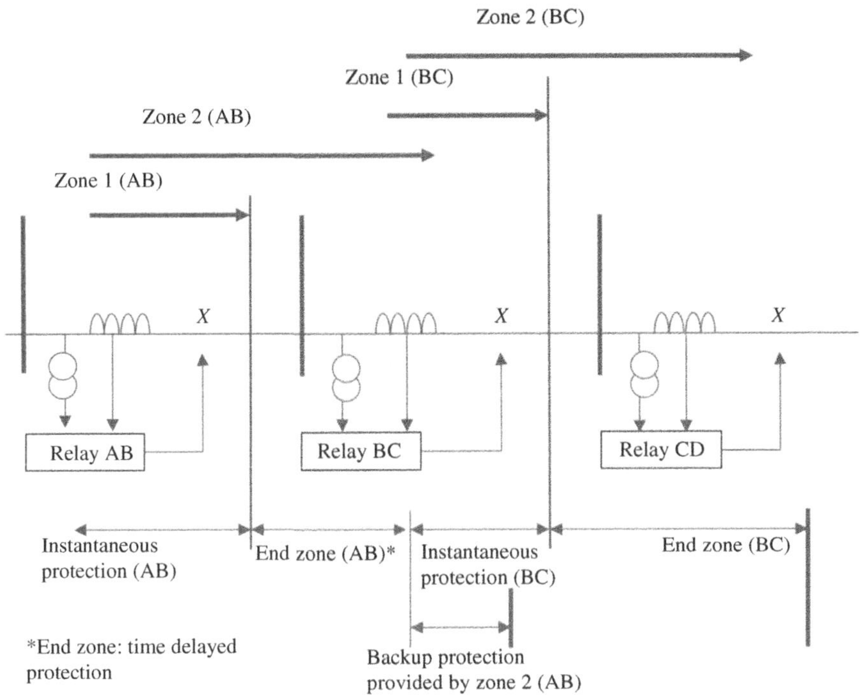

Figure 2.2 Nonunit protection with constrained reach setting.

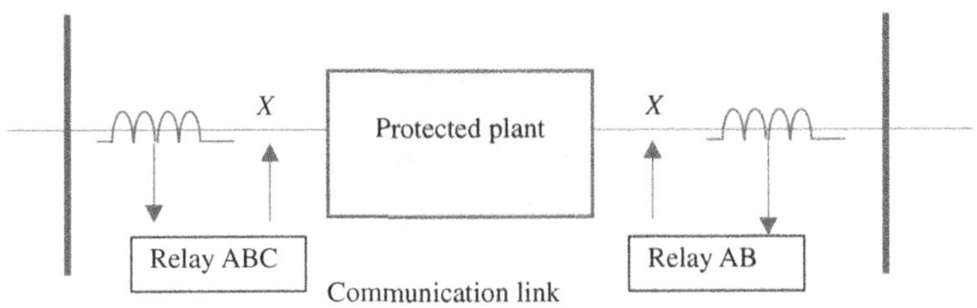

Figure 2.3 Unit protection.

Unit protection uses information obtained from more than one location in the power supply network and hence has multiple relaying points. When used to protect large units of plant, such as feeder circuits, this leads to the need for communication systems. Since the protection will only operate for faults or abnormal operating conditions affecting adjacent plant, unit protection provides inherent and absolute discrimination in protecting the adjacent plant.

Therefore, there is no need to delay the operation of the protective gear, and the unit protection is instantaneous in operations. But since it will not respond to faults affecting the adjacent plant, it cannot provide backup protection for the protective gear protecting that adjacent plant. The operation of unit protection schemes is covered in more detail in section five of this book.

2.4 Nonunit Protection and Unit Protection

In order to provide comprehensive protection, an overlay for a power supply system, the system is divided into manageable units, which can be properly protected individually. These units normally consist of a single unit of power system plant together with the switchgear used to connect it or disconnect it from the rest of the power system. Typically, these units of plant may consist of, for example, a transformer, a busbar, and a feeder circuit. The choice of what equipment constitutes a protected unit can become complicated where different pieces of plant are inherently operated together. The classical example of this is the generator and transformer combination, where the transformer provides the link for the generator to the power system and is therefore dedicated to that generator. In this case, the generator and its transformer may constitute a self-contained unit of plant and can be treated as a single unit for protection.

The division of a power system into manageable units is illustrated by the network shown in Figure 2.4. This represents a part of a power supply system being supplied by a local generator and incoming transmission lines, together with a distribution system fed by a step-down transformer, including subtransmission feeders and local distribution feeders. In order to protect this network, the system has been divided into manageable units that are connected to each other using circuit breakers. In this example, the manageable units of the power system that can be protected individually consist of:

- a generator and its associated transformer
- an incoming transmission line
- one half of the EHV busbar
- a transformer
- a HV busbar
- a subtransmission line
- a distribution busbar
- a distribution feeder

Each of these manageable units represents a single zone of protection. Taken together, the zones of protection provide a complete overlay for the network, and none of the network is left unprotected. As presented in Figure 2.4, each of the

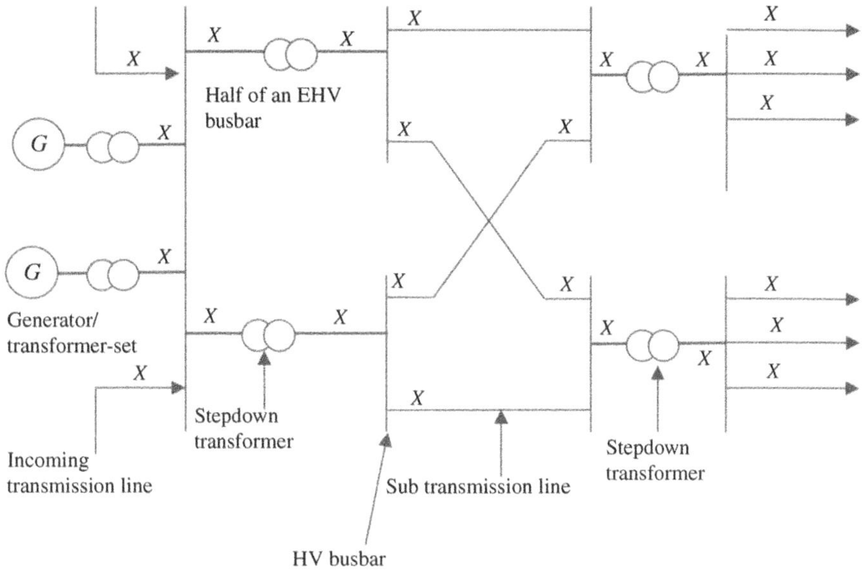

Figure 2.4 Zones of protection.

zones overlaps with the two zones of protection. The position of the boundaries of the zones of protection depends on the type of protection is used, the positions of the primary transducers, and the circuit breakers. Where the operation is inherently restricted to operate for faults occurring in the zone of protection, that is, the protection is a unit protection, the boundaries are generally defined by the location of current transducers used to provide the protective devices with current information. The boundary case for an overlapping zone is shown in Figure 2.5;

Figure 2.5 Overlapping zones of protection.

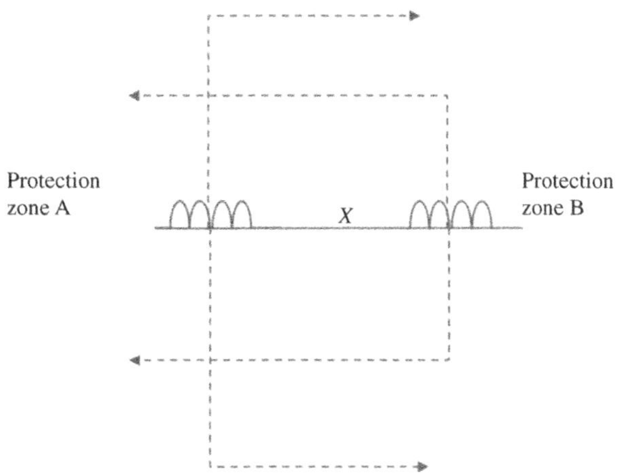

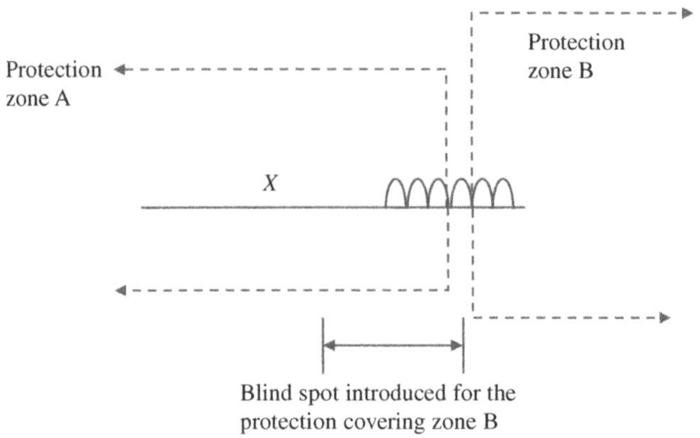

Figure 2.6 Nonoverlapping zones of protection.

current transducers are located on each side of the circuit breaker, the first providing information to the protective gear for zone A and the second providing information for the protective gear for zone B. This represents the ideal type of protection scheme since there are no inherent weaknesses in the protection.

An immediate constraint, frequently applied to protection systems, is the cost of the equipment that can be used. Figure 2.6 shows two current transducers with two outputs, where a cost constraint is applied. Under these conditions, the transducer output provides information to the protective gear for zones A and B. The two zones of protection do not overlap, and an inherent blind spot is introduced for the protection zone B. Should a fault occur in this blind spot between the current transducer and the circuit breaker, protective gear B will not trip. The fault could continue to be fed from the power system connected to the plant contained in zone B after the protective gear for zone A has operated and opened the breaker. The conditions described above are not uncommon, and since they produce a weakness in the protection system's coverage of the network, either additional protection is required to detect faults in this blind spot, or different types of protection are required that overcome this weakness. Where additional protection is used to detect faults not cleared by the operation of the protective gear associated with zone A of Figure 2.6, the protection scheme will generally include breaker failure protection. This provides cover for the blind spot and also protects against the scenario if the circuit breaker fails to clear the fault. An alternative approach to the protection is to use a nonunit type of protection where the boundary is defined by the poles of the circuit breaker, and time is used to discriminate between the protection system's response to faults in zones A and B. The boundaries of the zones of protection when nonunit protection is used are defined by the relaying

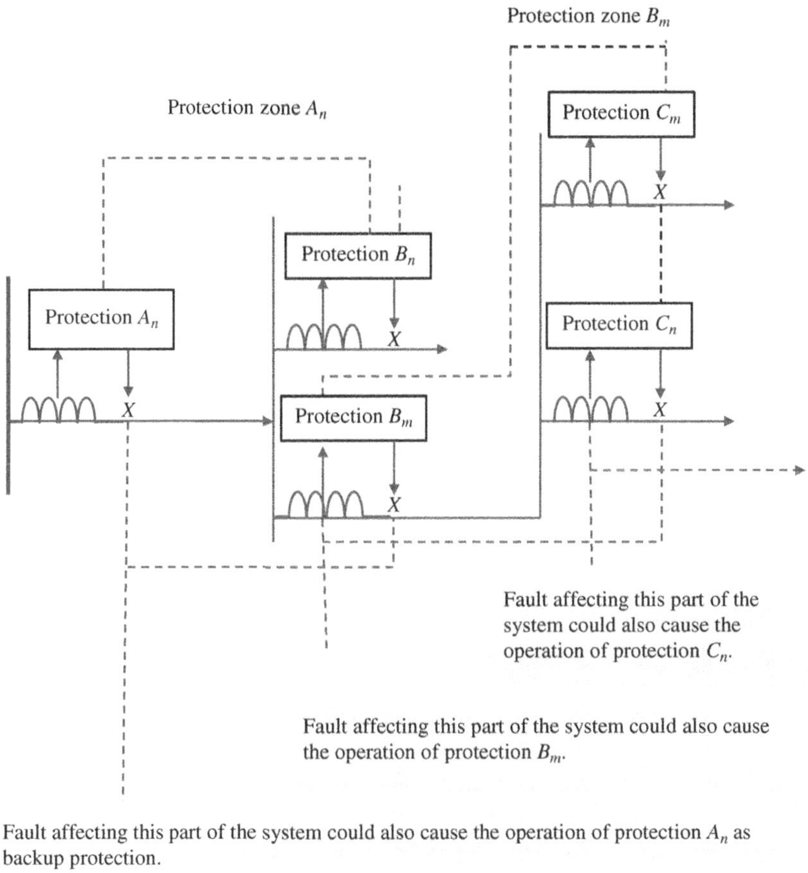

Figure 2.7 Zones of protection associated with nonunit protection using time for discrimination.

point of the protection scheme and the location of the circuit breakers controlled by the protection that protects the adjacent plant. This is shown in Figure 2.7.

Since the operation of the protection is not restricted to faults within its principal zone of protection, the protective gear associated with zone A_n of Figure 2.7 could operate for faults on the following feeder, which is covered by zone B_m. With this type of protection, discrimination between the two sets of protective gear is provided by ensuring that for faults in zone B_m, the protective gear associated with zone A_n is time delayed with respect to that for zone B_m. The protective gear associated with zone B_m, therefore, has time to clear the fault before the protective gear associated with zone A_n operates.

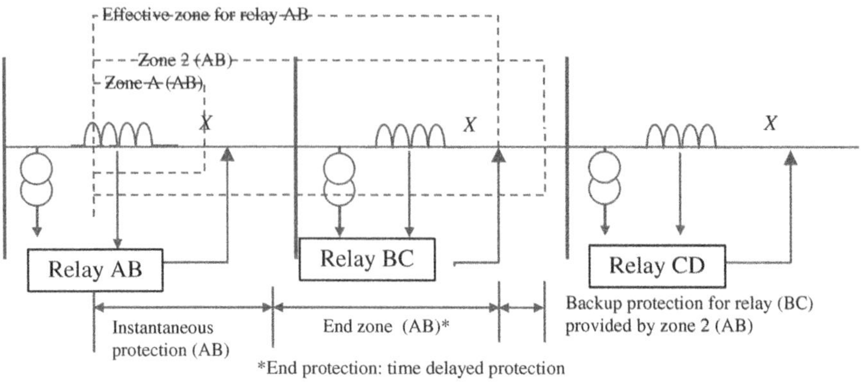

Figure 2.8 Zones of protection associated with nonunit protection with constrained reach.

The third type of protection zone is provided by the second type of nonunit protection, where the boundaries of the protection zone are defined by the reach setting of the protective system. This type of nonunit protection is discussed in Section 2.2.1.

The boundaries for the zones of protection associated with a classical stand-alone distance protection are shown in Figure 2.8 to ensure that the protective gear associated with zone 1 of Figure 2.8 does not operate for faults on the following feeder; its reach setting is chosen such that, despite all possible errors, it will not operate for faults affecting the immediate feeder. This reach setting is therefore not protected by the zone 1 elements, and a second zone is required. This second zone reach is chosen such that it will operate for all major faults affecting the immediate feeder, and as such, it may well operate for faults affecting the following feeder circuit. To provide discrimination between protective gear protecting the immediate feeders and protecting the following feeder, this second zone, zone 2, is time delayed. As a result, faults affecting the end zone of the immediate feeder are also cleared after the time delay. Should the protective gear associated with the following feeder fail to clear a fault in the region covered by zone 2, the zone 2 elements will provide backup protection.

2.5 Backup Protection

The protection system is a fundamental part of the power supply system, and as such, the power system needs to be protected from failure or abnormal operating conditions in the protective gear as well as the main power system plant.

The inherent backup protection capabilities described above can be used to clear fault or abnormal operating conditions affecting a following unit of the plant if they are not cleared by their own protective gear. An unfortunate weakness of this approach is that in order to disconnect the faulted plant from the network, the normal plant is also disconnected.

Failures in the protective gear can arise from a variety of causes. The failure could be caused by failure in the protection relaying, the primary transducers, the trip control scheme, or the circuit breaker itself. An alternative to depending on the inherent backup protection capabilities of the protective gear associated with the adjacent feeder is to use a second protective system to support the existing protection.

Where the greatest risk of a failure in the protective gear is attributed to the protection relaying, a second relaying system can be used, which operates in parallel with the main system. This is frequently referred to as a primary and secondary protection scheme. If the risk of failure of the protection relaying, the primary transducer, or the trip control scheme is considered unacceptable, a parallel protection system can be justified. This type of scheme is referred to as a dual main scheme of a main A/main B scheme.

Other type of backup protection schemes include trip supervision schemes, which monitor the trip control scheme, and fuse failure protection schemes that are used to monitor three-phase voltage transformer and either inhibit the protection scheme should one phase of the voltage transformer be lost or switch the protection scheme to a second set of voltage transformers.

2.5.1 Primary and Secondary Protection Schemes

A typical primary and secondary protection scheme is shown in Figure 2.9. This shows a subtransmission feeder system protected by a distance protection as the primary system and an inverse definite minimum time delay overcurrent protection as the secondary protection. Both sets of protection share the same primary transducers, namely the current transformers.

Under normal conditions, the primary protection is expected to detect the presence of faults on the protected feeder and to disconnect it from the rest of the network by tripping the circuit breaker. If, however, the primary protection fails

Figure 2.9 Primary and secondary protection scheme.

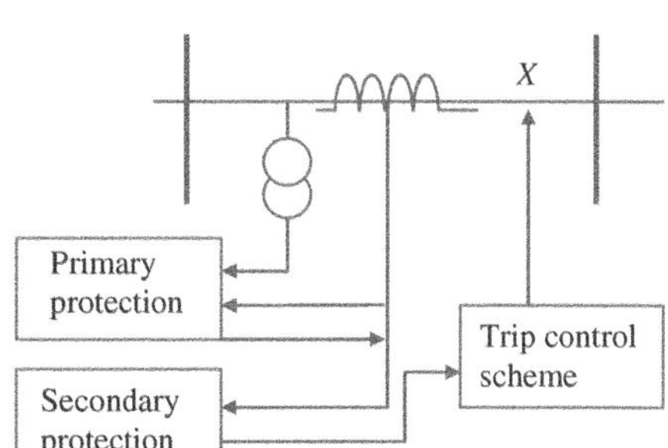

to operate due to an internal failure, the overcurrent relay will carry the same fault current and **albeit** time delay, it will operate and trip the circuit breaker.

In this case, both the primary and the secondary protection will provide inherent backup protection for faults on the system fed by this feeder.

2.5.2 Dual Main Protection Schemes

With a dual main scheme, the primary transducers, the protection relays, and the trip control schemes are duplicated, and both protection systems are chosen to provide the tripping capabilities required by the main protection. This reduces the risk of protection nonoperation to the probability of the circuit breaker only failing. In several cases, the expense of separate transducers is avoided by using a separate outputs from a common transducer.

A typical dual main protection scheme is illustrated in Figure 2.10. This shows a scheme used to protect a subtransmission scheme, which consists of a distance protection scheme and a current differential scheme. Both schemes provide a comparable level of protection for the feeder, with similar tripping times to initiate circuit breaker operation.

Different utilities operate different philosophies for dual main protection, and the variations include:

- Parallel sets of similar relays
- Parallel sets of relays using different operating principles
- Parallel sets of relays form different suppliers
- Parallel sets of relays form different suppliers using different operating principles

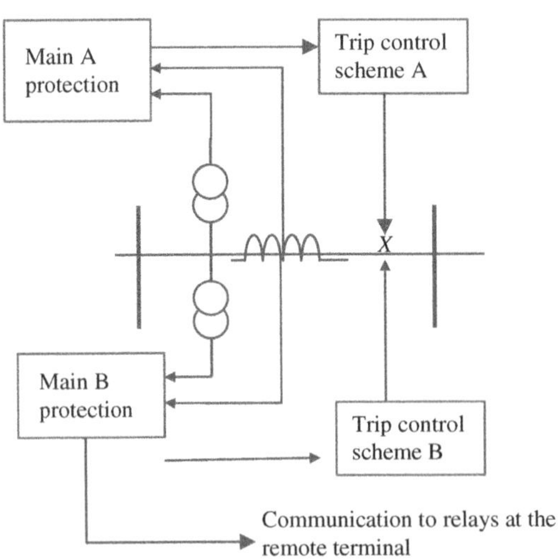

Figure 2.10 Dual main protection schemes.

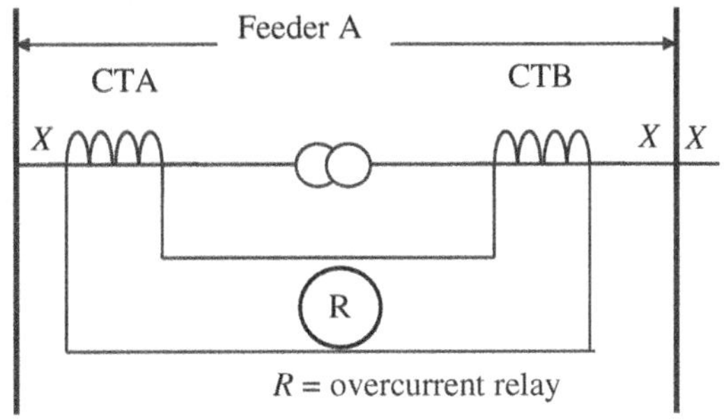

Figure 2.11 Protection scheme associated with a transformer and feeder circuit.

2.6 Self-Assessments Questions

Self-Assessment Question 2.1

Considering the protection scheme for the transformer and feeder circuit shown in Figure 2.11, describe how the scheme would respond to a fault on the transformer and a fault on the feeder given that:

2.1a) Current Transformer A (CTA) and Current Transformer B (CTB) have the same ratio.
2.1b) CTA has a ratio which is slightly higher than that of CTB, a small mismatch of, say, 10%.
2.1c) CTB is omitted.

Bibliography

Jamil Abdo, Tariq Masood “New Technique to develop Intelligent, highly reliable & Efficient Unmanned Power Generation Modular” published in 22nd International Conference on Renewable Energies and Power Quality (ICREPQ’24) Bilbao (Spain), 26–28 June 2024. https://www.icrepq.com/icrepq24/392-24-abdo.pdf.

Jamil Abdo, Tariq Masood “Operational Reliability- Production Best Practice Pragmatic Case Studies & Improvement” published in Proceedings of the 9th North American Conference on Industrial Engineering and Operations Management, Washington D.C., USA, 4–6 June 2024.

Jamil Abdo, Tariq Masood “Techno Economics Analysis to Reshape Global Clean and Green Energy” published in 22nd International Conference on Renewable Energies and Power Quality (ICREPQ’24) Bilbao (Spain), 26–28 June 2024. https://www.icrepq.com/icrepq24/415-24-abdo.pdf.

Qingtao Cao, Yun Jiang, Chunxiu Liu, Xinzhe Ma, Pengpeng Wang “Research on Fault Analysis of Multi source Distribution Network Based on Adaptive

Protection" published in 7th Asia Conference on Power and Electrical Engineering (ACPEE) 2022.

Qingfa Chai, Jiayin Jiang, Yi Tang, Anning Wang, Congcong Li "High Reliability Relay Protection Setting Scheme of Distribution Network" published in 2023 International Conference on Power System Technology (PowerCon) 2023.

Dong Chen, Bing Hu, Mingjun Xue, Yinian Qi, Changsong Oin "Research and Application of the System of Relay Protection and Self-Healing for Distribution Network with Mixed Use of Circuit Breaker and Load Switch" published in 2023 International Conference on Power System Technology (PowerCon) 2023.

Jane Courtney, Aaron McDonnell "Impact on Distribution System Protection with the Integration of EG on the Distribution Network" published in 2019 54th International Universities Power Engineering Conference (UPEC) 2019.

Feng Deng, Xiang Ji, Guo-Fang Huang "A Centralized Distribution Protection System" published in 2012 China International Conference on Electricity Distribution 2012.

Bruno Fazio, Sanja Bajic, Serge St-Antoine "Relay Protection Coordination Analysis using Fault Current Measurements: A Case Study of Hydro Quebec Distribution Network" published in 2022 IEEE Power & Energy Society General Meeting (PESGM) 2022.

Hana A Halim, Melaty Amirruddin, Nur S Noorpi, Nurhakimah M Mukhtar "An Improved Protection Scheme for Smart Distribution Grid" published in 2013 1st International Conference on Artificial Intelligence, Modelling and Simulation 2013.

IEEE Power and Energy Society "IEEE Recommended Practice for Fault Diagnosis and Protection in Smart Distribution System" IEEE Std 2748-2023 2024.

Hang Ji, Dawei Wang, Qiang Xu, Shangqing Yuan "DC Power Distribution System and Its Protection Configuration" published in 6th International Conference on Power and Renewable Energy (ICPRE) 2021.

Rani Kumari, Bhukya K Naick "Protection Coordination Impact on Reliability of Distribution Systems With Distributed Energy Resources" published in 2024 Third International Conference on Power, Control and Computing Technologies (ICPC2T) 2024.

Bin Li, Chao Li, Fengrui Guo, Ying Xin "Overcurrent Protection Coordination in a Power Distribution Network With the Active Superconductive Fault Current Limiter" published in IEEE Transactions on Applied Superconductivity 2014.

Qin Lijun, Wang Ying, Zheng Shenrui, Guo Qing "Situation of Study on Distribution Network Protection Containing Distributed Generation and Application Scheme of Intelligent Protection" published in 2011 4th International Conference on Electric Utility Deregulation and Restructuring and Power Technologies (DRPT) 2011.

Zhichao Lin, Yingmin Wang, Riping Zhong, Busheng Luo, Qiujie Ma, Chuankun Yu "Coordinated Optimization Method of Distributed Protection for Distribution

Network Based on Multi-Agent" published in 2023 8th Asia Conference on Power and Electrical Engineering (ACPEE) 2023.

HanYu Liu, LongHua Mu "Integrated Protection on Shipboard Integrated Power System" published in 2011 International Conference on Advanced Power System Automation and Protection 2011.

Tariq Masood, Jamil Abdo "Maintenance Strategy Production Best Practices – Pragmatic Case Studies and Operations Improvement" published in Proceedings of the 9th North American Conference on Industrial Engineering and Operations Management Washington D.C., USA, 4–6 June 2024.

Tariq Masood, Jamil Abdo "New Technique to Enhance Hybrid Renewable Energy Resources' Operational Flexibility and Reliability" published in 22nd International Conference on Renewable Energies and Power Quality (ICREPQ'24) Bilbao (Spain), 26–28 June 2024. https://www.icrepq.com/icrepq24/396-24-masood.pdf.

Tariq Masood, Jamil Abdo "Review: Uninterrupted Liquid Hydrogen (E-Methanol) Production framework to deliver Clean & Green Energy" published in 22nd International Conference on Renewable Energies and Power Quality (ICREPQ'24) Bilbao (Spain), 26–28 June 2024. https://www.icrepq.com/icrepq24/414-24-masood.pdf.

Tariq Masood, N Al-Emadi, SA Qureshi, YY Almulla, AJ Khan, SK Shah "FACTS Control Devices (Statcom, SSSC and UPFC) Re-Configuration Techniques By PSIM/MATLAB" published in 2007 International Conference on Electrical Engineering 2007. http://ieeexplore.ieee.org/abstract/document/4287295.

Tariq Masood, Raj K Aggarwal, SA Qureshi, RAJ Khan "STATCOM and SVC Control Operations and Optimization During Network Fault Conditions" published in 2010 IEEE International Symposium on Industrial Electronics 2010. https://ieeexplore.ieee.org/document/5636868.

Tariq Masood, Raj K Aggarwal, Suhail Aftab Qureshi, RAJ Khan "STATCOM Model Against SVC Control Model Performance Analyses Technique by MATLAB" published in International Conference on Renewable Energies and Power Quality (ICREPQ'10) 2010. https://icrepq.com/icrepq'10/655-Masood.pdf.

Tariq Masood, Suhail A Qureshi, RK Aggarwal, DP Kothari "Multiple FACTS Devices Integration Through Distributed Control System and Its Application on the GCC Interconnection" published in 2013 7th IEEE GCC Conference and Exhibition (GCC) 2013. https://ieeexplore.ieee.org/abstract/document/6705750.

Tariq Masood, Abdel Aty Edris, Suhail Aftab Qureshi, Muhammad Tajammal, Murtaza Hashmi, DP Kothari, Samer Kareem "SSSC Tailored to Optimization Performance in Between United Arab Emirates and Oman on the GCC Power Grid" published in the 13th IET International Conference on AC and DC Power Transmission Conference on 14–16 February 2017. http://ieeexplore.ieee.org/document/7934934.

Tariq Masood, Abdel Aty Edris, Suhail Aftab Qureshi, Muhammad Tajammal, Murtaza Hashmi, DP Kothari, Samer Kareem "Smart Grid Operations and Control Challenge by Implementing STATCOM Tailored to Optimization Performance at the GCC Power Grid" published in the 13th IET International Conference on AC and DC Power Transmission Conference on 14–16 February 2017. http://ieeexplore.ieee.org/document/7934988.

Tariq Masood, Muhammad Tajammal, Samer Kareem, Suhail Aftab Qureshi, Murtaza Hashmi, DP Kothari "Smart Grid Operational Functions and Control Challenges by Implementing SSSC Tailored to Optimize Performance in Between Qatar and KSA on the GCC Electrical-Power Grid" published in the ASTESJ Journal Advances in Science, Technology and Engineering Systems Journal Vol. 2, No. 1, 2017. https://astesj.com/special-issue-4.

Tariq Masood, Muhammad Tajammal, Samer Kareem, Suhail Aftab Qureshi, Abdul Jabbar Khan, DP Kothari, Nair Jameel "Unified Power Flow Controller Modeling and Analysis Technique on the GCC Power Grid Operations and Control" published in 2018 IEEE International Energy Conference, Limassol, Cyprus, 7–9 June 2018. https://ieeexplore.ieee.org/document/8398834.

Tariq Masood, Muhammad Tajammal, Samer Kareem, Suhail Aftab Qureshi, Murtaza Hashmi, Nasir Jameel "Unified Power Flow Controller (UPFC) Modeling and Analysis Technique by Using Matlab and EMTP" published in International Conference on Renewable Energies and Power Quality (ICREQP'18) Salamanca (Spain) 21–23 March 2018. http://www.icrepq.com/icrepq18/250-18-masood.pdf.

Tariq Masood, Muhammad Tajammal, Samer Kareem, Suhail Aftab Qureshi, Murtaza Hashmi, Naser Jameel "STATCOM Controller Performance Enhancement and Optimization Technique" published in International Conference on Renewable Energies and Power Quality (ICREQP'18) Salamanca (Spain) 21–23 March 2018. http://www.icrepq.com/icrepq18/249-18-masood.pdf.

Tariq Masood, Samer Karim Shah, Muhammad Tajammal, "Unified Power Flow Controller Modeling and Analysis Technique on the GCC Power Grid" published in 2018 IEEE International Energy Conference (ENERGYCON) 2018. https://ieeexplore.ieee.org/abstract/document/8398834.

Vassilis C Nikolaidis, Evangelos Papanikolaou, Anastasia S Safigianni "A Communication-Assisted Overcurrent Protection Scheme for Radial Distribution Systems With Distributed Generation" IEEE Transactions on Smart Grid Vol. 7, No. 1, 2015.

LiJun Qin, Ying Wang, Cuijuan Hao, "Multi-Agent System Wide Area Protection Considering Distributed Generation Impact" published in International Conference on Advanced Power System Automation and Protection 2011.

VS Sazanov, AI Kovalenko, AA Voloshin, SA Danilov, EA Voloshin, DA Degtyarev "The Development of System for Automatic Adaptive Change of Relay Protection Settings in Distribution Networks" published in 2022 5th International Youth

Scientific and Technical Conference on Relay Protection and Automation (RPA) 2022.

VS Sazanov, AI Kovalenko, AA Voloshin "The Development of System for Automatic Adaptive Change of Relay Protection Settings in Distribution Networks" published in 5th International Youth Scientific and Technical Conference on Relay Protection and Automation (RPA) 2022.

Xujiang Shi, Chong Tong, Xiaoming Pan "Dynamic Lightning Protection of PET Embedded AC/DC Hybrid Distribution Network" published in 36th International Conference on Lightning Protection (ICLP) 2022.

Xiaohui Song, Yu Zhang, Song Zhang, Shizhan Song, Jing Ma, Weibo Zhang "Active Distribution Network Protection Mode Based on Coordination of Distributed and Centralized Protection" published in 2017 China International Electrical and Energy Conference (CIEEC) 2017.

M Szablicki, P Rzepka, A Halinka, P Sowa "Diagnosis of Challenges for Power System Protection – Selected Aspects of Transformation of Power Systems" published in Modern Electric Power Systems (MEPS) 2019.

Tariq Masood, RK Aggarwal, SA Qureshi, RAJ Khan "Statcom Control Reconfiguration Technique for Steady State and Dynamic Performance Optimization During Network Fault Conditions" published in International Conference on Renewable Energies and Power Quality (ICREPQ'10) Granada (Spain) 2010. https://www.icrepq.com/icrepq'10/656-Masood.pdf.

Yuwei Xiong, Wei Chen "Positive Sequence Current Differential Protection Based on Least Square Method for Distribution Network with Distribution Generation" published in 2021 IEEE 4th International Electrical and Energy Conference (CIEEC) 2021.

Kun Yang, Zhen Xiang, Ruixiong Yang "Decentralized Protection Coordination Based on multi-agent system for Active Distribution Networks" published in 2022 7th Asia Conference on Power and Electrical Engineering (ACPEE) 2022.

Yu Zou, Minchuan Liao, Ruiqing Lian "Discussion on Lightning Protection of Distribution Network and Its Differentiation" published in 2020 International Conference on Intelligent Computing, Automation and Systems (ICICAS) 2020.

3

Nonunit Protection of Distribution Feeders

3.1 Objectives

At the end of this section, you will have had an introduction to the application of nonunit protection to protect distribution feeders, and you will be able to:

- Explain the operation of the fuse for protecting distribution feeders
- Choose suitable values to protect a radial distribution feeder system
- Appreciate the use of fuses together with circuit breakers for high-voltage feeders
- List the advantages offered by relays compared with fuses
- Choose a suitable setting for inverse definite minimum time-delayed overcurrent relays for the protection of a radial distribution feeder system
- Appreciate the limitations of these relays when protecting interconnected feeder circuits
- Describe the application of directional overcurrent relays for the protection of interconnected distribution feeders

3.2 Introduction

The major part of a power system consists of distribution feeders operating at a voltage range of 3.3–38.5 kV. Most of these feeders are protected either by fuses or by stand-alone overcurrent relays, both used as nonunit protection. The first reported application of protection system was in 1864 when fuses were used to protect submarine cables. Until the early 1910s, fuses provided the only viable method of protecting power supply system and most of the leading names in electrical engineering were involved with their development.

About the turn of the century, the limitations of fuses were becoming apparent, and other types of protection systems were being investigated, and, in 1910, the first circuit breakers and protection relays were introduced.

Protection of Electrical Power Distribution Systems: Smart grid, Microgrid, AI, and Cyber Security,
First Edition. Tariq Masood, Jamil Abdo, and Atif Iqbal.

These early protection relays were basic overcurrent devices, and their design was based on the attracted armature relay and later on the kilowatt-hour meter. The similarity of these later designs to the kilowatt-hour meter and their saturating characteristic earned them the title of "bad wattmeters." Following the success of these early electromechanical relays and their potential to solve a wide variety of power system protection problems, new relaying concepts were developed, and many of these still provided the basis of modern protection systems. Most of the principles of protection were derived between 1910 and 1940. Most of the current range of electrotechnical relays are based on designs from this era and was first introduced from 1940 to 1960. In many of today's modern power supply networks, the majority of relays are still electromechanical.

The advantages of electronic relays over electromechanical relays were appreciated in the 1940s, showing that they only became viable with the introduction of "reliable" semiconductor devices. The first commercially available "static" relay was introduced in the early 1960s. The expenses and complexity of these relays generally limit their application to the protection of the extra high voltage (EHV) feeder circuit.

Developments in semiconductor electronics were closely reflected in the new relay designs. In the 1970s, integrated circuits were first used in relays, and in the early 1980s, the first microprocessor-based relays became commercially available. As with other electronics technologies, the first application of microprocessor-based relays was to protect EHV feeders, although the manufacturing potential of high-volume microprocessor relays was quickly realized.

Arguably, the greater impact of microprocessors on protection has been the introduction of competitive distribution protection relays, notably the inverse definite minimum time delay (IDMT) overcurrent relay. This relay is widely used to provide nonunit protection for distribution feeders.

3.3 Fuses

The fuse perceived to be the simplest protection device is used to protect power supply networks. It is connected in series with the protected circuit and carries the same current. When this current exceeds the fusing current, the "fuse" melts, thus breaking the circuit supplying power to the load. The fuse, therefore, both detects the fault condition and directly isolates the faulted plant from the supply network.

By definition, the fuse comprises all of the parts that form the unit and not just the conductor, which melts under fault conditions. The fuse consists of the "fuse link" and its holder. The fuse link, the part which has to be replaced after operation, contains the fuse element, the conductors, which melt during operation, its

Figure 3.1 A low-voltage fuse.

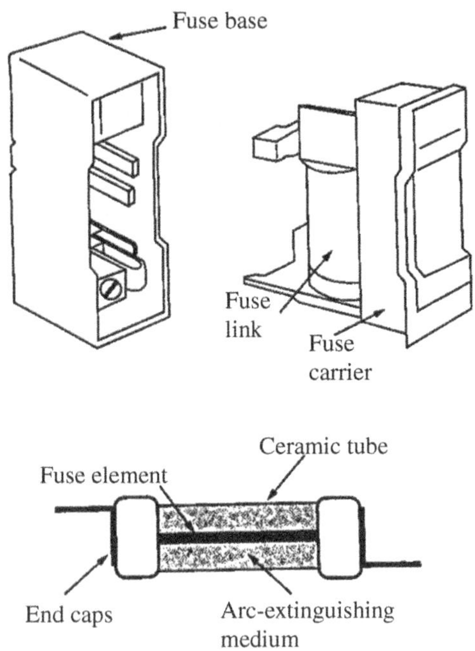

case, and an arc extinguishing filling. The case may comprise a glass or ceramic tube with suitable end connectors, and the filling may be of high-purity silicon oxide. The components of a typical industrial fuse are shown in Figure 3.1.

The principal advantages of the fuse are as follows:

- perceived to be a simple device,
- simple to use,
- direct acting,
- inexpensive,
- fail safe,
- very fast at high fault currents, and
- the operating characteristics can be graded.

The principal disadvantages of the fuse are as follows:

- requires replacement after operation,
- provides only limited protection for low-fault currents,
- is nondirectional, and
- has limited voltage breaking capacity.

The great success of the fuse has led to its use in a wide range of applications, and a large variety of specialist devices are now available. Fuses are extensively used on low-voltage systems rated at 415 V with a current rating of 2–1600 A. High-voltage fuses are available for 11 and 33 kV systems with a current rating of 5–60 A, and specialist devices are available for 66 kV applications.

High-voltage fuses, with three-phase voltage ratings in the range 7.2–34.5 kV, are frequently fitted with striker pins which can be used to operate a flag to indicate that either the fuse has ruptured or trigger a trip mechanism to disconnect all three phases. These strike pins are activated by a small pyrotechnic charge, which fires the pin out of the end of the fuse to operate a mechanical mechanism. The pin also prevents reclosure of the switch until the ruptured fuse has been replaced.

These fuses are long to support the voltage associated with high-voltage applications. A 12 kV fuse mounted in air has to be typically 300 mm long, and a 34.5 kV rated fuse, again mounted in air, is more than 500 mm long. Fuses that are designed to be immersed in oil can be shorter.

An alternative approach to protect high-voltage circuits is to use fuses with circuit breakers and current transformers. The circuit breaker breaks the fault current and provides high-voltage isolation, and the current transformers provide the fusing current, which ruptures the fuse elements. These schemes are described in more detail in the following sections.

3.3.1 Fuse Operating Characteristics

The operating characteristics of a fuse are presented either as a curve relating the pre-arcing time against prospective current, similar to that shown in Figure 3.2, or as the let-through capacity (I^2t) similar to that shown in Figure 3.3.

The concepts of pre-arcing times and prospective currents enable manufacturers to provide standard operating characteristics for fuses that avoid the variation in

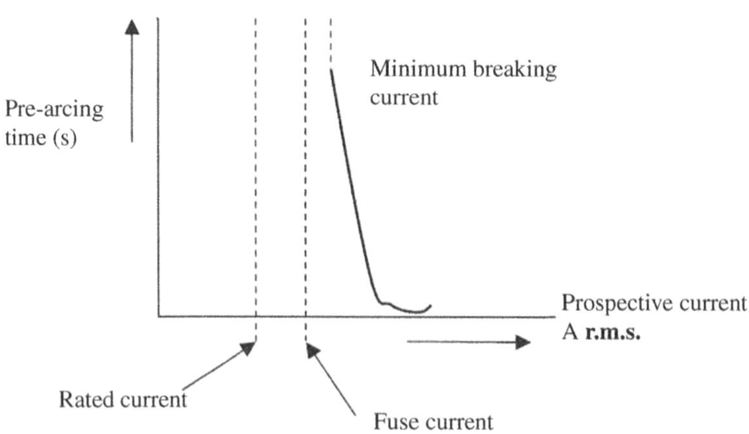

Figure 3.2 Fuse characteristics pre-arcing time against prospective current.

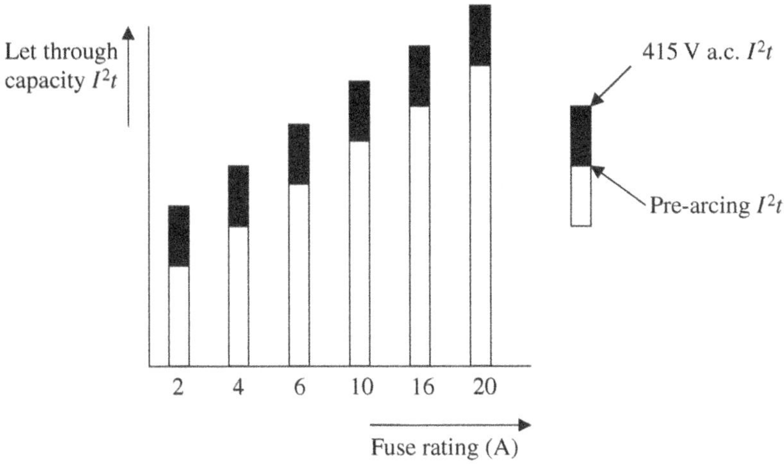

Figure 3.3 Fuse characteristics let-through capacity (I^2t).

performance resulting from different application circuits in which fuses are used. These are defined as:

pre-arcing time equal to the time required for the fuse to melt when carrying a specified **r.m.s.** current and prospective current equal to the **r.m.s.** current that would flow in the circuit if the fuse element were replaced by a link of zero impedance. The operating time for the fuse is the time required to break the circuit supplying power to the fault, which comprises the pre-arcing time plus the arcing time. The arcing time is the time required for the area to extinguish and depends on the characteristics of the circuit carrying the fault current and the voltage applied across it. The use of the prospective current avoids the complication caused by the voltage drop across the fuse during its melting and arcing periods.

Other terms used to define the operation of fuses include:

rated current = the current that the fuse can carry continuously without damage or degradation to the fuse element.

minimum fusing current = that current that will cause the fuse elements to rupture in an "infinite time."

fusing factor = the ratio between the minimum fusing current and the rated current.

minimum breaking current = the current which the fuse manufacturer will guarantee to cause the fuse element to rupture in a "reasonable time"; typically, the fusing factors of fuse designs for general application are in the range 1.25–1.75, and the minimum breaking currents vary between two and a half and five times the fuse's rated current.

In practice, it is impossible to determine the current at which the fuse element will rupture in an "infinite" time and therefore the period used for measuring the minimum fusing current is set at 1.67 or 4 h, depending on which specification is used. The "reasonable" times used for the minimum breaking currents are between 10 and 100 s. The typical accuracy of these curves is within a tolerance of $\pm 10\%$. An alternative method for defining a fuse's operating characteristic is to use its let-through capacity (I^2t). This technique is also valuable for assessing the fuse's performance when the fault currents are very high, and the operating times can be less than half a power system frequency cycle.

The (I^2t) value indicates the energy that the fuse will allow to pass into the faulted circuit before it operates. As shown in Figure 3.3, a value is given for the pre-arcing time and for the current breaking time with a variety of source supply voltages.

One advantage of using the fuse's let-through capacity is that it enables different types of fuses to be graded together, since the (I^2t) capacity relates directly to its rupturing requirements.

3.3.2 Protecting Radial Feeder Circuits with Fuses

The operating characteristics of fuses are predictable within reasonable tolerance bands. This enables fuses to be graded such that for a supply network protected by fuses, when there is a fault, only the fuse closest to the fault will operate. Therefore, although other fuses will have carried the fault current, they will not rupture, and the supply to the rest of the network will not be broken. For most straightforward applications, the grading exercise is based on the fuses' rated currents. When a more rigorous analysis is required or when other protection devices are involved, the operating characteristics or let-through capacities are used.

Considering the radial network shown in Figure 3.4, this can be protected using fuses at the supply end of each feeder. The value of these fuses is chosen so that

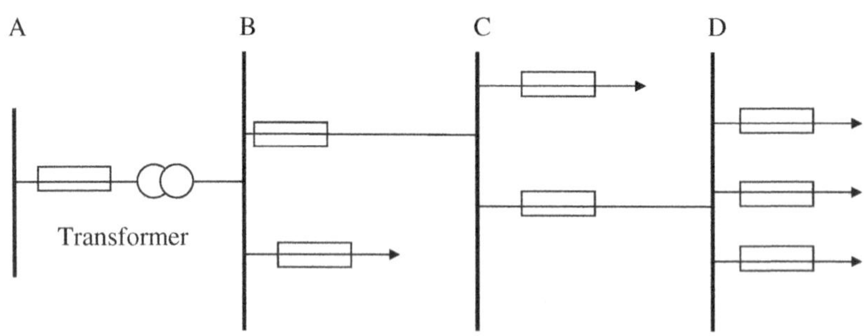

Figure 3.4 Radial distribution feeder system protected by fuses.

only the fuse closest to the fault will operate and hence isolate the faulted plant from the source supply. The value of the other fuses, which will also carry the fault current, is such that they will not operate before the fuse that is required to rupture. Supplies to other circuits fed via those fuses will therefore not be broken.

The rated current of the fuse closest to the load is chosen so that it is marginally greater than the maximum load current anticipated. Typically, this will be 1.1 or 1.2 times the maximum load current anticipated. The ratings of the fuses feeding this fuse from the supply the upstream fuses, are chosen such that they will not operate until the fuse closest to the fault has operated. The standard approach is to use a fuse. Naturally, this may be limited by the availability of a particular fuse rating.

Whenever there is a conflict between the 1.1 and 1.2 times, the maximum load carrying capacity of the fuse and its value being twice that of the next downstream fuse, the higher value is always chosen.

The fuse's let-through capacity also provides a basis for grading exercise. The general grading rule based on the let-through capacity is that the (I^2t) upstream fuse should be 40% higher than that of the next downstream fuse. This roughly equates to the upstream fuse having twice the current rating of the downstream fuse.

Exercise 3.1 The system shown in Figure 3.5 is protected by fuses. Suggest suitable values of fuses that should be installed to protect each of the feeders shown, given that the available fuse ratings are:

Low-voltage fuses: 35 A, 63 A, 125 A, 160 A, 250 A, 315 A, 500 A, 630 A, and 1000 A.

High-voltage fuses:

16 A, 25 A, 40 A, 50 A, 63 A, and 80 A

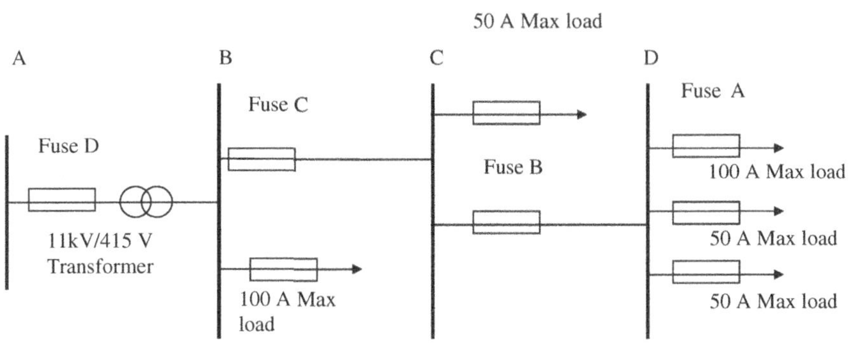

Figure 3.5 Distribution system protected by fuses.

Fuse A

The maximum load passing through this fuse is 100 A, therefore, the fuse should be rated at 1.1 to 1.2 times this value.

i.e.	110 A or 120 A
Available LV fuse rating	125 A
Therefore suggested fuse rating	125 A

Fuse B

The first approach is to determine the value of the fuse that would accommodate the load requirements, and the second is to determine the value of the fuse that would grade with the downstream fuse.

i) The maximum load passing through this fuse is 200 A, therefore, the fuse should be rated at 1.1 to 1.2 times this value.

i.e.	220 A or 240 A
Available LV fuse rating	250 A
Therefore suggested fuse rating	250 A

ii) The maximum value of a downstream this fuse is 125 A, **(fuse A)**; therefore for grading, **fuse B** should have a value of at least twice this.

i.e.	250 A
Available LV fuse rating	250 A
Therefore suggested fuse rating	250 A

Since both approaches give a suggested value of 250 A, this is the suggested value for this fuse.

Fuse C

i) The maximum load passing through this fuse is 250 A, therefore, the fuse should be rated at 1.1 to 1.2 times this value.

i.e.	275 A or 300 A
Available LV fuse rating	315 A
Therefore suggested fuse rating	315 A

ii) The maximum value of a downstream fuse is 250 A, **(fuse B)**; therefore, for grading, **fuse C** should have a value of at least twice this.

i.e.	500 A
Available LV fuse rating	500 A
Therefore, suggested fuse rating	500 A

Since the higher of these is a suggested value of 500 A, this is the suggested value for this fuse.

Fuse D: This fuse is on the high-voltage side of the 11 kV/415 V transformer, and its suggested rating is modified by the presence of that transformer.

At this stage, assume that the transformer is an ideal transformer. To determine the required fuse rating for an equivalent LV fuse and using an approach similar to that used for **fuse B** above:

- The maximum load passing through this fuse is 350 A, therefore, the fuse should be rated at 1.1 to 1.2 times this value, that is, 385 A or 420 A.
- The maximum value of a downstream fuse is 500 A **(fuse C)**; therefore, for grading, fuse (LV equivalent) should have a value of at least twice this, that is, 1000 A.

Therefore, the suggested rating for this fuse will be equivalent to 1000 A on the LV system. Referring this to the HV system, the equivalent fuse rating is:

$$100 \times \frac{415}{11000} = 37.73 \text{ A}$$

The next available fuse rating is 40 A.

Therefore, the suggested value for the HV fuse is 40 A.

3.3.3 Expulsion Fuse for Protecting Radial Feeder Circuits

The major limitation of the fuse for protecting radial feeders is its ability to rupture fault current and support the voltage required to provide the isolation between the supply and the faulted plant. This is a particular problem with highly inductive, high-voltage feeders. Where space allows, a simple solution to this problem is to use a special fuse link and housing that, when the fuse base element ruptures, allows the fuse carrier to fall away from the fuse base and therefore mechanically breaks the circuit. This system, the expulsion or drop-out fuse, is illustrated in Figure 3.6 and is widely used on overhead line feeders. After the operation, the fuse carrier has to be lifted from the housing, and the fuse element is replaced. The fuse carrier is then lifted back into position and clipped into the housing to complete the electrical circuit. During this operation, the supply side of the housing is invariably live, so a long-insulated pole is used to replace the fuse carrier.

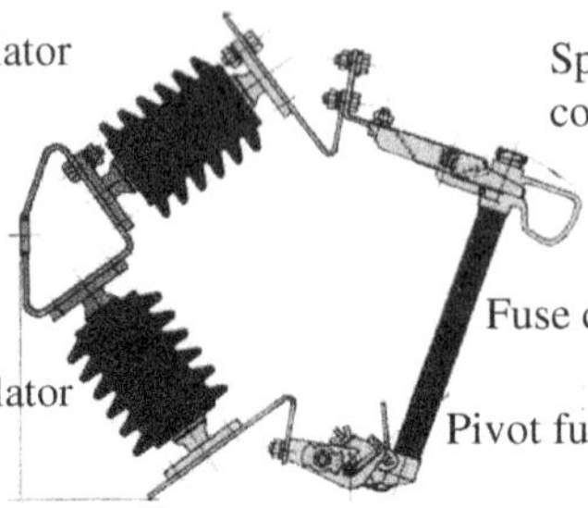

Figure 3.6 Expulsion fuses for protection of overhead feeders.

3.3.4 Circuit Breaker-Assisted Fuse Protection Schemes

The fuse's difficulty in supporting the voltage required to provide isolation between the supply and the faulted plant can also be circumvented by using a circuit breaker to break the fault current. In this scheme, the fuse is used to control the operation of the circuit breaker. The fuse monitors the high-voltage circuit's current by using a current transformer, and therefore, a low-voltage fuse can be used to protect a high-voltage circuit. This scheme is illustrated in Figure 3.7.

Since current transformers are used to provide the currents passing through the fuse, this scheme can be extended to provide increased sensitivity for earth faults. This is illustrated in Figure 3.8 and uses the summation current from the three-phase connected current transformer to provide the earth current through the earth fault fuse.

Although the fuse is no longer required to break the fault current, it is still a direct-acting technique, and the fault current provides the energy required to operate the circuit breaker. A station battery is not required.

This scheme can be graded with standard fuses using the fuse's rated current together with the ratio of the current transformer.

Exercise 3.2 Referring to the distribution system shown in Figure 3.5, suggest a suitable value for a low-voltage fuse to be used in place of fuse D if the incoming feeder were protected by breaker-assisted fuse scheme with a current transformer ratio of 80/5 A.

Available fuse ratings are:

0.5 A, 1 A, 2 A, 3 A, 5 A, 7 A, and 10 A

From the calculation in the first exercise, the required HV fuse rating is 37.73 A.

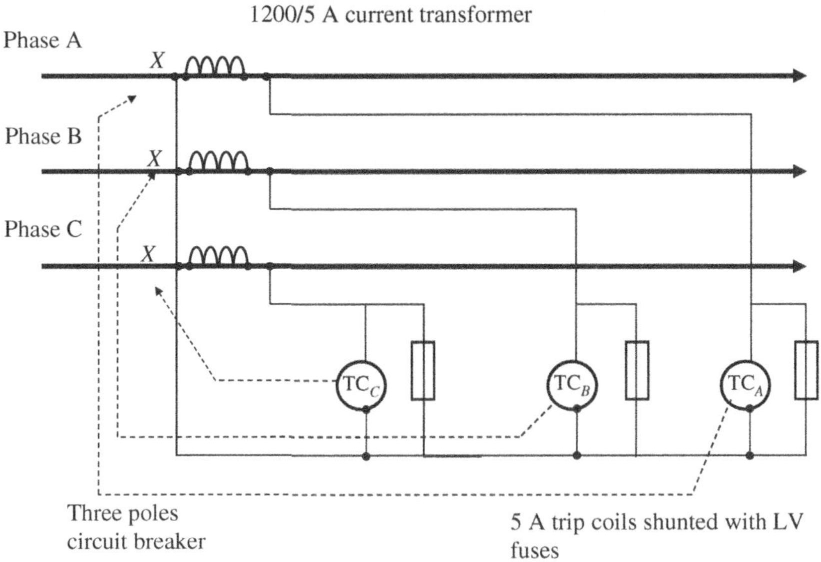

Figure 3.7 Fuse protection used with circuit breaker.

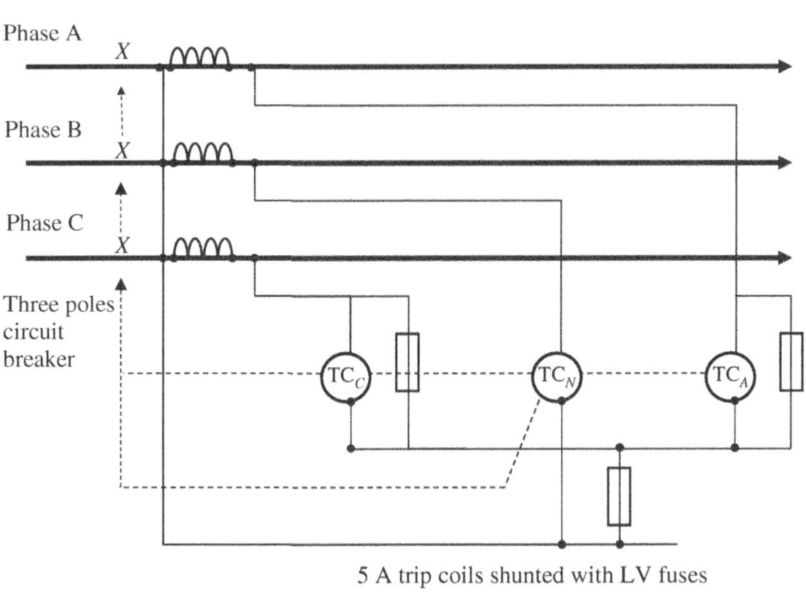

Figure 3.8 Fuse protection providing both phase and earth fault protection.

Using an 80/5 A current transformer, this is equivalent to:

$$37.73 \times \frac{5}{80} = 2.36 \text{ A}$$

On the secondary of the current transformer
The next available fuse rating from those available is 3 A.

3.4 Overcurrent Relays

The most frequently used type of protection for distribution feeders is the overcurrent. These relays provide nonunit, single-point relaying, and they monitor the measured power system current at a particular point in the network but cannot determine where the fault is on the network. Their settings are coordinated with other relays located at other points in the network so that, under fault conditions, only the relay that is required to trip actually trips.

The protection relay is a device designed to initiate the disconnection of a part of an electrical installation or to operate an alarm signal whenever a fault or other abnormal condition is detected. In common with all relays, a protection relay monitors one or more input signals and, when these satisfy predetermined conditions, the relay operates by closing a pair of output electrical contacts. When used for protection, closing of these output contacts causes the associated circuit breaker to trip and open the circuit.

Overcurrent relays monitor the secondary current obtained from a current transformer whose primary winding is connected in the high-voltage current-carrying circuit. Typically, they are rated at a nominal current of either 1 or 5 A. The current transformer transforms the primary current to a manageable level and also provides isolation between the high voltage of the power system and the low voltage of the relay and the protection equipment.

The basic types of overcurrent relays are:

- the instantaneous overcurrent relay,
- the definite time overcurrent relay, and
- the inverse characteristic overcurrent relay.

Although all three types are used for protecting distribution feeders together with a suitable combination, the inverse characteristic overcurrent relays are the most common. Since the tripping times of the inverse characteristics' relays become shorter as the fault current increases, they provide advantages over both instantaneous and definite time relays. The inverse characteristic relays are generally referred to as inverse definite minimum time overcurrent, or IDMT relays.

The current/time characteristic of IDMT relays enables them to be graded with other similar relays positioned at other points in the network so that they provide discrimination for faults on adjacent plant. An example of a grading exercise based on IDMT overcurrent relays is included in the book. This exercise considers a simplified power system because this is to be done by hand. For more involved networks, computer-based coordination systems are available. Instantaneous overcurrent relays are used in conjunction with IDMT relays and autoreclose schemes to protect feeder circuits. They are also used as high-set elements for a great variety of applications where their setting corresponds to an excessive fault current. Definite time-delayed overcurrent relays are used in conjunction with IDMT relays and autoreclose schemes for feeder protection, as well as being used alone for earth fault protection.

The first overcurrent relays were introduced in the early 1900s, and they have proved to be very effective and are now available in electromechanical, static, and digital designs. Because of their significance to the UK supply chain, **British standard BS 14** has established their current–time characteristic curves.

3.5 IDMT Overcurrent Relays

3.5.1 Operating Characteristics

The three principal characteristics defined by BS 142 are the STANDARD INVERSE, the VER INVERSE, and the EXTREMELY INVERSE characteristics as shown in Figure 3.17. These current–time curves, shown in Figure 3.1, were initially derived from the operating characteristics of electromechanical relays and are hence defined empirically. They are generally represented by plotting the operating time in seconds along the y-axis and input current measured as a multiple of the setting current along the x-axis. The principal area of interest is forming input currents of twice the setting current to 20 times the setting current relays; these curves have been characterized by equations. The equations used for one series of relays to define the British Standard specification are:

STANDARD INVERSE

$$T = \frac{0.14\,T}{\left[\frac{I_f}{\mathrm{PI}_N}\right]^{-0.02} - 1} \tag{3.1}$$

VERY INVERSE

$$T = \frac{13.5\,T}{\left[\frac{I_f}{\mathrm{PI}_N}\right] - 1} \tag{3.2}$$

EXTREMELY INVERSE

$$T = \frac{120\,T}{\left[\frac{I_f}{\mathrm{PI}_N}\right]^{-2} - 1} \tag{3.3}$$

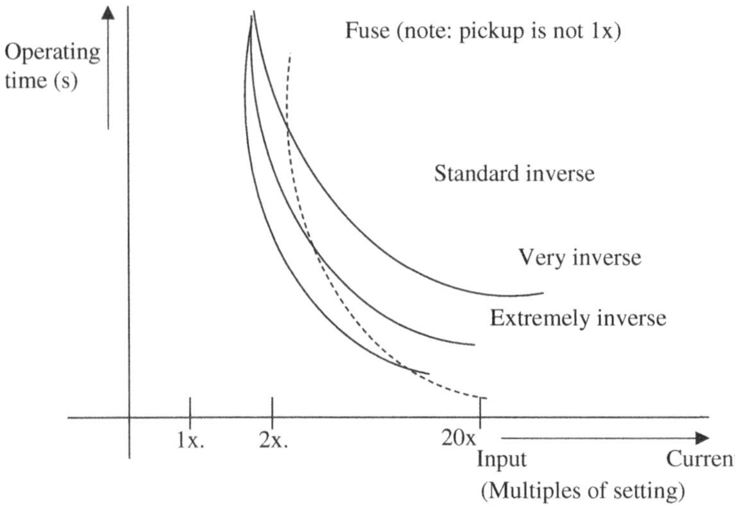

Figure 3.9 IDMT overcurrent relay operating characteristics.

where

- I_f is the fault current as seen by the relay.
- I_N is the nominal relay current.
- T is the time multiplier setting – the relay's time setting.
- P is the plug scale multiple – the relay's current setting.

Exercise 3.3 The operating characteristic of the protection system depends on the characteristics of the relay and its primary transducers. Assuming that the current transformer associated with an electromechanical overcurrent relay saturates at 10 times the relay current setting, sketch the resulting characteristics for the standard, very and extremely inverse IDMT relays. For simplicity, assume that the current transformer provides a linear output until it saturates and thereafter it provides a constant output current.

More ambitious students may wish to refer to the subsection on current transformer saturation in the following section noting that energy is required to rotate the disc.

The standard inverse, because of its tolerance to current transformer saturation, is the most widely used characteristic in the United Kingdom. The very inverse characteristic is an inverse linear characteristic, and in most feeder protection applications, it provides an operating time proportional to the distance to the fault position along the feeder. The extremely inverse characteristic, defined by

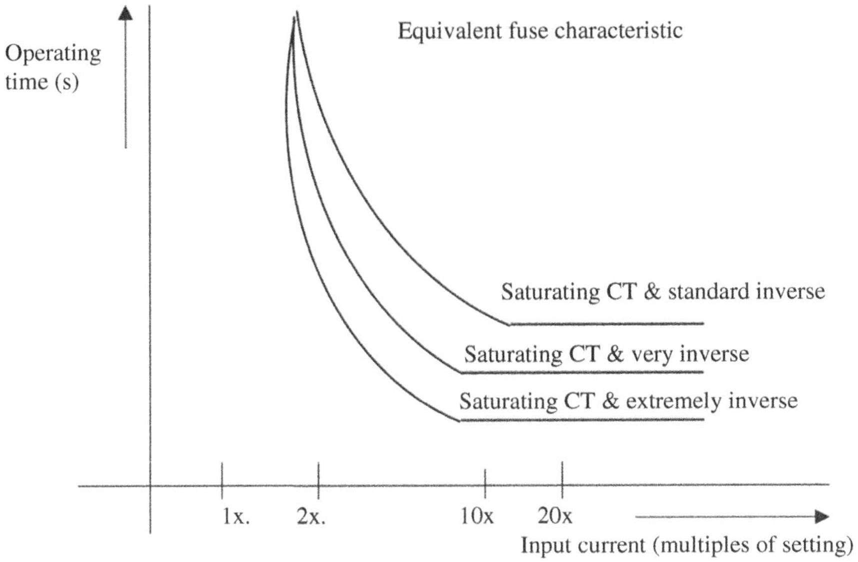

Figure 3.10 IDMT operating times due to primary CT saturation for Exercise 3.3.

an inverse square law, is similar to the characteristic of a fuse and is the most suitable for applications where part of the network is protected by fuses.

Other characteristics exist for specialist applications. These include overload protection and the protection of rectifier circuits.

Manufacturers need not comply with British Standard, and several leading international manufacturers supply their own preferred curves. These curves are similar to the British Standard curves. A wide range of IDMT characteristics is available, and where a system may contain relays from more than one manufacturer, system analysis is needed. With the introduction of digital overcurrent relays, customer-defined characteristic curves are also available.

3.5.2 Relay Types

3.5.2.1 Electromechanical Relays

The first IDMT overcurrent relays were developed from single-phase rotating disc watt-hour meters. Since, in the United Kingdom, the saturating standard inverse characteristic has always been the most widely used, these relays were often referred to as bad wattmeters (wattmeter should never saturate!). The majority of electrotechnical designs manufactured today were developed between 1940 and 1960. The main improvement that has been made to these designs has been to reduce costs and size, and to extend the setting range and to take advantage of modern methods of manufacture.

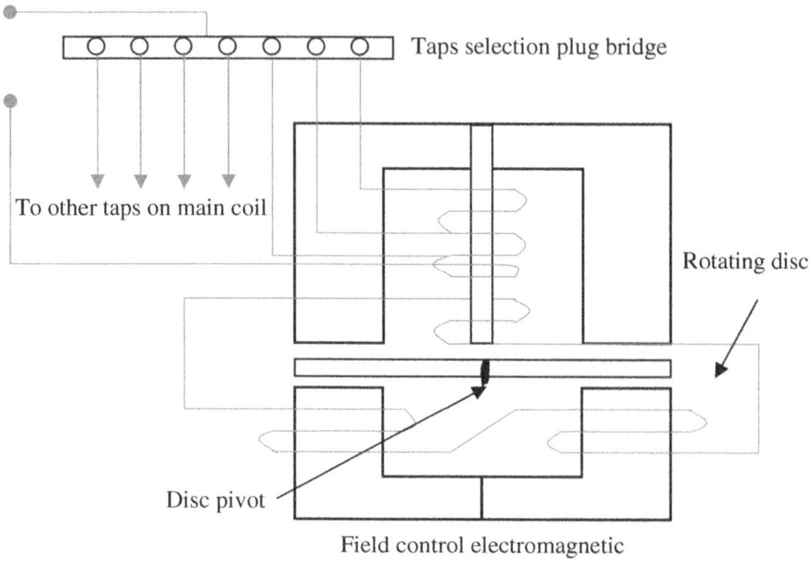

Figure 3.11 Principal components of an electromechanical IDMT overcurrent relay.

The principal components of an electrotechnical IDMT relay are shown in Figure 3.11 and the characteristic curve shaping is achieved through the magnetic characteristics of the electromagnet and the interaction of the magnetic field produced from the input current and the eddy currents in the disc. When the relay is deenergized, the disc is normally held against the backstop by the restraining spring. The input current generates a force to counter this, and when the current exceeds the pickup setting (i.e. exceeds the strength of the spring), the disc rotates. A moving contact is attached to the disc and rotates with it. This moving contact reaches a fixed contact and completes an electrical circuit, thus tripping the relay. The operating time corresponds to the time required for the disc to rotate from the backstop and reach the fixed contact, thus completing the electrical circuit.

The relay's two settings are the current setting, controlled by varying the number of turns on the electromagnet, and the time setting, controlled by adjusting the position of the backstop for the rotating disk. The current setting is referred to as the plug scale (or setting) multiplier, PSM, and the time setting is the time multiplier setting, TMS.

Following the British Standard definition, the relay "starting current," the input current that will cause the disk to move off the backstop, is set between 3 and 5% above the relay setting current, that is, nominal current times PSM. Its "closing current" that will result in the relay contacts closing is set between 15 and 30% above its setting current. If the input current falls below 90% of its setting current, the relay will reset, and the disk will return to the backstop under the force of the restraining spring.

Since the power required to rotate the disk is derived from the input current, these relays have a comparatively high burden. With modern designs, this can vary between 1 and 3 VA at the nominal current. As the input current increases, the burden also increases.

Exercise 3.4 Assuming that an overcurrent relay has a burden of 1 VA at its setting current, and that its input impedance is constant, calculate the burden at 10 times its setting current.

Letting

$$I_s = \text{setting current}$$

$$Z_s = \text{input imedance at setting current}$$

Then,

$$I_s^2 \times Z_s = 1\text{ VA} \tag{3.4}$$

At the 10 times setting current,

$$I_{\text{IN}} = 10 \times I_s$$

Z_s remain constant

$$\text{Burden} = I_N^2 \times Z_s = (10\,I_s)^2 \times Z_s = 100\text{ VA}$$

This relay burden of 100 VA at 10 times the setting current places heavy demands not only on the primary current transformer but also on the relay's input circuit. The relay must be able to withstand these input currents and hence power level for the maximum period associated with fault clearance, together with possible autoreclose on to permanent faults.

3.5.2.2 Static (Analog) Relays

The first static relays, using an analog electronic circuit, were introduced in the early 1970s. These designs used transistor circuits but were later replaced by operational amplifiers. Compared to electrotechnical relays, they offered the advantages of low burden, improved curve repeatability, reduced frequency errors, reduced temperature errors, and could be housed in smaller cases. The principal disadvantages of these electronic relays were that they were invariably more expensive, they required a power source, generally obtained from a d.c. station battery, and early design gained a reputation for poor reliability.

A simplified block diagram for a static overcurrent relay is shown in Figure 3.12. Input isolation is achieved using a current transformer (CT), which, together with a resistance across its secondary, provides the curve-shaping circuit with an

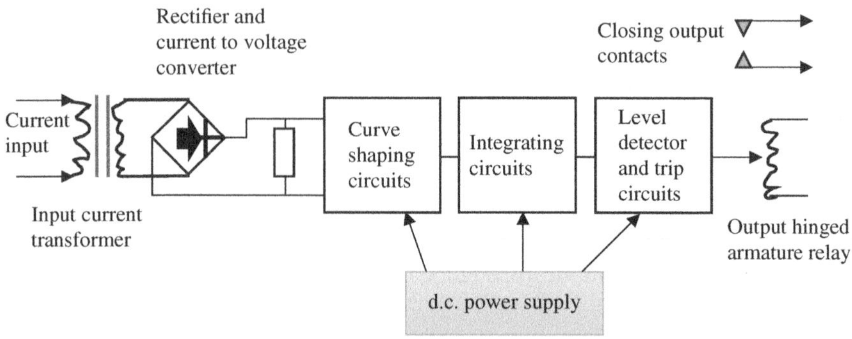

Figure 3.12 Block diagram for a static IDMT overcurrent relay.

input voltage proportional to the input current. The current setting, PSM, can be adjusted either by adjusting this CT turn ratio or by varying the value of the loading resistor. The time setting, TMS, is an internal setting in the electronic decision-making circuit.

The relay output generally uses a hinged armature electromechanical relay, which provides isolated output contacts. The harsh electrical environment of the power system substation is hazardous for the well-being of relay electronics, and hence, this method of isolation is very attractive. Solid-state, thyristor-switched outputs are available from some suppliers, but care is required to avoid malfunctioning because of the electrical environment.

In general, although static overcurrent relays have enjoyed some success in the industrial market, they have not been successful in the utility market except for certain specialist applications that are difficult to provide using electromechanical technology.

3.5.2.3 Digital Relays

Commercial digital overcurrent relays were first introduced in the late 1970s and early 1980s. The early designs used dedicated digital electronics and Erasable Programmable Read-Only Memory (EPROMs) to hold the curve-shaping data. These were soon superseded by microprocessor-based relays. The availability of single-chip microcomputers provided designers with few components and therefore a high tolerance to the harsh electrical environment of the power system.

A typical block diagram of a microprocessor overcurrent relay is shown in Figure 3.13. As with the static design, a current transformer provides input isolation, and a hinged armature relay provides the output isolation. In this design, the current setting is provided by switching the value of the CT loading resistance, whereas the time setting uses digital inputs into the microprocessor.

An immediate advantage of digital technology was the opportunity to provide switch-selectable characteristic curves. This was a particular advantage for the

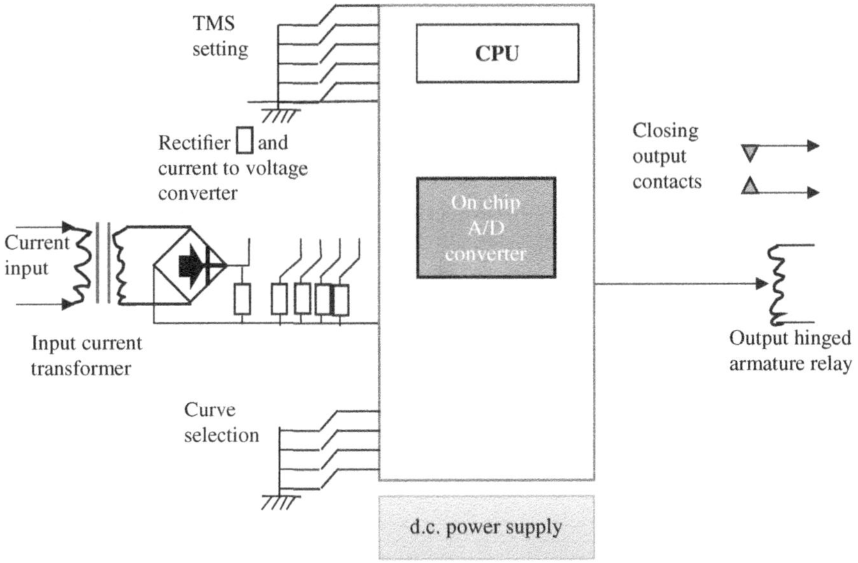

Figure 3.13 Typical block diagram of a microprocess IDMT overcurrent relay.

manufacturer since one design could contain all the standard curves and hence replace a range of electromechanical or static equivalents. Improved manufacturing techniques also reduced the production costs of these designs.

Low-input burden enabled relays to have wide current setting ranges. These can vary between **20:1, 24:1, and 32:1** depending on relay type. Thus, again a single digital design could cover a wide range of electromechanical variants.

Opportunities to reduce the component count and the physical size of the relays eased environmental problems despite the increased sensitivity of the microelectronics to interference.

Although these relays, as with static relays, have the disadvantage of requiring a power supply, systems have been developed that can provide power from the protected circuit's load and fault current. These units, when used with the overcurrent relays, effectively provide self-powered relays. More recently, the use of digital designs has enabled complex multi-characteristic curves to be provided and for the overcurrent relaying functions to be complemented by other protection functions and/or control functions, such as an autoreclose scheme.

3.5.3 Overcurrent Relay Setting

When protecting any power system, the operations of the relays should be such that they operate for faults on the plant they are protecting and not for faults on another plant, however, close. This discrimination is the principal requirement of any protection system. Relay operations for faults elsewhere on the network and subsequent circuit breaker opening will disconnect the healthy plant from the network.

Naturally, should the protection system for one unit of plant fail for any reason, further protection should be included to isolate that "faulted plant" from the network. So, there is a need for backup protection and the operation of circuit breakers disconnecting a larger part of the system. The protection system for a unit of plant is an integral part of that protected plant; hence, if the protection system fails, then the unit is considered faulty.

The second priority of the protection scheme is the need to break the flow of fault current as quickly as is prudent. The protection system should require the minimum operating time for high fault currents while operating satisfactorily for minimum fault currents. Overload protection is generally covered by other schemes and is not, therefore, the concern of the main protection system.

3.5.3.1 Protecting Feeder Circuits with IDMT Overcurrent Relays

In order to protect a network of feeder circuits with IDMT relays located in each substation, their settings are chosen so that they correctly discriminate between each other, such that only the relay required to trip will do so. The simplified radial network shown in Figure 3.14 will be protected by relays located at the input for each feeder at locations A, B, C, and D. The analysis of the protection system is referred to as a grading exercise. Grading exercises are applied to all types of network configurations; however, this radial network of plain feeders provides a useful introduction to the procedures. For an effective power system analysis, a one-line diagram of the power system network is essential, together with source capacities and plant impedances, which are generally expressed in MVA and symmetrical component impedances, respectively. If the system includes motors, and

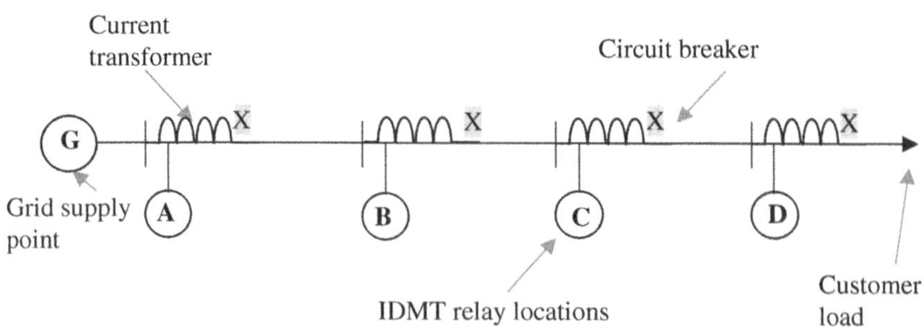

Figure 3.14 A radial transmission network.

or small or medium-sized generator, their starting currents and decrement curves under fault conditions will also be required.

To evaluate the effectiveness of the protection systems and their settings, details of the initial choices of relays will be required, together with the position and ratios of the primary current transformer's excitation curves, their secondary winding impedance, cabling impedances, and details of other loads.

When protecting a network with IDMT relays, it is advisable that the relays used should be of the same type and with similar operating characteristics. This simplifies the matching of curve shapes and ensures that variation in power system frequency and ambient temperatures will introduce minimum complications. At each substation, different relays are used to protect the feeder against phase faults and earth faults. The settings of the phase fault relays are based on three-phase faults and hence consider the system voltages and positive phase sequence impedances. Earth fault relay settings are based on single-phase-to-ground faults and, hence, require details of the positive, negative, and zero phases' sequence impedance components.

3.5.3.2 The Relay Current Settings

The first part of the analysis is to select the relay current settings. These are chosen so that the pickup current is well above the maximum load current expected in the circuit while at the same time, it is well below the minimum fault current for which the relay may be required to operate. Invariably, this will lead to the relays furthest from the grid supply point and closest to the ultimate load having the most sensitive settings.

As a general rule, the current setting of the relay at a point A in the power system I_{RA} is chosen such that:

$$I_{RA} > 2(I_{load})_{max}$$

and

$$I_{RA} < \frac{1}{2}(I_{fault})_{min}$$

where $(I_{load})_{max}$ is the maximum load current that the relay will experience and $(I_{fault})_{min}$ is the minimum fault current that it will encounter. The actual choice of the relay setting is a compromise between these limits. Occasionally, these limits conflict, and the choice of settings has to be examined in more detail.

Exercise 3.5 Considering the 11 kV overhead feeder network shown in Figure 3.15. Choose the current setting for IDMT phase fault relays at locations A, B, C, and D, noting that the source capacity can vary between 75 and 150 MVA and the line impedances are as shown in the diagram. Assume that the relays have a nominal rating of 5 A and PSM settings of 50, 75, 100, 125, 150, 175, and 200% are available.

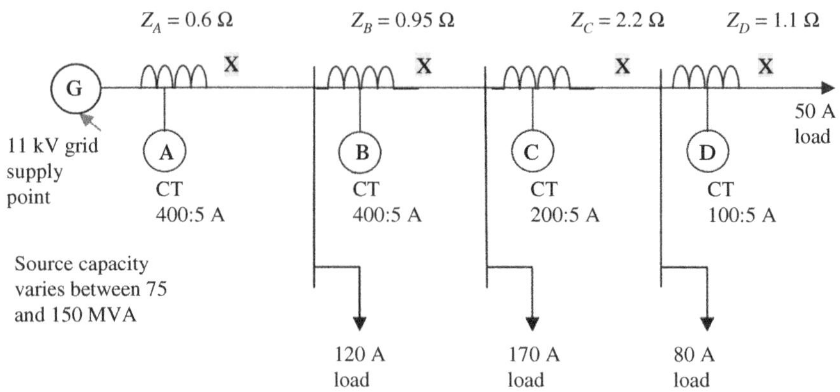

Figure 3.15 A radial feeder network to be analyzed.

To choose the current settings for the relays at locations A, B, C, and D in Figure 3.15, you must first calculate the load and fault currents encountered by the relays. The relay setting is then chosen to be a value that can be set on the relay, which satisfies the relationship:

$$I_{\text{SET}} > 2(I_{\text{LOAD}})_{\text{MAX}}$$

and

$$I_{\text{SET}} < \frac{1}{2}(I_{\text{FAULT}})_{\text{MIN}}$$

$(I_{\text{FAULT}})_{\text{MIN}}$ being the lowest fault current for which the relay must operate.

Since these relays are phase fault relays, their settings are chosen using the fault currents that will flow for balanced three-phase faults.

Considering the load currents:

$$(I_{\text{LOAD}})_D = 50\text{ A}$$

$$\begin{aligned}(I_{\text{LOAD}})_C &= 80 + (I_{\text{LOAD}})_D\text{ A}\\ &= 80 + 50\text{ A} = 130\text{ A}\end{aligned}$$

$$\begin{aligned}(I_{\text{LOAD}})_B &= 170 + (I_{\text{LOAD}})_C\text{ A}\\ &= 170 + 130\text{ A} = 300\text{ A}\end{aligned}$$

$$\begin{aligned}(I_{\text{LOAD}})_A &= 120 + (I_{\text{LOAD}})_B\text{ A}\\ &= 120 + 300\text{ A} = 420\text{ A}\end{aligned}$$

The analysis of the fault current requires the circuit impedances to be calculated: instead of defining the source as an ideal voltage with a source impedance, it has been defined by its voltage, 11 kV, and a range of source capacities, which in this case can vary from 75 to 150 MVA. This can be converted into source impedance using:

$$Z_{\text{eq}} = \frac{V^2}{M}\ \Omega \tag{3.5}$$

where
Z_{eq} is the equivalent source impedance (Ω);
V is the source voltage (kV);
M is the source capacity (MVA).
The 75 MVA source can be expressed as:

$$(Z_s)_{75} = \frac{V^2}{M} = \frac{11 \times 11}{75} 1.6133\ \Omega$$

$$(Z_s)_{75} = 1.61\ \Omega.$$

And the 150 MVA source can be expressed as:

$$(Z_s)_{150} = \frac{V^2}{M} = \frac{11 \times 11}{150} 0.8066\ \Omega$$

$$(Z_s)_{150} = 0.81\ \Omega$$

Returning to the fault currents, the greatest fault current that will be encountered by the relay at location A will be for a fault at the relaying position when the source is at its strongest, that is, 150 MVA. The circuit impedance will only consist of the source impedance $(Z_s)_{150}$

$$\therefore (I_{\text{FAULT}})_{\text{MAX}} = \frac{\frac{V}{\sqrt{3}}}{Z}$$

$$(I_{\text{FAULT}})_{\text{MAX}} = \frac{\frac{11000}{\sqrt{3}}}{0.81}\ \text{A}$$

$$(I_{\text{FAULT}})_{\text{MAX}} = 7840\ \text{A}$$

The minimum fault current that the relay at location A will encounter for a fault on its protected feeder corresponds to a fault at the far end of feeder A when the source capacity is at its weakest, that is, 75 MVA. This will correspond to the maximum circuit impedance for an in-zone fault, $Z_{\text{MAX},}$ where

$$Z_{\text{MAX}} = (Z_s)_{75} + Z_A$$

$$= 1.61 + 0.6\ \Omega = 2.21\ \Omega$$

The fault current will be derived from:

$$(I_{\text{FAULT}})_{\text{MIN}} = \frac{\frac{V}{\sqrt{3}}}{Z_{\text{MAX}}} \tag{3.6}$$

Table 3.1 Different location results.

Relay location	Minimum in-zone fault impedance	Maximum in-zone fault impedance	Maximum in-zone fault current	Minimum in-zone fault current	Maximum load current	CT ratio
A	0.81	2.21	7840	2874	420	400/5
B	1.41	3.16	4504	2010	300	400/5
C	2.36	5.36	2691	1185	130	200/5
D	4.56	6.46	1393	983	50	100/5

$$(I_{\text{FAULT}})_{\text{MIN}} = \frac{\frac{11{,}000}{\sqrt{3}}}{2.21}\text{ A} = 2874\text{ A}$$

Similar exercises can be undertaken for the relays at locations B, C, and D, and the results are tabulated as shown in Table 3.1.

For the relay at location A, its setting should be at least twice the maximum load current that it will encounter; therefore, its setting I_{RA}

$$I_{\text{RA}} > 2(I_{\text{LOAD}})_{\text{MAX}\,A}$$

$$I_{\text{RA}} > 2 \times 420\text{ A}$$

$$I_{\text{RA}} > 840\text{ A}$$

Its setting must also be at most half of the minimum fault current for which it must operate. The relay at location A must operate for all in-zone faults; therefore, its setting I_{RA}

$$I_{\text{RA}} < \frac{1}{2}(I_{\text{FAULT}})_{\text{Min}\,A}$$

$$I_{\text{RA}} < \frac{1}{2}(2874)\text{ A}$$

$$I_{\text{RA}} < 1437\text{ A}$$

Therefore, to operate correctly for all faults on feeder A,

$$840 > I_{\text{RA}} > 1437\text{ A}$$

Since this is a generous range, the opportunity for providing backup protection for feeder B can also be considered. For this, the relay at location A must operate for all in-zone faults for feeders A and B and its setting I_{RA} must satisfy

$$I_{\text{RA}} < \frac{1}{2}(I_{\text{FAULT}})_{\text{Min}\,B}$$

$$I_{RA} < \frac{1}{2}(2010)\text{ A}$$

$$I_{RA} < 1005\text{ A}$$

Therefore, to operate correctly for all faults on feeder A and to provide backup for faults on feeder B

$$840 > I_{RA} > 1005\text{ A}$$

The possible settings on the relay are provided by the nominal rated current, its PSM setting, and the current transformer ratio.

$$\therefore I_{SET} = I_N \times \text{PSM} \times \text{CT}$$

$$I_{SET} = 5 \times \text{PSM} \times \frac{500}{5}\text{ A}$$

Using a PSM of 150%

$$(I_{SET})_{150\%} = 5 \times \frac{150}{100} \times \frac{500}{5}\text{ A}$$

$$(I_{SET})_{150\%} = 875\text{ A}$$

Using a PSM of 200%

$$(I_{SET})_{200\%} = 5 \times \frac{200}{100} \times \frac{500}{5}\text{ A}$$

$$(I_{SET})_{200\%} = 1000\text{ A}$$

The preferred setting to satisfy the range of acceptable settings and the settings available is therefore

$$(I_{SET})_{175\%} = 875\text{ A}$$

This is expressed in primary current and can also be expressed in relay current, where

$$(I_{SET})_{175\%} = PSM\, I_N$$

$$(I_{SET})_{175\%} = \frac{175}{100} \times 5\text{ A}$$

$$(I_{SET})_{175\%} = 8.75\text{ A}$$

Similar exercises can be undertaken for the relays at locations B and C, and the results are tabulated as shown in Table 3.2.

The choice of the settings for the relays at A and B illustrates compromises that are often required. The lack of information about the ultimate load system leads to the choice of the lowest current setting for relay D, which accommodates the maximum load expected.

Table 3.2 Different location results.

Relay location	Maximum relay setting (A) inc. backup	Minimum relay setting (A) load	Chosen relay setting (A)	CT Ratio	PSM setting (% PSM)
A	1001.5	820	800	500:5	200
B	591	600	600	400:5	150
C	490.5	260	300	200:5	150
D	*	100	100	100:5	100

3.5.3.3 Relay Time Settings

The relay time settings are chosen to ensure that they operate as quickly as possible with the proviso that the relay closest to the fault, and which sees the fault current, opens its circuit breaker before any other relays have time to trip. It is not sufficient that the correct relay trips before the others in the scheme. To prevent possible malfunctioning, fault clearance times, relay inertia, and relay, transducer, and system errors all must be considered.

To ensure that only the correct relay trips, the minimum time separation required between adjacent relays for the largest fault current that they are likely to experience is set to the grading margin. This is the total time required for the correct circuit breaker to clear the fault current, for overshoot in the relay, which is required not to trip, to accommodate relay and transducer errors, and to provide a safety margin to cover other effects. A breakdown of the grading margin is shown in Figure 3.16.

The time allowed for the circuit breaker to clear the fault depends on the type of breaker used. Modern vacuum and SF6 breakers can clear the fault current under three cycles (60 ms); other types of modern circuit breakers require a maximum time of about five cycles (100 ms), whereas older switchgear could require up to seven cycles (140 ms).

Relay overshoot is caused by the relay inertia preventing it from stopping rotation immediately; the fault current is removed, and applies to both electromechanical and electronic relays, the typical overshoot allowance for electromechanical relays is with a time of 50 ms, and for the electronic relay is 30 ms.

Relay errors depend on the class of relay used. These are defined by specification **BS-142** and international standard **IEC 255-4**. Typically, the majority of electromechanical relays have a timing error specification of ±7.5% and electronic relays have a specification of ±5.0 %. Transducer errors are treated in a manner similar to relay errors and depend on the class of transducers used. Typical current transformers used for IDMT relays are assumed to have an error equivalent to a relay timing error of ±10.0%. In addition to these, a safety margin is added to ensure that

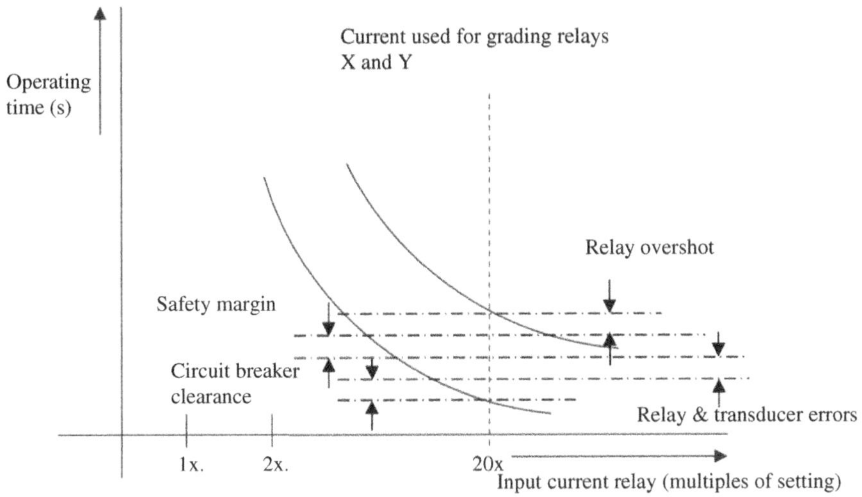

Figure 3.16 Timings used to form the grading margin.

any unforeseen complications do not result in malfunctioning. In general, 100 ms provides a suitable margin.

The equation for the grading margin, t_G, is:

$$t_G = \frac{(2E_R + E_{\text{CT}})}{100} t_R + t_{\text{CB}} + t_O + t_{\text{SM}} \tag{3.7}$$

where

E_R is the relay error (in%);

E_{CT} is the current transformer error (in%);

t_R is the relay operating time;

t_{CB} is the circuit breaker fault clearance time;

t_O is the relay overshoot time;

t_{SM} is the safety margin.

For most applications, this is too involved, and a grading margin of 400 ms is used, consisting of 150 ms for the circuit braker clearance, 50 ms of that relay overshoot, 100 ms for errors, and 100 ms safety margin.

Exercise 3.6 Determine the relay TMS settings for Figure 3.7 considered above given that the relay at D has a TMS setting of 0.1. Given also that the relay has a very inverse characteristics, that the TMS settings range from 0.1 to 1.0 in steps of 0.1, and that a grading margin of 0.4 s should be used.

The relay characteristic curve is given by:

$$t = \frac{13.5\,T}{\left[\frac{I_f}{\text{PI}_N}\right] - 1} \tag{3.8}$$

Table 3.3 TMS and PMS settings.

	D	C	B	A
PSM setting (A)	150	300	600	875
Max. fault current for downstream relay		1393	2691	4504
Trip time for downstream relay		0.163	0.339	0.415
Required trip time for relay, including grading margin		0.563	0.739	0.815
Calculated TMS		0.152	0.191	0.25
Chosen TMS	0.1	0.2	0.2	0.3

The first grading exercise considers the response of the relays at C and D to the largest fault current that both relays will experience. This corresponds to a fault close to the relaying point on the feeder taken from D. From Table 3.1, the maximum fault current seen by relay D is 1393 A. Using its PSM setting of 150 A and TMS setting of 0.1, the operating time can be derived from the characteristic equation:

$$t_D = 0.163\ \text{s}$$

For this fault, the operating time of relay C, t_C, should be the operating time of relay D plus the grading margin, that is,

$$t_C = 0.163 + 0.4 = 0.563\ \text{s}$$

The current setting for relay C has been chosen to be 300 A, therefore, from the characteristic equation, the TMS setting, T, can be found, that is,

$$T = 0.152$$

This setting is not available, and the relay must be set to 0.2. The settings for relay C are

$$\text{PSM}_C = 150\% \text{ and } \text{TMS}_C = 0.2$$

The procedure is tabulated for ease of presentation.

The chosen PSM and TMS settings define the relay characteristic curves, which can be presented as time–current curves as shown in Figure 3.9.

3.6 Instantaneous Overcurrent Relays

Instantaneous overcurrent relays provide high-speed tripping whenever the fault current exceeds the pickup current and are most frequently used as high-set units for detecting excessive fault currents. They have limited discrimination capabilities and are generally used in conjunction with other types of protection.

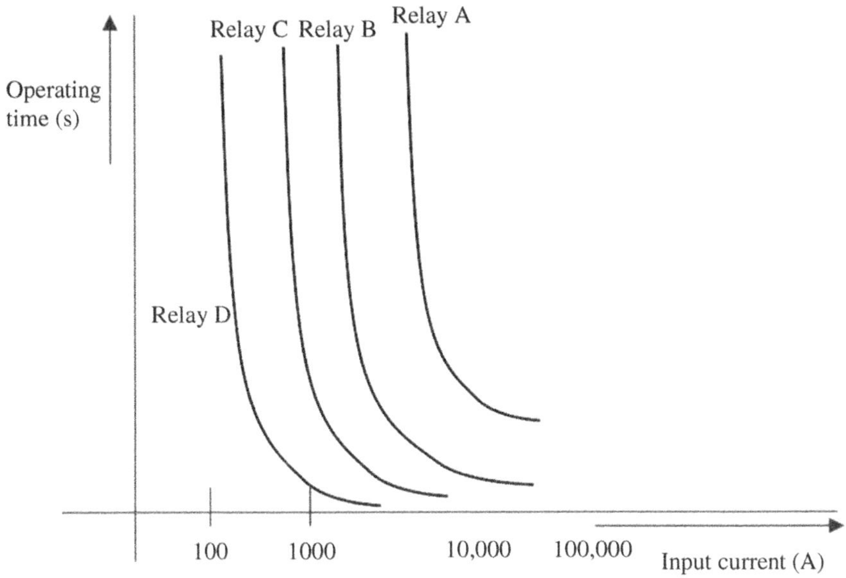

Figure 3.17 Characteristic curves for IDMT relays to protect radial feeder in Figure 3.6.

For feeder circuits, they can improve discrimination by providing high-speed response for close-in faults or where the protected system impedance is greater than the source impedance, such as with long feeders or circuits containing transformers.

Care must be taken with the settings of instantaneous relays, and consideration is given to their overreach under transient conditions. General practice is to see that the instantaneous relays are 20 or 30% higher than the largest fault current that would be produced by a fault at the far end of the protected feeder. This provides a margin to accommodate transducer errors, impedance errors, and transient overreach problems.

3.7 Conclusion

Overcurrent relaying provides a comparatively simple nonunit protection for feeder circuits and hence is the most widely used protection for electricity supply systems. The main disadvantages are that discrimination is difficult with instantaneous relays, and that IDMT relays, which overcome this problem, do so at the expense of tripping times. This is particularly undesirable for faults close to the point of generation or bulk supply.

As power supply networks become more interconnected, other difficulties arise in the use of overcurrent relaying, and other types of protection have to be used.

3.8 Self-Assessment Questions

Self-Assessment Question 3.1

Determine the optimum settings for the very inverse characteristic IDMT relays positioned at A, B, C, and D used to protect the feeder network shown in Figure 3.10, given the power system data shown in the diagram and that the relays have very inverse characteristic, are nominally rated at 5 A, and have PSM settings of 50, 75, 100, 125, 150, 175, and 200% and TMS settings from 0.1. Use a grading margin of 0.4 s.

Figure 3.18 Feeder network model for SAQ 3.1.

Overcurrent relays are commonly used for earth fault protection. Since in a three-phase circuit the earth current can rarely be measured directly, the summation of the three-phase current is used to provide the residual current, and this provides the input to the relay. The typical current transformer and relay connections are shown in Figure 3.19. Figure 3.19a shows the full scheme configuration with three separate phase fault relays and an earth fault relay. Since for any fault combination, two relays respond to the fault, this scheme can be reduced to that shown in Figure 3.19b, where the B phase relay is omitted and the scheme contains the A phase, C phase, and earth fault relays. Where only earth fault protection is required, the three current transformers are connected in parallel, as shown in Figure 3.19c.

The settings of the overcurrent relays are determined like those for the phase fault relays. Naturally, since the relays do not normally experience load current, their minimum settings are chosen such that the relays are tolerant to any normal imbalance in the three-phase system and mismatch in the current transformers.

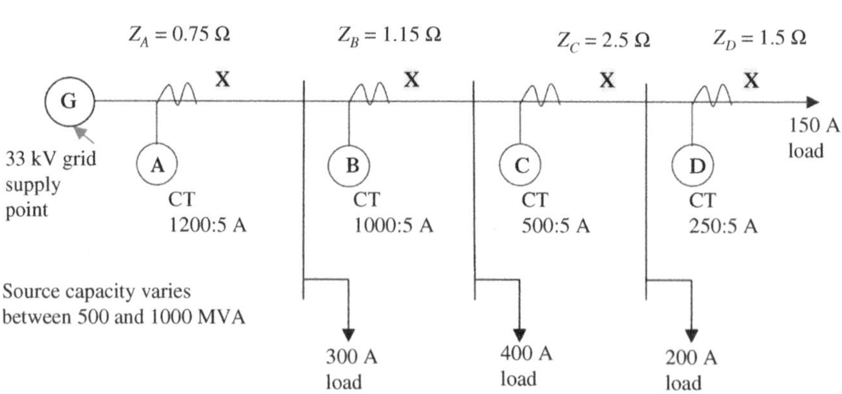

Figure 3.18 Feeder network model for SAQ 3.1.

Figure 3.19 Earth fault protection using overcurrent relays. (a) Full scheme configuration, (b) phase fault relay configuration, and (c) earth fault relay only configuration.

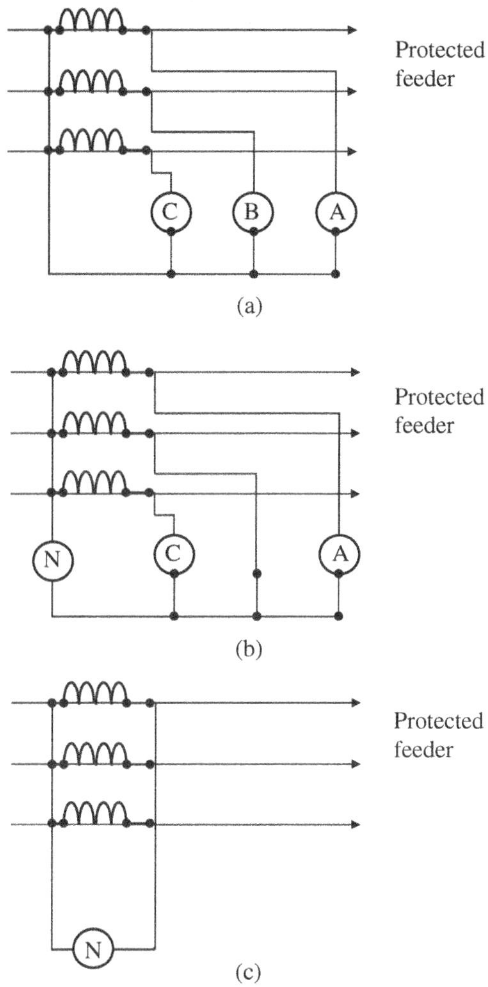

The use of delta/star transformers in the power supply system effectively breaks the power system into sections and constrains the flow of earth fault currents. They therefore provide boundaries at which the grading exercise starts again, and the relay close to the delta winding can be set to have a low time setting.

3.8.1 Protecting Feeder Circuits Using IDMT Overcurrent Relays with Instantaneous Elements

Instantaneous overcurrent relays provide high-speed tripping whenever the fault current exceeds the pickup current and are mostly frequently used as high-set units for detecting excessive fault current. They have limited discrimination capabilities and are generally used to complement other types of protection. For feeder circuit protection, they can improve the performance of an IDMT protection scheme by providing high-speed response for close-in faults or where the protected system impedance is large in comparison to theta of the protected system, such as with long feeders or circuits containing transformers.

Care must be taken with the settings of instantaneous relays, and consideration is given to their overreach under transient conditions. General practice is to set the instantaneous relays to 20 or 30% higher than the largest fault current that would be produced by a fault at the far end of the protected feeder. This provides a margin to accommodate transducer errors, impedance errors, and transient overreach problems. Transient overreach is caused by exponentially decaying d.c. components in the current waveform, and it is generated by the fault. It is defined as:

$$\text{Transient overreach} = \frac{I_1 - I_2}{I_2} \tag{3.9}$$

where

I_1 is the relay pickup current with steady-state **r.m.s.** current;

I_2 is therelay pickup current with a fully offset current waveform measured as steady state amperes.

A typical value for the transient overreach of a relay designed for low overreach is between 1 and 5%. For simpler relays, the transient overreach can be as high as 30%

Although described as instantaneous elements, instantaneous overcurrent relays require time to operate. Their operating times vary from a quarter to half a power system cycle for the simpler relays, to between three-quarters and two cycles for those relays designed to have a low transient overreach. The immediate advantage of using instantaneous overcurrent elements with IDMT relays is reduced tripping times for faults close to a relaying point. This is illustrated in Figure 3.20, which compares the operating curves for a protection scheme using IDMT relays and those for a scheme using both IDMT and instantaneous overcurrent elements. A further advantage of using instantaneous overcurrent elements with IDMT relays is that the TMS settings of the upstream relays can be reduced since the grading exercise is undertaken between the setting of the instantaneous element and not at the maximum current seen by the downstream IDMT relay. This improvement in operating times is illustrated in Figure 3.21.

3.8.2 Protecting Feeder Circuits Using Directional Overcurrent Relays

All of the above have considered radially fed feeder circuits where there is a single point of generation, and the currents always flow in one direction. Multiple generations, or the interconnection of feeder circuits, often lead to the situation

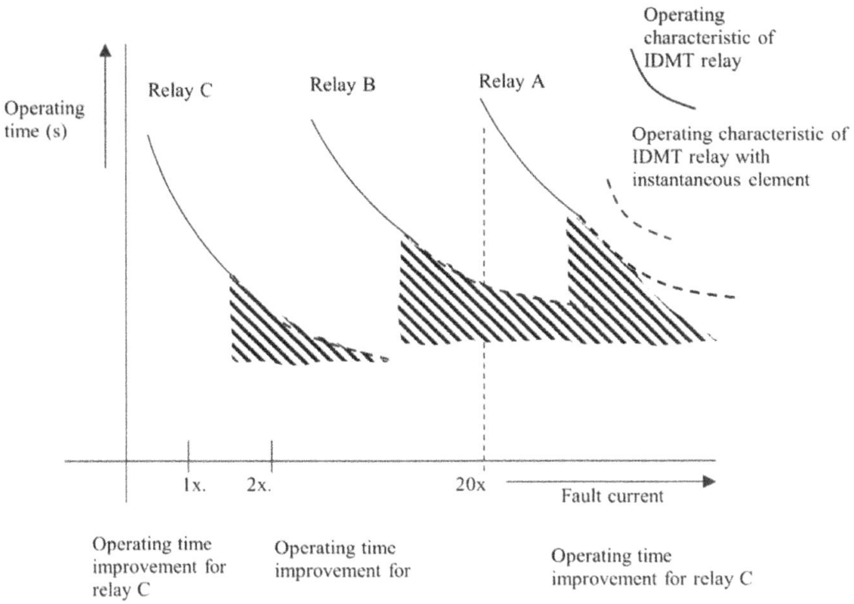

Figure 3.20 Improved protection performance provided by adding instantaneous elements.

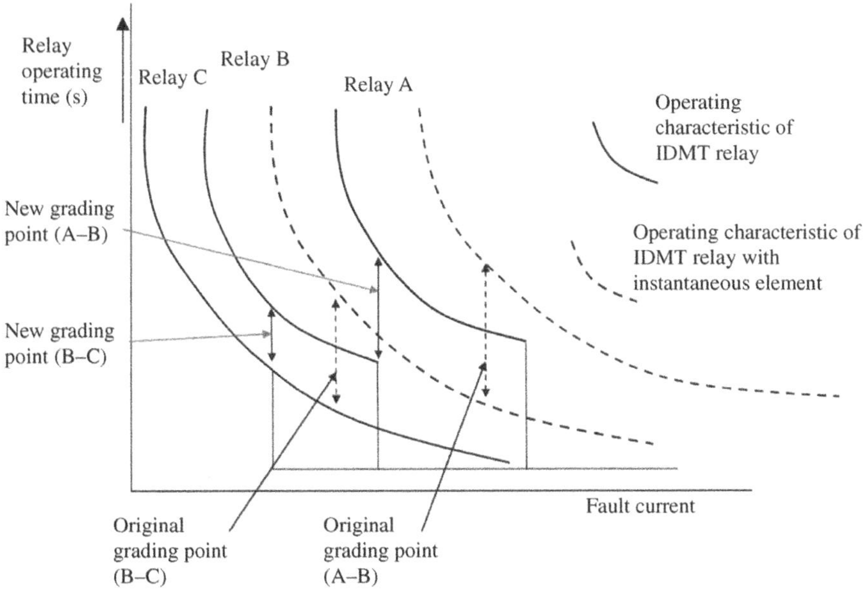

Figure 3.21 Further improvement to protection performance by grading.

where fault currents can flow in either direction through the relaying point and hence the relay. In themselves, overcurrent relays are nondirectional and hence they cannot differentiate between faults on different sides of their relaying point. This compromises the grading exercise used to determine relay settings and can lead to protection maloperations. Therefore, there is a need to use a directional element in combination with overcurrent relays, which allows the relay to operate when the fault current is flowing in one direction but inhibits it when the current flows in the reverse direction.

The directional element is generally a single-phase power direction relay or a cosØ relay that uses voltage and current input signals to determine the direction of power flow during the fault. A typical electrotechnical directional element uses an induction cup unit that is subjected to a positive torque and hence operates when the power flow is in the designed forward direction and a negative torque, which restrains the element when the power flow is in the reverse direction. The typical static relay uses a phase angle between the current flowing and the input voltage that is within the range of $-90°$ to $+90°$. A relay setting can be introduced by rotating the input voltage phase by the required angle.

3.8.3 Phase Fault Protection Using Directional Overcurrent Relays

One complication with voltage-polarized directional elements is the potential for voltage collapse on the faulted phase. This is frequently avoided by polarizing the directional element for one specific phase with a voltage input derived from the other two phases. Thus the "a" phase element is polarized by the $V_{bc,}$ the "b" phase element by $V_{ca,}$ and the "c" phase element by V_{ab}. This is often referred to as the 90 or quadrature connection. Under fault conditions, the phase current normally lags the voltage by an angle equal to the system impedance angle. Since most power system plants are inductive, this can frequently vary from 30° to 85° lagging. To accommodate this, the directional element's setting is set at either 30° or 45°. With a setting of 30° the quadrature connected element will operate when the fault current is between 30° leading and 150° lagging of the phase voltage, and with 45° setting, the element will operate when the fault current is between 45° leading and 135° lagging. The 30°setting is generally recommended for the protection of plain feeders where the feeding transformer or generator and its star point are behind the relaying point, and 45° setting is recommended for transformer feeders where the transformer is included in the feeder. The 45° setting should be used whenever the protected circuit could experience 2–1 current distribution.

3.8.4 Phase Fault Protection of Parallel Feeders

A typical parallel feeder system is illustrated in Figure 3.22 and shows overcurrent relays located at the terminal of the two feeders. For both feeders, there is no risk

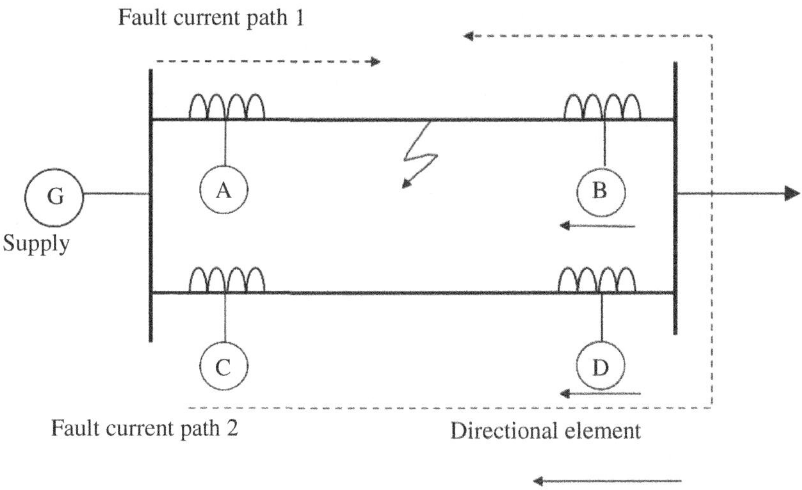

Figure 3.22 Parallel feeder protection.

of reverse current flow in the relays at the terminal connected close to the source, and nondirectional overcurrent will provide satisfactory protection. Directional elements are required at the feeder's remote end, otherwise, both feeders could be tripped out of service for an internal fault on either one.

For a fault on either feeder, the parallel feeder will provide a path for fault current, and hence the fault will be fed from both ends and through both feeders. Considering the fault F, shown in Figure 3.22, the relay B must trip the fault before relay C has time to operate, otherwise, feeder CD will be tripped unnecessarily. The operation of relay B must therefore be graded with that of relay C. Relay D should be restrained by its directional element.

Since the normal load current flow for relays B and D is in the reverse direction to the fault current for which they are required to trip, these relays will be inhibited under normal conditions. Also, since they are only required to operate for faults on the feeders AB and CD, respectively, and are not required to grade with other relays, they can be set to sensitive settings. Typically, these relays use a current setting of half the maximum load current and use a minimal TMS setting, usually 0.1.

Relays A and C must be graded with the setting used for relays D and B, respectively, as well as other relays used on the downstream circuit. In general, the operation of downstream relays dominates the grading exercise and therefore dictates the settings of these relays.

3.8.5 Phase Fault Protection of Ring Main Circuits

Ring main circuits are a common feature of power supply systems since they enable the load to be supplied from either or both directions around the ring.

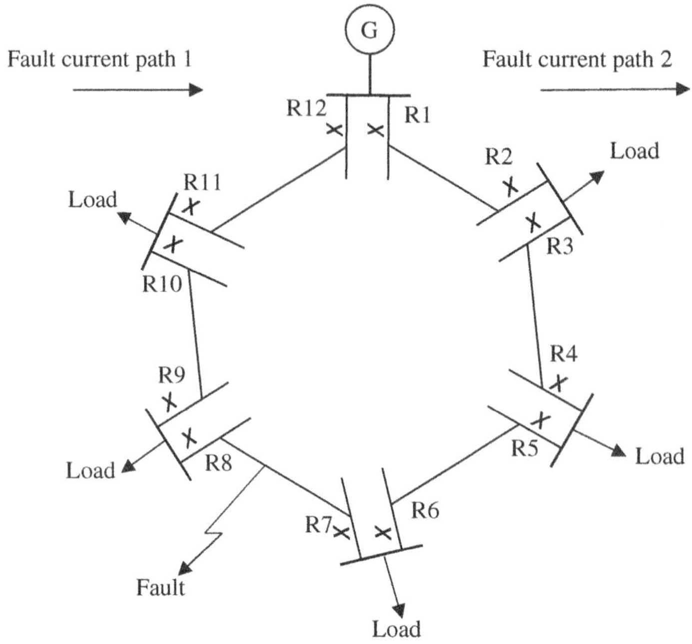

Figure 3.23 Ring main protection.

A typical ring-main circuit is shown in Figure 3.23 to simplify the protection of this configuration, the ring is frequently operated as an "open-ring" with one of the circuit breakers left open. This is not necessary, though, when directional overcurrent relays are employed rather than standard overcurrent relays. The main disadvantage of this approach is the high cost of the additional relays.

Since a fault on the ring or on a load supplied by the ring will be fed from both sides, separate relays are required to detect clockwise and anticlockwise fault current. Their setting is chosen such that only the relays closest to the fault operate, and hence, clockwise and anticlockwise grading exercises are required.

Consider the ring main circuit shown in Figure 3.23 and the fault between relays R7 and R8, the fault is fed clockwise via relays R1, R2, R3, R4, R5, and R7 and anticlockwise via relays R12, r11, R10, R9, and R8. The relay settings are chosen such that only relays R7 and R8 operate, hence disconnecting the faulted feeder but maintaining the supply to all of the loads.

All of the relays from R2 to R11 are directional relays, therefore, for the above fault in the clockwise fault current path, relays R1 R3, R5, and R7 will respond to the fault current and relays R2, R4, and R6 will be inhibited. In the anticlockwise fault current path, relays R12, R10, and R8 will respond to the fault, and relays R11 and R9 will be inhibited.

The use of the directional elements allows separate grading exercises, first for those relays responding to clockwise current flow, and second to anticlockwise current flow. The settings for relays R1, R3, R5, R7, R9, and R11 are selected from the clockwise grading exercise, and those for relays R12, R10, R8, R6, R4, and R2 are selected from the anticlockwise grading exercise. The order of this grading exercise is in the order listed above. For both relays R1 and R12, the fault current will always pass into the ring circuit, and hence, directional elements are not required.

The use of multiple infeeds, or when small generators are connected to the ring, complicates the choice of relay settings, and proper discrimination may not be possible. Under these conditions, the ring may need to be broken or unit-type protection employed.

3.8.6 Earth Fault Protection Using Directional Overcurrent Relays

The greatest proportion of faults involve earth, and directional earth fault protection is frequently employed to provide discrimination between faults on the incoming and outgoing circuits, where the system can be fed from either direction.

The difficulty with earth faults is that the fault could affect any of the three phases or a combination of them; therefore, the directional element must be polarized with an input that is phase independent and only governed by the direction to the fault. The earth fault current is derived from the summation of the three-phase currents to provide the residual current. The most frequently used method of polarization is to use an open delta transformer. This scheme is illustrated in Figure 3.24, where the directional earth fault relay is used to protect against earth faults on the outgoing feeder.

The current setting of the overcurrent element is chosen such that it is above the expected imbalance in the primary system and the possible mismatch in the current transformers. The directional element has an angular setting depending upon the type of system being protected. When used with a solidly earthed system, the angular setting is generally set at either 45° or 60° and when used with a resistively earthed system, the angular setting is generally set 0° or 15°.

In systems where it is not possible to provide a residual voltage signal, current polarization is used. The polarizing current signal is generally obtained from the earth current flowing from a feeding transformer's star point or the earth current flowing in an earthing transformer where the system is connected to a transformer's delta winding. The connections used with a transformer's star point earth are shown in Figure 3.25, for which the directional element is set so that the relay protects the feeder circuit and not the transformer.

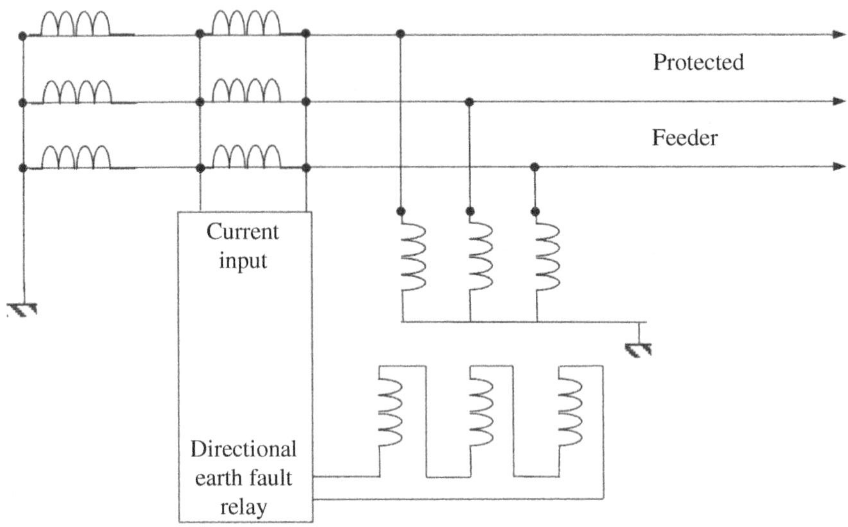

Figure 3.24 Directional earth fault overcurrent relay using voltage polarization.

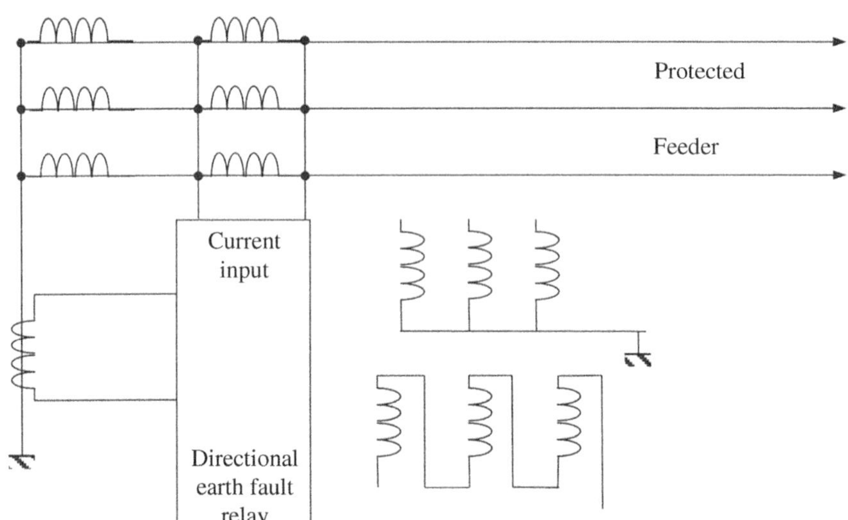

Figure 3.25 Directional earth fault overcurrent protection using current polarization.

3.8.7 Computer-Based Overcurrent Relay Grading

The above exercise has considered a radial network of plain feeders with a single grid supply point. Although these systems do exist, they are becoming the minority as supply companies are increasingly using an interconnected network to enhance the security of supply. These networks can be analyzed by hand, but with the advent of readily available computer systems and power system fault analysis software, this job is now best handled by computer.

Most relay coordination software has been developed from fault analysis packages where the characteristics of the appropriate protection relays are included. Details of the types of protection are input into the software along with the system parameters. Most programs accept suggested relay settings, and some can progress through an extensive grading exercise to determine the desired relay settings.

These packages are extremely useful with interconnected systems since they complicate the choice of appropriate settings. The first complication is that the fault current can be supplied from more than one point in the network and/or by more than one route. To protect these systems, there is a need for protection systems to be located at both ends of the feeder and for directional elements to assist in ensuring proper discrimination. Even with these, however, the grading exercise can highlight the impossibility of providing an effective protection scheme and the need to use other types of relaying.

Bibliography

Jamil Abdo, Tariq Masood "New Technique to Develop Intelligent, Highly Reliable & Efficient Unmanned Power Generation Modular" published in 22nd International Conference on Renewable Energies and Power Quality (ICREPQ'24) Bilbao (Spain), 26–28 June 2024. https://www.icrepq.com/icrepq24/392-24-abdo.pdf.

Jamil Abdo, Tariq Masood "Operational Reliability – Production Best Practice Pragmatic Case Studies & Improvement" published in Proceedings of the 9th North American Conference on Industrial Engineering and Operations Management Washington D.C., USA, 4–6 June 2024.

Jamil Abdo, Tariq Masood "Techno Economics Analysis to Reshape Global Clean and Green Energy" published in 22nd International Conference on Renewable Energies and Power Quality (ICREPQ'24) Bilbao (Spain), 26–28 June 2024. https://www.icrepq.com/icrepq24/415-24-abdo.pdf.

Qingtao Cao, Yun Jiang, Chunxiu Liu, Xinzhe Ma, Pengpeng Wang "Research on Fault Analysis of Multi source Distribution Network Based on Adaptive Protection" published in 7th Asia Conference on Power and Electrical Engineering (ACPEE) 2022.

Qingfa Chai, Jiayin Jiang, Yi Tang, Anning Wang, Congcong Li "High Reliability Relay Protection Setting Scheme of Distribution Network" published in 2023 International Conference on Power System Technology (PowerCon) 2023.

Lekshmi R Chandran, Anju Parvathy VS, Ilango K, Manjula G Nair "Adaptive Over Current Relay Protection in a PV Penetrated Radial Distribution System With Fuzzy GA Optimization" published in IEEE 19th India Council International Conference (INDICON) 2022.

Dong Chen, Bing Hu, Mingjun Xue, Yinian Qi, Changsong Oin "Research and Application of the System of Relay Protection and Self-Healing for Distribution Network with Mixed Use of Circuit Breaker and Load Switch" published in 2023 International Conference on Power System Technology (PowerCon) 2023.

Jane Courtney, Aaron McDonnell "Impact on Distribution System Protection with the Integration of EG on the Distribution Network" published in 2019 54th International Universities Power Engineering Conference (UPEC) 2019.

Feng Deng, Xiang Ji, Guo-Fang Huang "A Centralized Distribution Protection System" published in China International Conference on Electricity Distribution 2012.

AA Voloshin, EA Voloshin, AI Kovalenko, SA Danilov, DA Degtyarev, VS Sazanov "System Of Automatic Change Of Relay Protection Operation Parameters In Distribution Networks" published in 2021 4th International Youth Scientific and Technical Conference on Relay 2021.

Qiang Liu, Shengwen Li, Runquan Meng, Jingchong Huo, Shaozhe Jiang "A Protection Scheme for Low Voltage DC Distribution System Based on Control and Protection Cooperation" published in 4th International Conference on Smart Power & Internet Energy Systems (SPIES) 2022.

Congbo Wang, Ke Jia, Bohan Liu, Jiankang Zhang "Coordination Control and Protection for Photovoltaic DC Distribution System" published in 2nd International Conference on Smart Power & Internet Energy Systems (SPIES) 2020.

Yuichi Yoshida, Makoto Setoguchi, Kazuyuki Ishimoto "A Study of Rational Lightning Protection Measures for Power Distribution Lines in the Chugoku Region" published in 33rd International Conference on Lightning Protection (ICLP) 2016.

Bin Li, Chao Li, Fengrui Guo Ying Xin "Overcurrent Protection Coordination in a Power Distribution Network with the Active Superconductive Fault Current Limiter" published in IEEE Transactions on Applied Superconductivity 2014.

Bruno Fazio, Sanja Bajic, Serge St-Antoine "Relay Protection Coordination Analysis Using Fault Current Measurements: A Case Study of Hydro Quebec Distribution Network" published in 2022 IEEE Power & Energy Society General Meeting (PESGM) 2022.

Hana A Halim, Melaty Amirruddin, Nur S Noorpi, Nurhakimah M. Mukhtar "An Improved Protection Scheme for Smart Distribution Grid" published in 2013 1st

International Conference on Artificial Intelligence, Modelling and Simulation 2013.

IEEE Power and Energy Society "IEEE Recommended Practice for Fault Diagnosis and Protection in Smart Distribution System" IEEE Std 2748-2023 2024.

Kun Yang, Zhen Xiang, Ruixiong Yang, Lin Jiang "Decentralized Protection Coordination Based on Multi-Agent System for Active Distribution Networks" published in 2022 7th Asia Conference on Power and Electrical Engineering (ACPEE) 2022.

Qin Lijun, Wang Ying Zheng Shenrui, Guo Qing "Situation of Study on Distribution Network Protection Containing Distributed Generation and Application Scheme of Intelligent Protection" published in 2011 4th International Conference on Electric Utility Deregulation and Restructuring and Power Technologies (DRPT) 2011.

Rani Kumari, Bhukya K Naick "Protection Coordination Impact on Reliability of Distribution Systems With Distributed Energy Resources" published in 2024 Third International Conference on Power, Control and Computing Technologies (ICPC2T) 2024.

Tariq Masood, Jamil Abdo "Maintenance Strategy Production Best Practices – Pragmatic Case Studies and Operations Improvement" published in Proceedings of the 9th North American Conference on Industrial Engineering and Operations Management Washington D.C., USA, 4–6 June 2024.

Tariq Masood, Jamil Abdo "New Technique to Enhance Hybrid Renewable Energy Resources' Operational Flexibility and Reliability" published in 22nd International Conference on Renewable Energies and Power Quality (ICREPQ'24) Bilbao (Spain), 26–28 June 2024. https://www.icrepq.com/icrepq24/396-24-masood.pdf.

Tariq Masood, Jamil Abdo "Review: Uninterrupted Liquid Hydrogen (E-Methanol) Production Framework to Deliver Clean & Green Energy" published in 22nd International Conference on Renewable Energies and Power Quality (ICREPQ'24) Bilbao (Spain), 26–28 June 2024. https://www.icrepq.com/icrepq24/414-24-masood.pdf.

Tariq Masood, Abdel Aty Edris, Suhail Aftab Qureshi, Muhammad Tajammal, Murtaza Hashmi, DP Kothari, Samer Kareem "SSSC Tailored to Optimization Performance in Between United Arab Emirates and Oman on the GCC Power Grid" published in the 13th IET International Conference on AC and DC Power Transmission Conference on 14–16 February 2017. http://ieeexplore.ieee.org/document/7934934.

Tariq Masood, Abdel Aty Edris, Suhail Aftab Qureshi, Muhammad Tajammal, Murtaza Hashmi, DP Kothari, Samer Kareem "Smart Grid Operations and Control Challenge by Implementing STATCOM Tailored to Optimization Performance at the GCC Power Grid" published in the 13th IET International Conference on AC and DC Power Transmission Conference on 14–16 February 2017. http://ieeexplore.ieee.org/document/7934988.

Tariq Masood, Muhammad Tajammal, Samer Kareem, Suhail Aftab Qureshi, Abdul Jabbar Khan, DP Kothari, Nair Jameel "Unified power Flow Controller Modeling and Analysis Technique on the GCC Power Grid Operations and Control" published in 2018 IEEE International Energy Conference, Limassol, Cyprus, 7–9 June 2018. https://ieeexplore.ieee.org/document/8398834.

Tariq Masood, Muhammad Tajammal, Samer Kareem, Suhail Aftab Qureshi, Murtaza Hashmi, Nasir Jameel "Unified Power Flow Controller (UPFC) Modeling and Analysis Technique by Using Matlab and EMTP" published in International Conference on Renewable Energies and Power Quality (ICREQP'18) Salamanca (Spain) 21–23 March 2018. http://www.icrepq.com/icrepq18/250-18-masood.pdf.

Tariq Masood, Muhammad Tajammal, Samer Kareem, Suhail Aftab Qureshi, Murtaza Hashmi, Naser Jameel "STATCOM Controller Performance Enhancement and Optimization Technique" published in International Conference on Renewable Energies and Power Quality (ICREQP'18) Salamanca (Spain) 21–23 March 2018. http://www.icrepq.com/icrepq18/249-18-masood.pdf.

Tariq Masood, Muhammad Tajammal, Samer Kareem, Suhail Aftab Qureshi, Murtaza Hashmi, DP Kothari "Smart Grid Operational Functions and Control Challenges by Implementing SSSC Tailored to Optimize Performance in Between Qatar and KSA on the GCC Electrical-Power Grid" published in the ASTESJ Journal Advances in Science, Technology and Engineering Systems Journal Vol. 2, No. 1, 2017. https://astesj.com/special-issue-4.

Tariq Masood, N Al-Emadi, SA Qureshi, YY Almulla, AJ Khan, SK Shah "FACTS Control Devices (Statcom, SSSC and UPFC) Re-Configuration Techniques By PSIM/MATLAB" published in 2007 International Conference on Electrical Engineering 2007. https://http://ieeexplore.ieee.org/abstract/document/4287295.

Tariq Masood, Raj K Aggarwal, SA Qureshi, RAJ Khan "STATCOM and SVC Control Operations and Optimization During Network Fault Conditions" published in 2010 IEEE International Symposium on Industrial Electronics 2010. https://ieeexplore.ieee.org/document/5636868.

Tariq Masood, Raj K Aggarwal, Suhail Aftab Qureshi, RAJ Khan "STATCOM Model Against SVC Control Model Performance Analyses Technique by MATLAB" published in International Conference on Renewable Energies and Power Quality (ICREPQ'10) 2010. https://icrepq.com/icrepq'10/655-Masood.pdf.

Tariq Masood, RK Aggarwal, SA Qureshi, RAJ Khan "Statcom Control Reconfiguration Technique for Steady State and Dynamic Performance Optimization During Network Fault Conditions" published in International Conference on Renewable Energies and Power Quality (ICREPQ'10) Granada (Spain) 2010. https://www.icrepq.com/icrepq'10/656-Masood.pdf.

Tariq Masood, Samer Karim Shah, Muhammad Tajammal "Unified Power Flow Controller Modeling and Analysis Technique on the GCC Power Grid" published in 2018 IEEE International Energy Conference (ENERGYCON) 2018. https://ieeexplore.ieee.org/abstract/document/8398834.

Tariq Masood, Suhail A Qureshi, RK Aggarwal, DP Kothari "Multiple FACTS Devices Integration Through Distributed Control System and Its Application on the GCC Interconnection" published in 2013 7th IEEE GCC Conference and Exhibition (GCC) 2013. https://ieeexplore.ieee.org/abstract/document/6705750.

VS Sazanov, AI Kovalenko, AA Voloshin "The Development of System for Automatic Adaptive Change of Relay Protection Settings in Distribution Networks" published in 2022 5th International Youth Scientific and Technical Conference on Relay Protection and Automation (RPA) 2022.

Vassilis C Nikolaidis, Evangelos Papanikolaou, Anastasia S Safigianni "A Communication-Assisted Overcurrent Protection Scheme for Radial Distribution Systems With Distributed Generation" IEEE Transactions on Smart Grid Vol. 7, No. 1, 2015.

Xiaohui Song, Yu Zhang, Song Zhang, Shizhan Song, Jing Ma, Weibo Zhang "Active Distribution Network Protection Mode Based on Coordination of Distributed and Centralized Protection" published in 2017 China International Electrical and Energy Conference (CIEEC) 2017.

Yu Zou, Minchuan Liao, Ruiqing Lian "Discussion on Lightning Protection of Distribution Network and Its Differentiation" published in 2020 International Conference on Intelligent Computing, Automation and Systems (ICICAS) 2020.

Yuwei Xiong, Wei Chen "Positive Sequence Current Differential Protection Based on Least Square Method for Distribution Network with Distribution Generation" published in 2021 IEEE 4th International Electrical and Energy Conference (CIEEC) 2021.

4

Protection Transducers

4.1 Objectives

When you have finished this section, you should be able to:

- Explain the operation and application of current and voltage transformers (also known as protection transducers) used in the protection of distribution systems
- Distinguish clearly between the basic differences in design criteria of the two types of transducers
- Define the accuracy requirements of the two transformers

4.2 Introduction

Protection relays require accurate reproduction of normal steady-state and abnormal (such as under-faults) conditions in a power system for correct sensing and operation. This information is input from the power system through current and voltage transformers.

Current and voltage transformers provide insulation for the protection relays from the high system voltage and a reduction of the primary current and voltage quantities. Thus, the transformer primary is connected to the power system with insulation compatible with that of the system. The secondary is standardized for the convenience of application and relay design. These relay input devices are a very important part of the overall protection scheme.

This book is concerned with the operation and application of current and voltage transformers in the protection of distribution system. Particular emphasis is placed on discussing specific aspects of the protection transducers to ensure their correct operation for different system and fault conditions, in particular, under a wide dynamic range of primary system current and/or voltage.

Protection of Electrical Power Distribution Systems: Smart grid, Microgrid, AI, and Cyber Security, First Edition. Tariq Masood, Jamil Abdo, and Atif Iqbal.

4.3 Current Transformer (CT)

A current transformer (CT) designed for measuring purposes (other than those for protection) operates over a range of current up to the normal current rating and has specific errors at that value. However, a protection CT is required to operate over a range of currents many times the circuit rating, albeit for short periods of time. The latter thus requires special design consideration to meet the accuracy requirements for protection purposes.

The normal design criterion is to limit the transformer core flux to the value where saturation commence – known as the knee-point flux – and therefore it is the maximum value of the magnetizing characteristics for a 100/1 A CT which is designed to give as high a value as possible of magnetizing impedance (this involves high-quality core material) so as to keep the magnetizing current I_e *(this is primarily responsible for both ratio and phase errors in the secondary output current, as shown in* Figure 4.1).

As small as possible up to the knee-point of the characteristic. The latter would ensure that both phase and ratio errors are very small, thus faithfully maintaining the primary/secondary current relationship (Figure 4.2).

The load of a CT is called the burden and can be expressed as VA or as an impedance. In the former case, the VA is taken to be at the CT nominal secondary

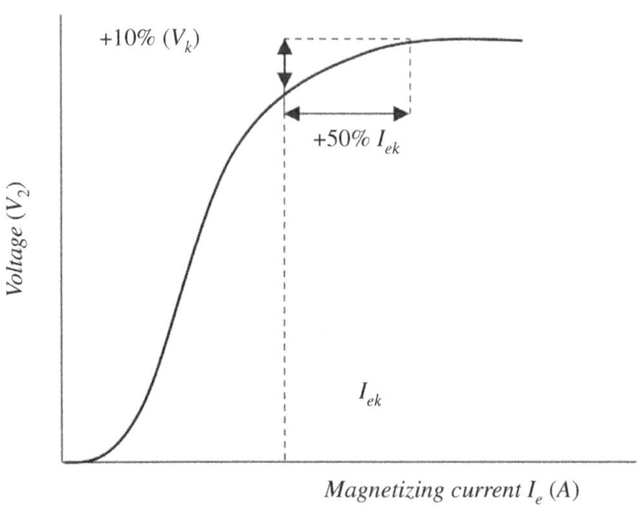

Figure 4.1 CT magnetizing characteristic.

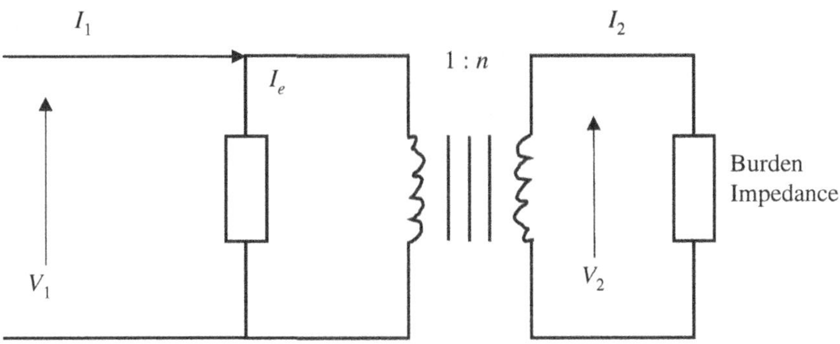

Figure 4.2 Equivalent circuit of current transformer.

current. For example, a 5VA burden on a 1A transformer would have an impedance given by:

Secondary voltage:

$$V_2 = \frac{5\ \text{VA}}{1\ \text{A}} = 5\ \text{V}$$

Impedance

$$Z = \frac{5\ \text{V}}{1\ \text{A}} = 5\ \Omega$$

All burdens are connected in series, and an increase in impedance increases the burden on the CT. The errors of transformation are also influenced by the magnitude and argument of the burden. The British Standard for specifying CTs for protection purposes is in terms of rated burden, accuracy class, and accuracy limit. Standard values of rated burden are: 2.5, 5, 7.5, 10, 15, and 35 VA.

There are essentially two accuracy classes of measuring CTs, and in British practice, these are referred to as 5P and 10P. As an example, a CT described as 15 VA Class 5P20 would mean that it is rated for a burden of 15 VA and will not have more than 5% error at 20 times rated current. It is often more convenient to refer directly to the maximum useful voltage that can be obtained from a CT. In this respect, the knee-point of the magnetizing current (as shown in Figure 4.1) is defined as that point beyond which an increase of 10% of secondary voltage will increase the magnetizing current by 50%. Design requirements of CTs for protection purposes are often specified in terms of the knee-point voltage and are generally known as Class X CTs. It should be noted that at voltages above the knee-point, saturation quickly occurs, and the device no longer behaves as a CT. In this respect, there are two primary causes of CT saturation: the magnitude of the current and a large offset in the current. However, in the case of distribution systems, the latter of the two is very small (this is due to the line/cable resistance being comparable to reactance), and saturation is therefore primarily governed by the magnitude of the currents. CTs are usually designed to have rated secondary currents of 1A and 5A.

Exercise 4.1 What is the significance of the knee-point voltage?

As can be seen from Figure 4.1, up to the knee-point voltage, the secondary voltage increases almost linearly with the exciting current. However, above this value, a small increase in secondary voltage requires a very large increase in exciting current, that is, about five times what is necessary below the knee-point voltage. It can thus be said that above the knee-point voltage, the CT goes into saturation and therefore no longer faithfully transforms the primary current. In CT design (used in protection application), it is thus important to ensure that the knee-point voltage is not exceeded for the worst-case fault.

In specifying the CTs, the connected burden and mode of operation must be taken into account, paying attention not only to the wide range of protection relays which may be connected but also to the variation in impedance over the range of any relay setting. As an example, consider the normal burden of an overcurrent relay to be 3 VA at 100% setting. If the normal setting range of the relay is 50% to 200% of normal current, then a 1A relay set to 50% will have a setting current of 0.5 A, and the voltage across the coil at this current setting would be:

$$V = \frac{3\ \text{VA}}{0.5\ \text{A}} = 6\ \text{V} \tag{4.1}$$

and the burden impedance would be:

$$Z = \frac{6\ \text{VA}}{0.5\ \text{A}} = 12\ \text{V} \tag{4.2}$$

Exercise 4.2 What will be the burden impedance at a relay setting of 200%?

For this case, the setting current will be 2 A, and the voltage across the coil will be

$$V = \frac{3\ \text{VA}}{2\ \text{A}} = 1.5\ \text{V}, \tag{4.3}$$

and the burden impedance would be:

$$Z = \frac{1.5\ \text{VA}}{2\ \text{A}} = 0.75\ \text{V} \tag{4.4}$$

If the characteristic of the relay is to be maintained at 20 times the relay setting, then a knee-point voltage not less than $(20 \times 6\,\text{V}) = 120\ \text{V}$ *for a 50% setting or* $(20 \times 1.5\ \text{V}) = 30\ \text{V}$ *for a* 200% setting would be required. It is apparent that the former of the two is more onerous and therefore the lowest relay setting must be taken into account when specifying the knee-point voltage. There is, however, an alleviating factor in that a relay operating at 20 times its setting will have saturated magnetically (i.e. if it is of the electromechanical type) and therefore the burden impedance will be reduced. In practice, the reduction for the overcurrent relay is about half the impedance at setting, which means that in the above example, a knee-point voltage of 60 V will suffice.

In many cases, the CTs associated with overcurrent protection must also cater for earth fault relays, which normally have a lower setting than overcurrent relays. An earth fault relay having a minimum setting of 20% will have a voltage setting of:

$$V = \frac{3\ \text{VA}}{0.2\ \text{A}} = 15\ \text{V}$$

and an impedance of:

$$V = \frac{15\ \text{V}}{0.2\ \text{A}} = 75\ \Omega$$

The maximum earth fault level may be restricted to, say, 10 times the relay setting. The knee-point voltage should therefore be greater than (10 × 15) V = 150 V or allowing for relay saturation, about 75 V. In this case, the CT specification is dictated by the earth faulted relay. A suitable CT would be 7.5 VA class 510. This will produce a voltage of 7.5 V at 1A rated current when connected to a 75 burden and will have only 5% error at 10 times rated current, that is, at a voltage of (10 × 7.5) V = 75 V. It is apparent from the CT specification that the knee-point voltage can be estimated.

Exercise 4.3 A CT with a 5A secondary winding is specified as 30 VA class 5P20. Estimate the transformer knee-point voltage.

Secondary voltage V (30 VA/5 A) = 6 V, that is, it will produce 6 V across the rated burden. The knee-point voltage is 20 times this value, that is, 120 V.

4.4 Voltage Transformer (VT)

A VT for use with protection has to fulfill an important requirement, which is that the secondary voltage must be an accurate representation of the primary voltage in magnitude and phase.

To meet the requirement, it is designed to operate at fairly low flux densities so that the magnetizing current and the ratio and phase angle errors are small. This effectively means that the core area for a given output is larger than that of a conventional power transformer, and this, in turn, results in an increase in the overall size of the transformer. The nominal secondary output is 63.5 V per phase to produce a line (phase-phase) voltage of 110 V.

With regard to accuracy, only in a few of the many applications are the phase and ratio errors likely to be of much significance. However, in practice, the likelihood of a VT being provided solely for protection purposes is small, and, therefore, the more stringent accuracies of instrumentation and metering are usually required.

By British Standards, all VTs are required to have ratio and phase errors within a prescribed 80 → 120% range of voltage and a range of burden from 25 → 100%.

For protection purposes, the accuracy of measurement is important during fault conditions, particularly when the voltage is greatly suppressed. Thus, a protection VT must also meet the extended range of requirement over a range of 5–80% rated voltage and for certain application, between 120 and 190% rated voltage.

4.5 Self-Assessment Questions

Self-Assessment Question 4.1a

Outline the fundamental difference between an ordinary CT and a CT used in protection. In the case of the latter, what measures are taken in the design procedures to ensure that a high accuracy is maintained under a wide range of fault conditions

Self-Assessment Question 4.2b

A 100/1 A CT is to be used in conjunction with an earth fault relay which has a normal burden of 5 VA at 100% setting and is set to 50% for detecting earth faults. If the worst-case primary fault current is 1 kA, what class of CT would be suitable for this application, assuming that a CT error of <5% is required and that the relay does not saturate?

Bibliography

Juan Noh, Wookyu Chae, Woohyun Kim, Sungyun Choi "A Study on Mesh Distribution system and Protection Coordination Using HILS System" published in 13th International Conference on Information and Communication Technology Convergence (ICTC) in 2022.

Congbo Wang, Ke Jia, Bohan Liu, Jiankang Zhang "Coordination Control and Protection for Photovoltaic DC Distribution System" published in 2nd International Conference on Smart Power & Internet Energy Systems (SPIES) in 2020.

Yuichi Yoshida, Makoto Setoguchi, Kazuyuki Ishimoto "A study of Rational Lightning Protection Measures for Power Distribution Lines in the Chugoku Region" published in 33rd International Conference on Lightning Protection (ICLP) in 2016.

Tara Khayyamim, Bita Nouparvar, Arman Safaei "A Novel Scheme for Grid Interconnection and Protection of EV Charging stations" published in 15th International Conference on Protection and Automation of Power Systems (IPAPS) in 2020.

Chong Tong, Yunfeng Cai, Ziyang Zhang "Artificial intelligence-Based Lightning Protection of Smart Grid Distribution system" published in International Symposium on Lightning Protection (XIV SIPDA) in 2017.

Harsha Ravindra, M. Omar Faruque "Impact of PV on Distribution Protection System" published in North American Power Symposium (NAPS) in 2012.

Hesam-Edin Hayati Soloot, Edris Agheb, Amir Hayati Soloot, Shahab Moghadam "A SWOT Analysis of Two Protection Strategies Due to the Expansion of Renewable

Distributed Generation Distribution Network" in published in 15th International Conference on Protection and Automation of Power Systems (IPAPS) in 2020.
Bin Li, Chao Li, Fengrui Guo, Ying Xin "Overcurrent Protection Coordination in a Power Distribution Network With the Active Superconductive Fault Current Limiter" Published in IEEE Transactions on Applied Superconductivity 2014.
Bruno Fazio, Sanja Bajic, Serge St-Antoine "Relay Protection Coordination Analysis using Fault Current Measurements" published in 2022 IEEE Power & Energy Society General Meeting (PESGM) 2022.
Dong Chen, Bing Hu, Mingjun Xue, Yinian Qi, Changsong Oin "Research and Application of the System of Relay Protection and Self-Healing for Distribution Network with Mixed Use of Circuit Breaker and Load Switch" published in 2023 International Conference on Power System Technology (PowerCon) 2023.
Feng Deng, Xiang Ji, Guo-Fang Huang "A Centralized Distribution Protection System" published in 2012 China International Conference on Electricity Distribution 2012.
Hana A. Halim, Melaty Amirruddin, Nur S. Noorpi, Nurhakimah M. Mukhtar "An Improved Protection Scheme for Smart Distribution Grid" published in 2013 1st International Conference on Artificial Intelligence, Modelling and Simulation 2013.
Jamil Abdo, Tariq Masood "New Technique to Develop Intelligent, Highly Reliable & Efficient Unmanned Power Generation Modular" Published in 22nd International Conference on Renewable Energies and Power Quality (ICREPQ'24) Bilbao (Spain), 26–28 June 2024. https://www.icrepq.com/icrepq24/392-24-abdo.pdf.
Jamil Abdo, Tariq Masood "Operational Reliability-Production Best Practice Pragmatic Case Studies & Improvement" published in Proceedings of the 9th North American Conference on Industrial Engineering and Operations Management Washington D.C., USA, 4–6 June 2024.
Jamil Abdo, Tariq Masood "Techno Economics Analysis to Reshape Global Clean and Green Energy" in 22nd International Conference on Renewable Energies and Power Quality (ICREPQ'24) Bilbao (Spain), 26–28 June 2024. https://www.icrepq.com/icrepq24/415-24-abdo.pdf.
Jane Courtney, Aaron McDonnell "Impact on Distribution System Protection with the Integration of EG on the Distribution Network" published in 2019 54th International Universities Power Engineering Conference (UPEC) 2019.
Kun Yang, Zhen Xiang, Ruixiong Yang, Lin Jiang "Decentralized Protection International Journal of Automation and Power Engineering Coordination Based on Multi-agent system for Active Distribution Networks" published in 2022 7th Asia Conference on Power and Electrical Engineering (ACPEE) 2022.
M. Saeed Ullah, Tariq Masood, Umair Zahid, Hafiz Abuzar Ahsan Electrochemical Oxidation of Dissolved organic Dye in Aqueous Solution published in IMASCON-2019, 2019.

Nasser Alemadi, Tariq Masood, R Aggarwal, Suhail Aftab Qureshi, D Kothari "Novel Selection Approach of STATCOM Device Tailored and Optimized Engineering Configuration to Implement in Between Qatar and Bahrain" published in International Journal of Automation and Power Engineering 2014. https://ieeexplore.ieee.org/document/9952497.

Qin Lijun, Wang Ying, Zheng Shenrui, Guo Qing "Situation of Study on Distribution Network Protection Containing Distributed Generation and Application Scheme of Intelligent Protection" published in 2011 4th International Conference on Electric Utility Deregulation and Restructuring and Power Technologies (DRPT) 2011.

Qingfa Chai, Jiayin Jiang, Yi Tang, Anning Wang, Congcong Li "High Reliability Relay Protection Setting Scheme of Distribution Network" published in 2023 International Conference on Power System Technology (PowerCon) 2013.

Rani Kumari, Bhukya K. Naick "Protection Coordination Impact on Reliability of Distribution Systems With Distributed Energy Resources" published in 2024 Third International Conference on Power, Control and Computing Technologies (ICPC2T) 2014.

Tariq Masood, Jamil Abdo "Maintenance Strategy Production Best Practices – Pragmatic Case Studies and Operations Improvement" published in Proceedings of the 9th North American Conference on Industrial Engineering and Operations Management Washington D.C., USA, 4–6 June 2024.

Tariq Masood, Jamil Abdo "New Technique to Enhance Hybrid Renewable Energy Resources' Operational Flexibility and Reliability" published in 22nd International Conference on Renewable Energies and Power Quality (ICREPQ'24) Bilbao (Spain), 26–8 June 2024. https://www.icrepq.com/icrepq24/396-24-masood.pdf.

Tariq Masood, Jamil Abdo "Review: Uninterrupted Liquid Hydrogen (E-Methanol) Production framework to deliver Clean & Green Energy" published in 22nd International Conference on Renewable Energies and Power Quality (ICREPQ'24) Bilbao (Spain), 26–28 June 2024. https://www.icrepq.com/icrepq24/414-24-masood.pdf.

Tariq Masood, Abdel Aty Edris, Suhail Aftab Qureshi, Muhammad Tajammal, Murtaza Hashmi DP Kothari, Samer Kareem "SSC Tailored to Optimization Performance in Between United Arab Emirates and Oman on the GCC Power Grid" published in the 13th IET International conference on AC and DC power Transmission Conference on 14–16 February, 2017. https://ieeexplore.ieee.org/document/9952497.

Tariq Masood, Abdel Aty Edris, Suhail Aftab Qureshi, Muhammad Tajammal, Murtaza Hashmi DP Kothari, Samer Kareem "Smart Grid Operations and Control Challenge by Implementing STATCOM Tailored to Optimization Performance at the GCC Power Grid" published in the 13th IET International conference on AC and DC power Transmission Conference on 14–16 2017. https://ieeexplore.ieee.org/document/9952497.

Tariq Masood, Muhammad Tajammal, Samer Kareem, Suhail Aftab Qureshi, Abdul Jabbar Khan, DP Kothari, Nair Jameel "Unified power Flow Controller Modeling and Analysis Technique on the GCC Power Grid Operations and Control" published in IEEE Energy Con, 2018 7–9 June Limassol Cyprus 2018. https://ieeexplore.ieee.org/document/9952497.

Tariq Masood, Muhammad Tajammal, Samer Kareem, Suhail Aftab Qureshi, Murtaza Hashmi, Nasir Jameel "Unified Power Flow Controller (UPFC) Modeling and Analysis Technique by using Matlab and EMTP" published in (ICREQP),18 Salamanca Spain 21–23 March, 2018. http://www.icrepq.com/icrepq18/250-18-masood.pdf.

Tariq Masood, Muhammad Tajammal, Samer Kareem, Suhail Aftab Qureshi, Murtaza Hashmi Naser Jameel "STATCOM Controller Performance Enhancement and Optimization Technique" published in (ICREQP),18 Salamanca Spain 21–23 March, 2018. http://www.icrepq.com/icrepq18/249-18-masood.pdf.

Tariq Masood, Muhammad Tajammal, Samer Kareem, Suhail Aftab Qureshi, Murtaza Hashmi DP Kothari Smart Grid Operational functions and Control Challenges by Implementing SSSC Tailored to Optimize performance in between Qatar and KSA on the GCC Electrical-power grid published in the ASTESJ Advances in Science Technology and Engineering Systems Journal 2017. https://astesj.com/special-issue-4/.

Tariq Masood, N Al-Emadi, SA Qureshi, YY Almulla, AJ Khan, SK Shah "FACTS Control Devices (Statcom, SSSC and UPFC) Re-Configuration Techniques By PSIM/MATLAB" published in: 2007 International Conference on Electrical Engineering https://ieeexplore.ieee.org/document/9952497.

Tariq Masood, Raj K Aggarwal, SA Qureshi, RAJ Khan "STATCOM and SVC Control Operations and Optimization During Network Fault conditions" 2010. https://ieeexplore.ieee.org/document/9952497.

Tariq Masood, Raj K Aggarwal, Suhail Aftab Qureshi, RAJ Khan "STATCOM Model Against SVC Control Model Performance Analyses Technique by MATLAB" published in International Conference on Renewable Energies and Power Quality (ICREPQ'10) 2010. https://icrepq.com/icrepq'10/655-Masood.pdf.

Tariq Masood, RK Aggarwal, SA Qureshi, RAJ Khan "Statcom Control Reconfiguration Technique for Steady State and Dynamic Performance Optimization During Network Fault Conditions" published in International Conference on Renewable Energies and Power Quality (ICREPQ'10) Granada (Spain) 2010. https://www.icrepq.com/icrepq'10/656-Masood.pdf.

Tariq Masood, Samer Karim Shah, Muhammad Tajammal, "Unified Power Flow Controller Modeling and Analysis Technique on the GCC Power Grid" published in 2018 IEEE International Energy Conference (ENERGYCON) 2018. https://ieeexplore.ieee.org/document/9952497.

Tariq Masood, Suhail A Qureshi, RK Aggarwal, DP Kothari "Multiple FACTS Devices Integration Through Distributed Control System and its Application on the GCC Interconnection" published in 2013 7th IEEE GCC Conference and Exhibition (GCC) 2013. https://ieeexplore.ieee.org/document/9952497.

VS Sazanov, AI Kovalenko, AA Voloshin "The Development of System for Automatic Adaptive Change of Relay Protection Settings in Distribution Networks" published in 2022 5th International Youth Scientific and Technical Conference on Relay Protection and Automation (RPA) 2022.

Vassilis C Nikolaidis, Evangelos Papanikolaou, Anastasia S. Safigianni A Communication-Assisted Overcurrent Protection Scheme for Radial Distribution Systems With\Distributed Generation published in IEEE Transactions on Smart Grid 2016.

Xiaohui Song, Yu Zhang, Song Zhang, Shizhan Song, Jing Ma, Weibo Zhang Active Distribution Network Protection mode Based on Coordination of Distributed and Centralized Protection published in 2017 China International Electrical and Energy Conference (CIEEC) 2017.

5

Unit Protection of Distribution Feeders

5.1 Objectives

When you have completed this section, you should be able to

- Define the basic principle of unit protection.
- Identify the need for bias and be able to derive both bias and operating quantities
- Assess the advantages and disadvantages of different types of differential protection schemes (including pilot wire and digital) used in practice
- Appreciate directional earth fault and rough balance protection

5.2 Introduction

Any feeder protection must have adequate speed of operation under fault conditions to minimize damage and provide discrimination in operation so as to localize the loss of supplies. In many circumstances, these requirements are best met by using a unit scheme, such as differential protection, for a particular section of a feeder. For example, in a distribution system (which is primarily projected by time graded overcurrent relays), the minimum permissible operating time for a particular scheme may be unacceptably long. Unit protection, on the other hand, provides consistent high-speed protection for the entire feeder length.

However, since it is based on comparing quantities at the two ends of a protected zone, it does have the disadvantage that it requires a communication link for transfer of information from one end to the other and can therefore suffer from some degradation in performance due to the limitation imposed by the communication link.

The section deals with some of the basic principles of unit protection as applied to distribution feeders and considers the different types of techniques used in practice.

Protection of Electrical Power Distribution Systems: Smart grid, Microgrid, AI, and Cyber Security,
First Edition. Tariq Masood, Jamil Abdo, and Atif Iqbal.

5.3 Conventional Differential Protection

Differential protection is principally of the longitudinal type, in which measuring equipment situated at each end of a feeder section is interconnected by means of pilot wires to compare the magnitude and phase angle of the current entering and leaving the feeder. Most differential protection schemes for feeder protection are based on the *Merz-Price* circulating current principle, in which, if the current flowing into the protected zone is the same as the current flowing leaving it, then the fault is not in the protected zone, thus restraining tripping.

On the other hand, if there is a difference in either phase or magnitude between the two quantities, then the fault is within the protected zone, and the relay should give a trip decision. One such scheme is shown in Figure 5.1, in which identical CTs are connected at the two ends of the protected zone.

In practice, pilot connections in the role of CT leads would present an excessive burden and so a further stage of current transformation is introduced, and it is desirable, for the purpose of tripping, to have a relay at each line end. This means that the relay is not connected at an equipotential point, and so to prevent false operation and cater for any inequalities in CT performance, particularly under heavy external fault conditions, a bias feature is generally included. The latter increases the effective relay settings as primary current magnitude is increased (see Figure 5.2).

Most of the differential systems used today for feeders are derived from one of the basic arrangements shown in Figure 5.2, depicting the circulating current scheme and the balanced voltage scheme. Of the two, the former is only situated to very short feeders (about 1 km distance) because of the excessive voltage developed across the CT as the feeder length goes up. The balanced voltage scheme, on the other hand, can be applied to much longer feeders (up to about 30 km).

Exercise 5.1 *Why is the circulating current scheme rather limited in application? Suggest an alternative arrangement.*

In circulating current schemes, the voltage developed across the CT secondaries increases in proportion to the length of the feeder, and therefore, very large CTs are required when applying this principle to feeders of length above 1 km. A balanced

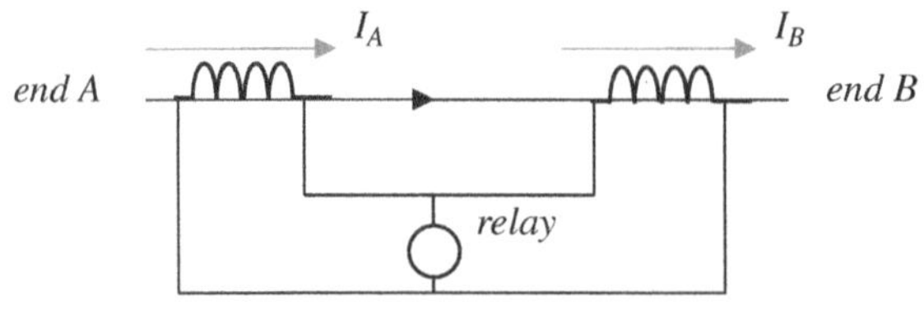

Figure 5.1 Circulating current system.

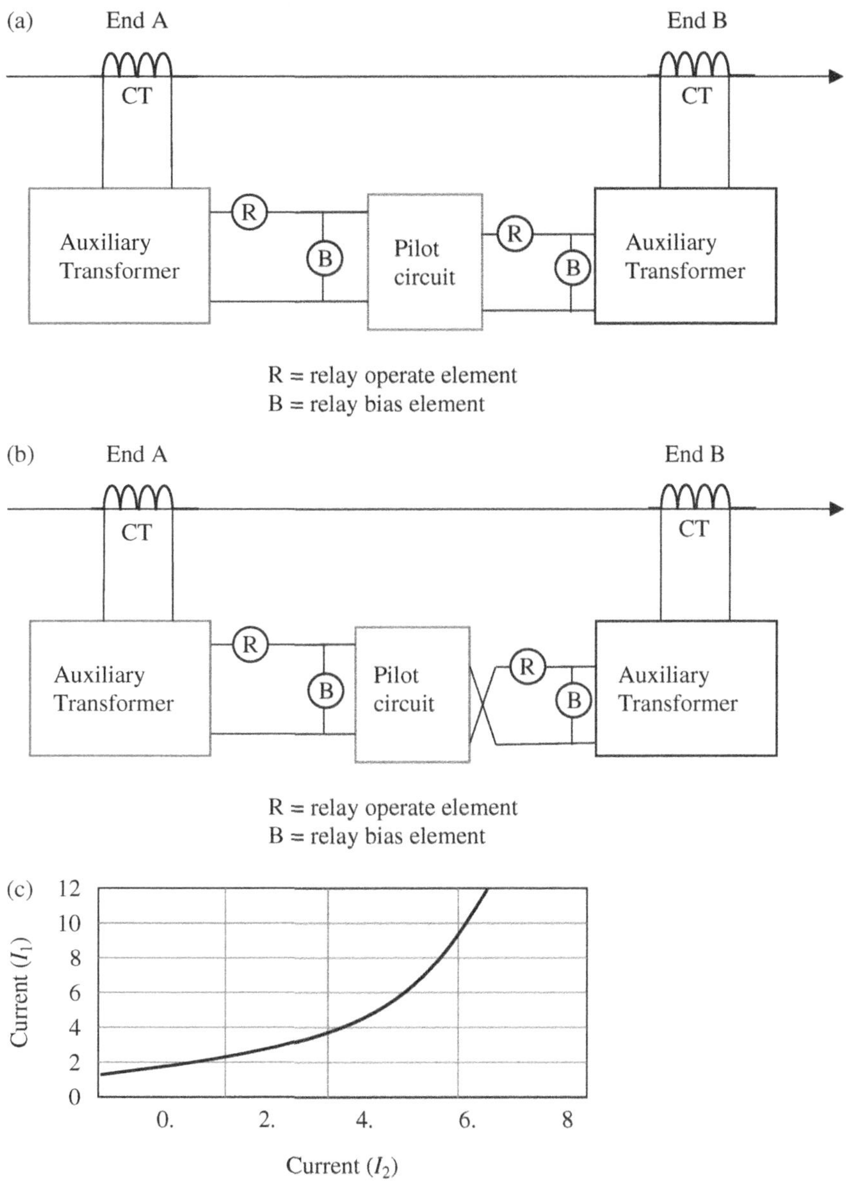

Figure 5.2 (a) Circulating current scheme, (b) balanced voltage scheme, and (c) typical bias curve.

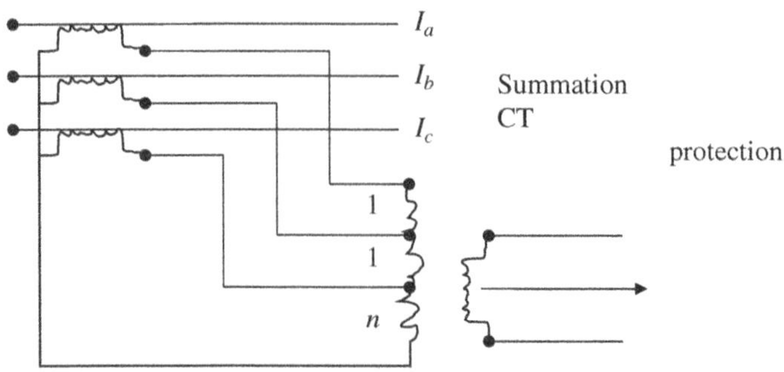

Figure 5.3 Summation transformer arrangement.

voltage protection is a better alternative because, in this arrangement, with the through current, the CTs are opposed and therefore open circuit. In this way, the voltage developed across the CTs is limited, and the protection can therefore be applied to much longer feeders up to about 30 km.

From an economic point of view, differential schemes for feeder protection should require only two pilot conductors for data transfer. In practice, this is achieved by reducing the three-phase quantities into a single signal for comparison in the protected circuit. The simplest method of producing a single-phase output is via a summation primary winding on an interposing auxiliary CT, as shown in Figure 5.3. The sections of the primary winding have turns in the ratio 1:1:n, where n may vary between 2 and 4.

With most protection systems employing pilot wires, we have to limit the output that can appear at the pilot terminal, in particular if the pilot wires are rented from a telephone company and are metallic. The voltage limitation may be controlled but the magnetizing characteristic of the summation CT and the addition of a nonlinear resistance across the transformer output terminals.

It follows that at voltage levels in excess of those at which the limitation is imposed, the current comparison achieved by the protection system will relate principally to phase angle only and not to the magnitude. This can often pose a protection problem, that is, the relay can run into blind spot problems whereby it can fail to detect an internal fault.

Exercise 5.2 *What problems can be caused by the imposition, by telephone authorities, if the maximum voltage that can be applied to rented pilot wires.*

Because of the voltage limitations imposed, at large fault currents involving large voltage at the pilot wire terminals, the comparison of the quantities would be based on single-phase only, rather than on both phases and magnitude. This can result in incorrect protection performance. The pilot wires used in distribution feeder protection are often of telephone-type cables, which have a high resistance/capacitance ratio, resulting in large attenuation and current imbalance under healthy conditions. In many differential schemes, the pilot resistance and capacitance are compensated for to minimize the error. With the simple, low-speed system, no attempt is made to offset variation in pilot resistance. Higher resistance merely increases the minimum operating level. In high-speed scheme, however, a variable resistor is usually included in each relay so that the complete pilot circuit may be padded up to the prescribed nominal value, thereby maintaining substantially constant operating level for a wide range of pilot resistance. The principal effect of capacitance between pilot conductors is to provide a shunt path for current, which would nominally be absent with ideally balanced voltage conditions on the pilot. The lowering of the shunt impedance by capacitance can be offset by the addition of a shunt reactor at each pilot terminal. When each reactor has been tuned to half the pilot capacitance at nominal power frequency, the high shunt impedance is substantially restored. Compensation is, however, only partially achieved because the phase relationship between pilot terminal currents is not fully corrected by this method. This arises because the distributive nature of the pilot resistance and capacitance limits the level of tuning achieved.

5.4 Digital Differential Protection

As mentioned in the previous section, the current differential protection as shown in Figure 5.2 has a limitation on the maximum voltages (and therefore the maximum currents) that can be applied at the pilot terminals. Voltages above the prescribed level are clipped through limiters, and this can cause problems with protection. Furthermore, because of the resistance and capacitance of the pilot wires, there are limitations on the maximum feeder length that can be satisfactorily protected. Another disadvantage is that the metallic wire is prone to electromagnetic interference, which can cause problems of nonlinearities when analog power frequency signals are being transmitted. A better arrangement is to modulate the power frequency signals before the transmission. One such signaling arrangement is as shown in Figure 5.4.

The instantaneous analog current waveforms at each terminal are modulated into instantaneous frequency signals have their frequencies normally within the bandwidth 500 – 2400 Hz for voice frequency channeling.

In practice, the secondary current of the CT is isolated and processed into an appropriate voltage through an auxiliary transformer. A voltage frequency converter is then employed to change the voltage signal to the frequency of the voice band. The main advantage of such a technique is that it is less liable to be affected by the nonlinearities of the transmission equipment.

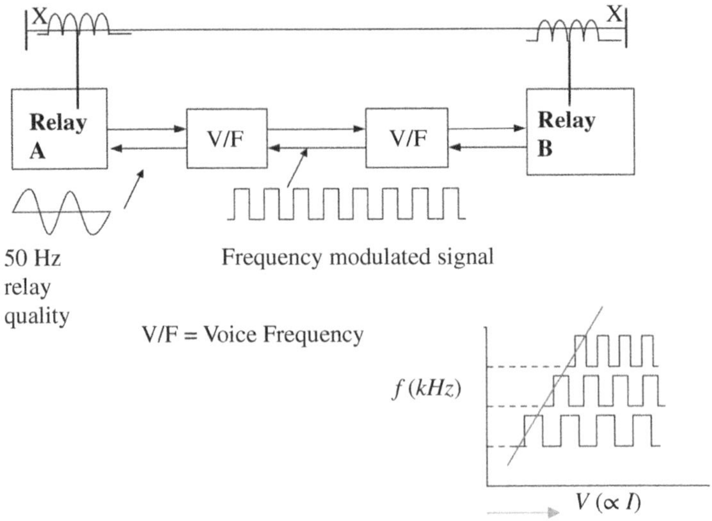

Figure 5.4 Differential protection based on frequency modulation.

Figure 5.5 shows the basic arrangement of digital current differential protection, the development of which has been made possible by the advent of high-capacity digital communication channels such as optical fiber. Fiber-optic links are now finding widespread usage in telecommunications and therefore can be rented from telephone companies for protection signaling.

In digital differential protection, since several pieces of data can be carried within a digital message at high speeds, it is possible to send the current signals of each phase to the remote and of the protected zone for comparison purposes, thus

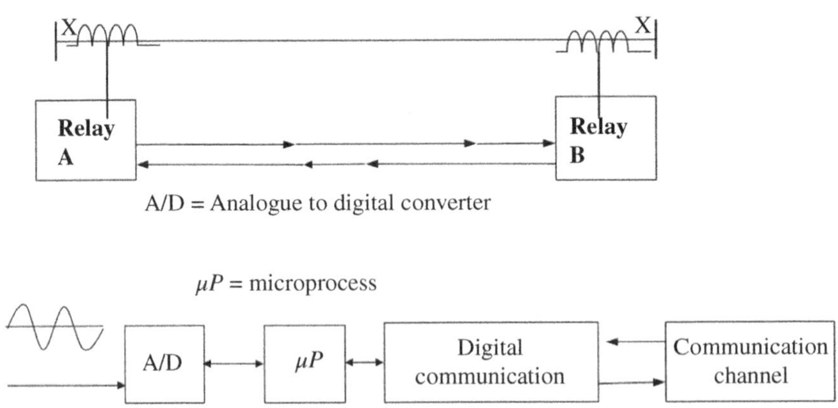

Figure 5.5 Digital differential protection.

obviating some of the protection problems associated with summated quantities. Furthermore, status information can also be sent within the digital message. It is normal practice to code the sampled power frequency current signals into binary data (on and off), and each signal is then sent as an 8-bit data stream. At the receiving end, the coded signals are converted back into original digital current signals by dividing the continuous bit stream into 8-bit words. It is apparent that since the transmitted data is simple in the form of on and off signals, it is far less prone to nonlinearities in the circuit and is much more tolerant to the presence of any circuit noise.

Exercise 5.3 *Outlines the advantage of digital differential protection over conventional analog technique.*

First, conventional analog techniques are prone to maloperation because of a requirement to summate three-phase quantities into a single-phase quantity (this can result in certain faults being undetected); and second, because of the nonlinearities in the metallic pilot wires due, for example, to electromagnetic interference. A digital technique, in which analog signals are coded into binary signals, is less prone to communication medium nonlinearities, and also if used in conjunction with a rented fiber optical link, then all three-phase quantities can be transmitted at high speeds, minimizing the problems associated with single summated quantities.

5.5 Directional Earth Fault Protection

Figure 5.6 shows how inverse definite minimum time (IDMT) relays are connected in practice for detecting phase and earth faults. As can be seen, phase fault overcurrent relays are often provided on only two phases since these will detect any interphase faults, the connection to the earth fault relay being unaffected by this consideration.

Figure 5.6 Typical IDMT relay arrangement.

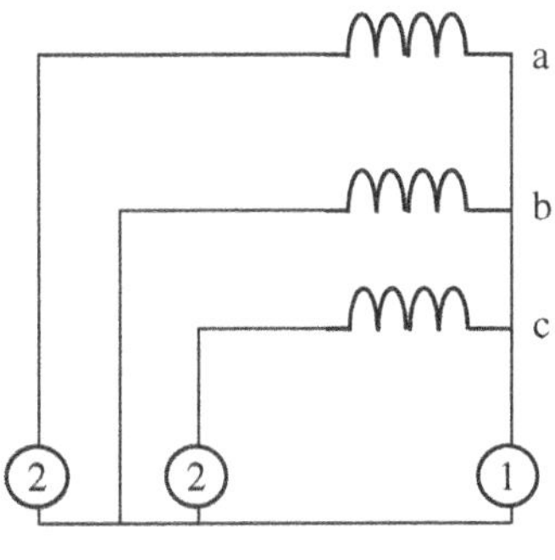

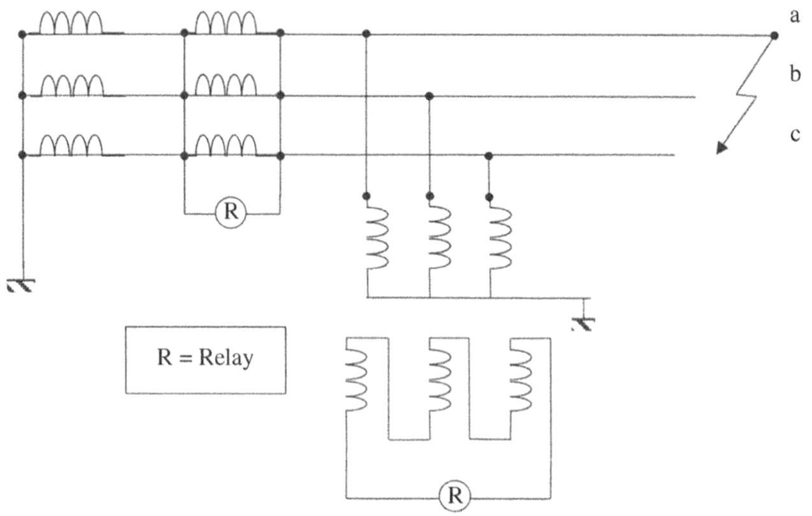

Figure 5.7 Voltage polarized directional earth fault relay.

However, with this type of arrangement, the phase fault relays can respond to the flow of earth fault current, and it is thus important that their directional response be correct for this condition (this is particularly so within IDMT relays are used for the protection of ring-distribution system). A related directional element is thus needed in this earth fault element. As can be seen, since the residual (or zero sequence) current may be derived from any phase, to obtain a directional response, it is necessary to obtain a related voltage. Such a voltage is the residual voltage of the system and is the vector sum of the individual phase voltages. An arrangement involving a five-limb voltage transformer (with a broken delta connection) as shown in Figure 5.7 provides the necessary residual voltage. This will be zero for balanced voltages, but for representing earth fault conditions, it will be equal to the depression of the faulted phase voltage. The residual quantities (both voltage and current) are thus applied to the directional element of the earth fault relay, which compares the phase angle between the two to ascertain the direction of the earth fault.

5.6 Rough Balance Protection

A variation of the simple differential protection provides both differential protection and generally system backup and overload protection using the same triple pole IDMT relay. The modified scheme is known as rough balance protection. These are primarily used for protecting either transformers or transformer feeders.

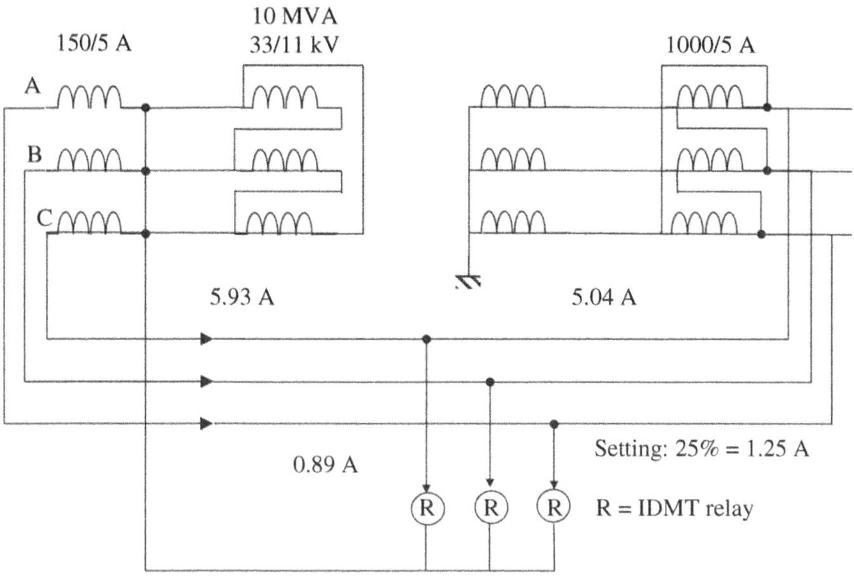

Figure 5.8 Rough balance protection.

A normal circulating current system is set up, but the ratio of the CTs (on either side of the protected zone) is deliberately unbalanced by a small amount. The relay is a standard IDMT type having a setting above the differential current during full load conditions. In the example illustrated in Figure 5.8 (which shows the technique applied to a transformer feeder), the transformer is rated 10 MVA, 33/11 kV connected delta-star.

Exercise 5.4 *Calculate the full load currents (I_{FL}) on the* 33 kV *and* 11 kV *sides.*

$$VA = \sqrt{3}V_{FL}I_{FL} \tag{5.1}$$

Thus, on the 33 kV side:

$$I_{FL} = \frac{MVA}{kV} \tag{5.2}$$

$$I_{FL} = \frac{10 \times 10^6}{\sqrt{3} \times 33 \times 10^3} = 175\ \text{A}$$

On the 11 kV side

$$I_{FL} = \frac{10 \times 10^6}{\sqrt{3} \times 11 \times 10^3} = 525\ \text{A}$$

CTs are 150/5 A star-connected on the 33 kV side and 900/5 A or 520/2.89 A delta-connected on the 11 kV side. Full load three-phase current will produce outputs from the CT groups as follows:

33 kV side:

$$\mathrm{CT}_{\mathrm{output}} = \frac{175 \times 5}{150} = 5.93\ \mathrm{A} \tag{5.3}$$

11 kV side:

$$\frac{525 \times 5}{900\sqrt{3}} = 5.04\ \mathrm{A}$$

A current of 0.89 A will therefore flow in the relay.

Exercise 5.5 *If the relay has a setting of 25% of 5 A, what will be the effective overload setting of the system.*

$$25\%\ \text{of}\ 5\ \mathrm{A} = 1.25\ \mathrm{A} \tag{5.4}$$

Thus, % overload setting

$$\text{Effective overload} = \frac{1.25}{0.89} \times 100 = 140\% \tag{5.5}$$

This is 140% of the rated current of the transformer. It can thus be seen that both differential protection with a nominal setting of 25% and the general overload feature are obtained with the same relay.

5.7 Self-Assessment Questions

Self-Assessment Question 5.1

Self-Assessment Question 5.1a

Explain why it is necessary to condense information into single-phase quantities at each end of a multiphase feeder employing a unit protection scheme of the current differential type. Outline the principles of the two main methods, that is, the outlines the principle of the two main methods, that is, the summation transformer and the phase sequence segregating network, used in practice and briefly discuss the reasons why the former can sometimes lead to "blind spots" under certain fault conditions.

Self-Assessment Question 5.1b

Give reasons for the need to condense information into a single-phase quantity at each end of a multiphase feeder employing a unit protection scheme of the current differential type. Outline the principle of a summation transformer used in practice.

Self-Assessment Question 5.1c

The summation transformer used for deriving relaying quantities for a differential feeder protection scheme is supplied with an unbalanced three-phase input of phase sequence a, b, and c, in the nominal manner. If the system conditions are such that the positive and negative phase sequence components in the "c" phase are identical and ratio positive and zero phase sequence components is 4. Show that a value of ($n = 3$) used with a 1:1:n primary turn ratio transformer will give a zero secondary output and will therefore be unsatisfactory.

Self-Assessment Question 5.2

Self-Assessment Question 5.2a

With the aid of circuit diagram, explain the principle of the *Merz-price* circulating current system used to protect a three-phase feeder. Give reasons why biasing is necessary in such a system.

Self-Assessment Question 5.2b

A 33-kV transformer feeder arrangement is protected with a Merz price circulating current protection scheme. The transformer on the high-voltage side is star connected and is delta connected to the low-voltage side. If the CT on the low-voltage side has a ratio 600/5 A, find the CT ratio on the high voltage.

Bibliography

Jamil Abdo, Tariq Masood "New Technique to Develop Intelligent, Highly Reliable & Efficient Unmanned Power Generation Modular" published in 22nd International Conference on Renewable Energies and Power Quality (ICREPQ'24) Bilbao (Spain), 26–28 June 2024. https://www.icrepq.com/icrepq24/392-24-abdo.pdf.

Jamil Abdo, Tariq Masood "Operational Reliability – Production Best Practice Pragmatic Case Studies & Improvement" published in Proceedings of the 9th North American Conference on Industrial Engineering and Operations Management Washington D.C., USA, 4–6 June 2024.

Jamil Abdo, Tariq Masood "Techno Economics Analysis to Reshape Global Clean and Green Energy" published in 22nd International Conference on Renewable Energies and Power Quality (ICREPQ'24) Bilbao (Spain), 26–28 June 2024. https://www.icrepq.com/icrepq24/415-24-abdo.pdf.

Nasser Alemadi, Tariq Masood, R Aggarwal, Suhail A Qureshi, D Kothari "Novel Selection Approach of STATCOM Device Tailored and Optimized Engineering Configuration to Implement in Between Qatar and Bahrain" published in International Journal of Automation and Power Engineering. 2014. https://doi.org/10.14355/ijape.2014.0302.01.

Qingfa Chai, Jiayin Jiang, Yi Tang, Anning Wang, Congcong Li "High Reliability Relay Protection Setting Scheme of Distribution Network" published in 2023 International Conference on Power System Technology (PowerCon) 2023.

Dong Chen, Bing Hu, Mingjun Xue, Yinian Qi, Changsong Oin "Research and Application of the System of Relay Protection and Self-Healing for Distribution Network with Mixed Use of Circuit Breaker and Load Switch" published in 2023 International Conference on Power System Technology (PowerCon) 2023.

Jane Courtney, Aaron McDonnell "Impact on Distribution System Protection with the Integration of EG on the Distribution Network" published in 2019 54th International Universities Power Engineering Conference (UPEC) 2019.

Bruno Fazio, Sanja Bajic, Serge St-Antoine "Relay Protection Coordination Analysis using Fault Current Measurements: A Case Study of Hydro Quebec Distribution Network" published in 2022 IEEE Power & Energy Society General Meeting (PESGM) 2022.

Renan H Furlan, Carlos H Beuter, Rodrigo P Bataglioli "Improvement of Overcurrent Protection Considering Distribution Systems with Distributed Generation" published in 18th International Conference on Harmonics and Quality of Power (ICHQP) 2018.

Hana A Halim, Melaty Amirruddin, Nur S Noorpi, Nurhakimah M Mukhtar "An Improved Protection Scheme for Smart Distribution Grid" published in 2013 1st International Conference on Artificial Intelligence, Modelling and Simulation 2013.

Rani Kumari, Bhukya K. Naick "Protection Coordination Impact on Reliability of Distribution Systems With Distributed Energy Resources" published in 2024 Third International Conference on Power, Control and Computing Technologies (ICPC2T) 2024.

Da Lei, Yunlei Yang, Mingjian Hu "Solutions for the Distributed Photovoltaic Access Distribution Network Relay Protection" published in 5th Asia Conference on Power and Electrical Engineering (ACPEE) 2020.

Bin Li, Chao Li, Fengrui Guo, Ying Xin "Overcurrent Protection Coordination in a Power Distribution Network With the Active Superconductive Fault Current Limiter" published in IEEE Transactions on Applied Superconductivity 2014.

Qin Lijun, Wang Ying, Zheng Shenrui, Guo Qing "Situation of Study on Distribution Network Protection Containing Distributed Generation and Application Scheme of Intelligent Protection" published in 2011 4th International Conference on Electric Utility Deregulation and Restructuring and Power Technologies (DRPT) 2011.

Tariq Masood, Jamil Abdo "Maintenance Strategy Production Best Practices – Pragmatic Case Studies and Operations Improvement" published in Proceedings of the 9th North American Conference on Industrial Engineering and Operations Management Washington D.C., USA, 4–6 June 2024.

Tariq Masood, Jamil Abdo "New Technique to Enhance Hybrid Renewable Energy Resources' Operational Flexibility and Reliability" published in 22nd International Conference on Renewable Energies and Power Quality (ICREPQ'24) Bilbao (Spain) 26–28 June 2024. https://www.icrepq.com/icrepq24/396-24-masood.pdf.

Tariq Masood, Jamil Abdo "Review: Uninterrupted Liquid Hydrogen (E-Methanol) Production Framework to Deliver Clean & Green Energy" published in 22nd International Conference on Renewable Energies and Power Quality (ICREPQ'24) Bilbao (Spain) 26–28 June 2024. https://www.icrepq.com/icrepq24/414-24-masood.pdf.

Tariq Masood, N Al-Emadi, SA Qureshi, YY Almulla, AJ Khan, SK Shah "FACTS Control Devices (Statcom, SSSC and UPFC) Re-Configuration Techniques By PSIM/MATLAB" published in 2007 International Conference on Electrical Engineering 2007. http://ieeexplore.ieee.org/abstract/document/4287295.

Tariq Masood, Raj K Aggarwal, SA Qureshi, RAJ Khan "STATCOM and SVC Control Operations and Optimization During Network Fault Conditions" published in 2010 IEEE International Symposium on Industrial Electronics 2010. https://ieeexplore.ieee.org/document/5636868.

Tariq Masood, Raj K Aggarwal, Suhail Aftab Qureshi, RAJ Khan "STATCOM Model Against SVC Control Model Performance Analyses Technique by MATLAB" published in International Conference on Renewable Energies and Power Quality (ICREPQ'10) 2010. https://icrepq.com/icrepq'10/655-Masood.pdf.

Tariq Masood, Suhail A Qureshi, RK Aggarwal, DP Kothari "Multiple FACTS Devices Integration Through Distributed Control System and Its Application on the GCC

Interconnection" published in 2013 7th IEEE GCC Conference and Exhibition (GCC) 2013. https://ieeexplore.ieee.org/abstract/document/6705750.

Tariq Masood, Abdel Aty Edris, Suhail Aftab Qureshi, Muhammad Tajammal, Murtaza Hashmi, DP Kothari, Samer Kareem "SSSC Tailored to Optimization Performance in Between United Arab Emirates and Oman on the GCC Power Grid" published in the 13th IET International Conference on AC and DC Power Transmission Conference on 14–16 February 2017. http://ieeexplore.ieee.org/document/7934934.

Tariq Masood, Abdel Aty Edris, Suhail Aftab Qureshi, Muhammad Tajammal, Murtaza Hashmi, DP Kothari, Samer Kareem "Smart Grid Operations and Control Challenge by Implementing STATCOM Tailored to Optimization Performance at the GCC Power Grid" published in the 13th IET International Conference on AC and DC Power Transmission Conference on 14–16 February 2017. http://ieeexplore.ieee.org/document/7934988.

Tariq Masood, Muhammad Tajammal, Samer Kareem, Suhail Aftab Qureshi, Murtaza Hashmi DP Kothari, "Smart Grid Operational Functions and Control Challenges by Implementing SSSC Tailored to Optimize Performance in Between Qatar and KSA on the GCC Electrical-Power Grid" published in the ASTESJ Journal Advances in Science, Technology and Engineering Systems Journal Vol. 2, No. 1, 2017. https://astesj.com/special-issue-4.

Tariq Masood, Muhammad Tajammal, Samer Kareem, Suhail Aftab Qureshi, Abdul Jabbar Khan, DP Kothari, Nair Jameel "Unified Power Flow Controller Modeling and Analysis Technique on the GCC Power Grid Operations and Control" published in 2018 IEEE International Energy Conference, Limassol, Cyprus, 7–9 June 2018. https://ieeexplore.ieee.org/document/8398834.

Tariq Masood, Muhammad Tajammal, Samer Kareem, Suhail Aftab Qureshi, Murtaza Hashmi, Nasir Jameel "Unified Power Flow Controller (UPFC) Modeling and Analysis Technique by Using Matlab and EMTP" published in International Conference on Renewable Energies and Power Quality (ICREQP'18) Salamanca (Spain) 21–23 March 2018. http://www.icrepq.com/icrepq18/250-18-masood.pdf.

Tariq Masood, Muhammad Tajammal, Samer Kareem, Suhail Aftab Qureshi, Murtaza Hashmi, Naser Jameel "STATCOM Controller Performance Enhancement and Optimization Technique" published in International Conference on Renewable Energies and Power Quality (ICREQP'18) Salamanca (Spain) 21–23 March 2018. http://www.icrepq.com/icrepq18/249-18-masood.pdf.

Tariq Masood, Samer Karim Shah, Muhammad Tajammal, "Unified Power Flow Controller Modeling and Analysis Technique on the GCC Power Grid" published in 2018 IEEE International Energy Conference (ENERGYCON) 2018. https://ieeexplore.ieee.org/abstract/document/8398834.

Vassilis C. Nikolaidis, Evangelos Papanikolaou, Anastasia S. Safigianni "A Communication-Assisted Overcurrent Protection Scheme for Radial Distribution

Systems With Distributed Generation" published in IEEE Transactions on Smart Grid Vol. 7, No. 1, 2015.

Harsha Ravindra, M Omar Faruque "Impact of PV on Distribution Protection System" published in North American Power Symposium (NAPS) 2012.

M Saeed Ullah, Tariq Masood, Umair Zahid, H Abuzar Ahsan "Electrochemical Oxidation of Dissolved Organic Dye in Aqueous Solution" published in IMASCON-2019.

VS Sazanov, AI Kovalenko, AA Voloshin "The Development of System for Automatic Adaptive Change of Relay Protection Settings in Distribution Networks" published in 2022 5th International Youth Scientific and Technical Conference on Relay Protection and Automation (RPA) 2022.

Hesam-Edin Hayati Soloot, Edris Agheb, Amir Hayati Soloot, Shahab Moghadam "A SWOT Analysis of Two Protection Strategies Due to the Expansion of Renewable Distributed Generation Distribution Network" published in 15th International Conference on Protection and Automation of Power Systems (IPAPS) 2020.

Xiaohui Song, Yu Zhang, Song Zhang, Shizhan Song, Jing Ma, Weibo Zhang "Active Distribution Network Protection Mode Based on Coordination of Distributed and Centralized Protection" published in 2017 China International Electrical and Energy Conference (CIEEC) 2017.

Tariq Masood, RK Aggarwal, SA Qureshi, RAJ Khan "Statcom Control Reconfiguration Technique for Steady State and Dynamic Performance Optimization During Network Fault Conditions" published in International Conference on Renewable Energies and Power Quality (ICREPQ'10) Granada (Spain) 2010. https://www.icrepq.com/icrepq'10/656-Masood.pdf.

Chong Tong, Yunfeng Cai, Ziyang Zhang "Artificial Intelligence-Based Lightning Protection of Smart Grid Distribution System" published in International Symposium on Lightning Protection (XIV SIPDA) 2017.

Kun Yang, Zhen Xiang, Ruixiong Yang "Decentralized Protection Coordination Based on Multi-Agent System for Active Distribution Networks" published in 7th Asia Conference on Power and Electrical Engineering (ACPEE) 2022.

Kun Yang, Zhen Xiang, Ruixiong Yang, Lin Jiang "Decentralized Protection Coordination Based on Multi-Agent System for Active Distribution Networks" published in 2022 7th Asia Conference on Power and Electrical Engineering (ACPEE) 2022.

Ping Zeng, Zhaoxin Du, Desheng Zhou, Xiaofeng Zou, Qian Feng "A Study on Adaptive Protection System of Distribution Network Based on Passive Protection Terminal of Ring Network Cabinet" published in IEEE 6th Conference on Energy Internet and Energy System Integration (EI2) 2022.

6

Transformer Protection

6.1 Objectives

When you have completed this section, you will be able to:

- List the main types of faults affecting the transformer
- Describe the different types of transformer protection
- Demonstrate some of the typical constraints and application considerations for different types of transformer protection

6.2 Introduction

In the context of the Power System Plant (Power Generation Plant), the transformer is probably the most versatile item since it is available in a bewildering selection of sizes, voltage levels, ratings, and connections. Distribution transformers range from a few kVA to large transformers exceeding 10 MVA, which require forced cooling and on-load tap changing. Almost without exception, all these transformers have one common feature: they are all constructed from copper windings wound on an iron core immersed in an oil tank. Another common feature is that, like all items of a power system plant, they are susceptible to faults. This section will examine the general area of transformer protection.

The main types of faults affecting transformers are as follows:

- Earth fault
- Phase fault
- Interturn fault
- Core fault

Protection of Electrical Power Distribution Systems: Smart grid, Microgrid, AI, and Cyber Security, First Edition. Tariq Masood, Jamil Abdo, and Atif Iqbal.

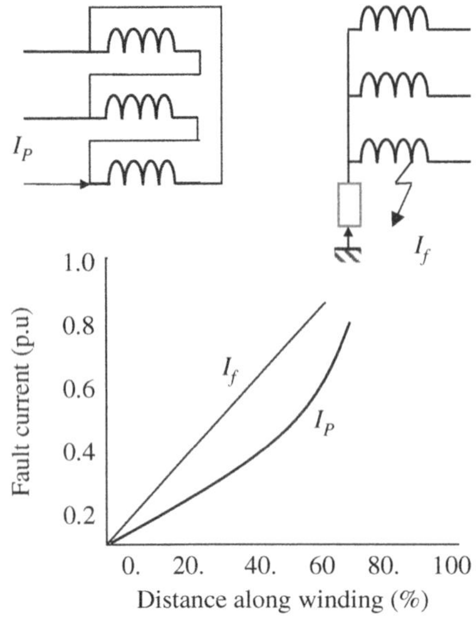

Figure 6.1 Typical transformer fault currents for a resistance earthed star winding.

6.2.1 Earth Fault

Faults between transformer windings and the earthed case will give rise to fault currents, which are highly dependent on the transformer earthing arrangement. For star-connected windings with resistance earthing, the fault current is influenced by the resistance value and the position of the fault. For a fault on the star winding, the distribution of fault currents, Figure 6.1, shows a virtually linear increase in fault current with position. If the star point is solidly earthed (Figure 6.2), the reactance in the fault path changes with position, being highly influenced by the leakage reactance of the winding.

This leads to a complex variation in fault current versus position. For delta-connected windings, there will be fault current generation at all positions of faults since the minimum phase to earth voltage along such a winding is half of the phase voltage value and occurs at the winding midpoint. A distribution of the fault current versus position is shown for a general case in Figure 6.3.

6.2.2 Phase Fault

Due to the physical separation of the phases of a three-phase transformer, phase-to-phase faults are rare. When they do occur, they lead to large fault currents.

Figure 6.2 Typical transformer fault currents for solid earthed star winding.

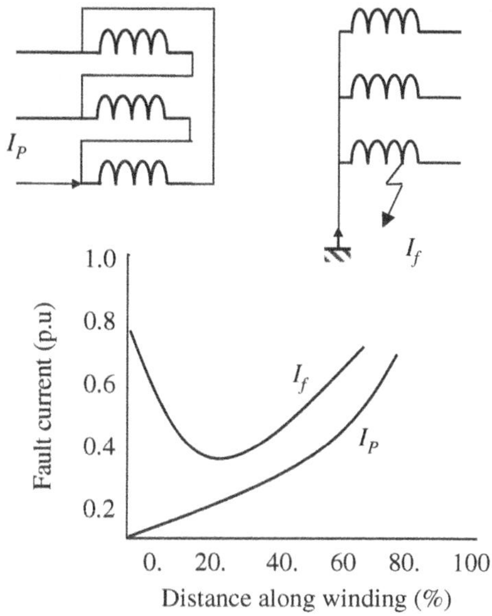

Figure 6.3 Typical transformer fault currents for delta winding: Note: general case only; I_f highly dependent on delta winding side earthing arrangements.

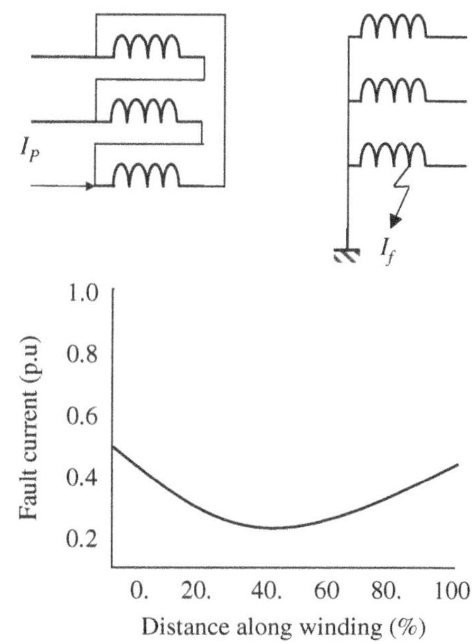

6.2.3 Interturn Fault

Interturn faults occur due to the breakdown of interwinding insulation, and such faults are not uncommon. On low-voltage transformers, these faults can follow incorrect transformer operation, such as high and prolonged through currents, or poor maintenance, such as loss of oil to allowing moisture ingress, which can cause deterioration of the insulation. On higher-voltage transformers, the presence of a steep-fronted impulse voltage will have a similar detrimental effect on the interwinding insulation. Note that earth faults are unlikely to be produced by this fault mechanism due to the higher winding of the earth insulation. Interturn faults give rise to heavy current flow in the shorted turns, which causes substantial heating of the transformer; externally, however, only small terminal currents will flow as a result.

6.2.4 Core Fault

Even if the integrity of the winding insulation is maintained in a transformer, the possibility of faults to the core causing heat generation exists, with a consequent risk of damage to the windings. Shorted lamination of the core can lead to substantial eddy current heating. A possible mechanism for this effect is degradation of the core clamping bolts, which can be caused by over-fluxing in abnormal operating conditions. In addition, failure of the cooling system or loss of tank oil can lead to core overheating.

6.3 Protection Overview

The protection best suited for a particular transformer will reflect the size and position of the transformer in the power system. For small distribution transformers that have no associated circuit breaker, the only option will be the provision of fuses – usually of the high rupturing capacity (HRC) type. For larger transformers, typically greater than 100 kVA, a circuit breaker will be fitted, allowing conventional overcurrent and earth fault relays to be applied. In view of the small fault currents for star winding faults close to the star point where resistive earthing is employed (Figure 6.1), restricted earth fault protection will give a better performance than conventional earth fault protection. For large distribution transformers employing forced cooling and on-load tap changers, typically in the range 5–10 MVA, some form of differential protection will be used. A further point regarding distribution transformer protection is that high-speed tripping is rarely required, since, unlike the transmission system, the stability of the distribution system is largely unaffected by faults.

In addition to these types of electrical protection, overheating protection will also be incorporated. Simple forms of overheating protection monitor the tank oil temperature and raise an alarm should it exceed some predefined limit. It should be noted that, owing to the relatively large thermal inertia of the transformer compared to, say, an overhead line, overheating need not dictate immediate action, and thus raising an alarm rather than tripping out the transformer may be sufficient. Severe localized heating can lead to the evolution of a gas bubble within the oil, which can be easily detected (Buchholz relay).

6.4 Differential Protection

6.4.1 Factors Affecting Differential Protection

Differential protection is commonly used on larger transformers. The following factors need to be considered when applying this form of protection:

- Transformation ratio
- Phase shifts in star/delta transformers
- Tap changing
- Magnetic inrush

6.4.2 Transformation Ratio

The basic application for an unbiased differential relay is shown in Figure 6.4 for a delta/delta transformer. Since the transformer has differing voltages on the primary and secondary, it can be rated with current transformer (CT) ratio. To specify the CT rating, the full load currents must be considered. Thus, if the transformer rating is M, then the full load currents will be:

$$\frac{M}{\sqrt{3} \times V_{lv}} = I_{lv} \tag{6.1}$$

and

$$\frac{M}{\sqrt{3} \times V_{hv}} = I_{hv} \tag{6.2}$$

The CTs must be chosen so that their rating is greater than the full load current. In Figure 6.4, let "K" represent the transformation ratio:

$$K = \frac{V_{hv}}{V_{lv}} = \frac{I_{lv}}{I_{hv}} \tag{6.3}$$

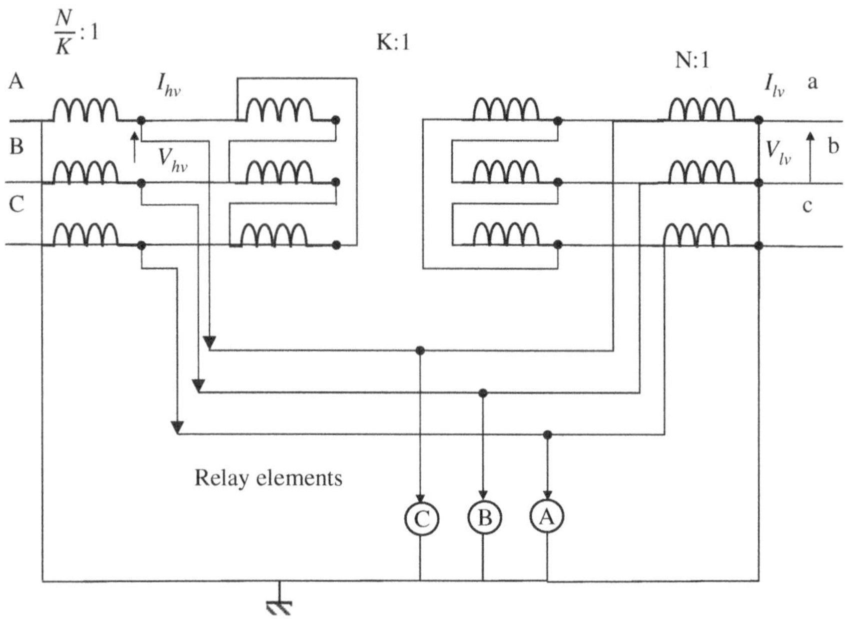

Figure 6.4 Unbiased differential scheme for delta/delta transformer.

If the CT secondaries are to be 1 A, then the CT ratio on the low-voltage winding will be the nearest ratio available to I_{lv}: 1. Now, taking into account the transformation ratio K, the CT ratio on the high-voltage winding will be the nearest ratio available to

$$I_{hv} : 1, \quad or \quad \frac{I_{lv}}{K} : 1.$$

Because the differential protection is influenced by the transformation ratio, it is capable of detecting an **interturn fault.** Although little change in line currents occurs for this fault, the transformation is affected. The relay elements shown in Figure 6.4 will be of the inverse time lag type. Occasionally, these will be used in conjunction with high-set instantaneous relays to give high-speed operation for high-current faults.

6.4.3 Phase Shifts in Delta/Star Windings

A delta/star transformer introduces a 30-phase shift across the terminals, for which it is necessary to compensate to ensure that load or external fault through current is measured in phase by both CTs. This is achieved by altering the CT connection on each side of the transformer. Figure 6.5 shows an unbiased

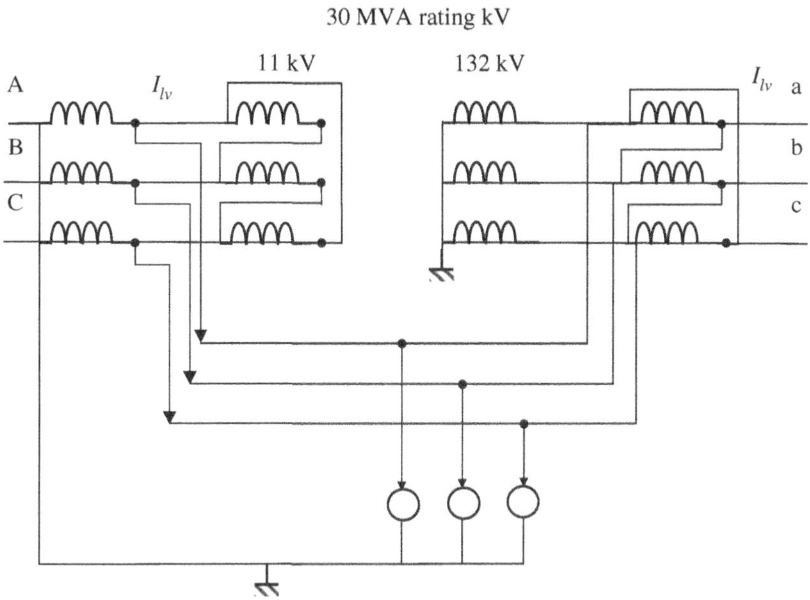

Figure 6.5 Unbiased differential scheme for delta/star transformer.

differential relay application where the delta side CTs are star connected and vice versa. This has two advantageous effects.

a) Elimination of phase shift. The delta/star transformer winding produces a +30° phase shift, whereas the delta star CT connection produces a −30° phase shift, thus giving a 0° phase shift to the transformed relaying currents.
b) Zero current elimination. In a transformer of this type with an earthed star point, zero sequence currents can flow through the star windings due to an external fault. This will produce transformed zero sequence currents circulating in the delta windings. If star-connected CTs were used on both sides of the transformer, then the zero sequence currents would be transformed on the high-voltage star-connected side, but not on the low-voltage delta-connected side, and could lead to the differential protection operating for external faults. However, by connecting the CTs attached to the star winding in delta formation, no zero-sequence current can pass through to the relay elements.

In applications where the CTs are connected in delta formation, it is necessary to derate the CT rating by $\frac{1}{\sqrt{3}}$.

Exercise 6.1 To calculate suitable CT ratios for the unbiased differential protection application in Figure 6.5, what is the spill current under full load for your choice of CT ratios?

First, we will calculate the full load currents.

$$I_{hv} = \frac{M}{\sqrt{3} \times V_{hv}} = \frac{30 \times 10^6}{\sqrt{3} \times 132 \times 10^3} = 131\ \text{A} \tag{6.4}$$

and

$$I_{lv} = \frac{M}{\sqrt{3} \times V_{lv}} = \frac{30 \times 10^6}{\sqrt{3} \times 11 \times 10^3} = 1575\ \text{A} \tag{6.5}$$

On the high-voltage winding, the CTs are of delta formation, and so the assumed full load current becomes $\frac{131}{\sqrt{3}}$ which is 76 A.

Thus, CTs must be chosen to transform currents of 1575 A and 76 A. Assuming that a secondary current of 1 A is required, we will choose a CT ratio of 1600 : 1 *and* 80 : 1, respectively. Note that this is a convenient choice of CT ratio; under some conditions, there may not be a closely suitable CT ratio available, which will consequently lead to a greater spill current. In order to calculate the spill, current I_{spill}, we must calculate the difference of the CT secondary current at full load:

$$I_{spill} = \frac{1575}{1600} - \frac{76}{80} = 0.034\text{A}$$

Exercise 6.2 Does the value of full load spill current allow the basic sensitivity of the relay to be set?

The answer to this question is no, and the relay is to remain stable under fault conditions. For through faults, the full load current may be exceeded many times. A typical factor considered for through-fault calculations is 10, thus, the spill current for the previous example would be 0.34 A, which would determine the basic relay sensitivity. Note that this would give a high sensitivity since the spill current is low.

6.4.4 Tap Changing

Most transformers allow variations in the transformation ratio by the use of taps. On some transformers, this may be varied automatically by the use of on-load tap-changing equipment. With unbiased differential protection, the relay operating current will need to be set high to avoid unwanted operation for high through currents, thus such protection will be relatively insensitive. Such relays are set with the tap assumed in the center position.

Exercise 6.3 In the first exercise, if the 132 kV winding can be tapped between +10% and −20% of its nominal voltage, calculate the spill current under the extremes of tapping position?

We will assume the CT ratios previously calculated. Under the maximum voltage, that is, 132 kV + 10%, the winding current will be:

$$I_{hv} = \frac{30 \times 10^6}{\sqrt{3} \times 1.1 \times 132 \times 10^3} = 119\text{A}$$

And the spill current is:

$$I_{spill} = \frac{1575}{1600} - \frac{119}{\sqrt{3} \times 80} = 0.126\text{A}$$

Under the minimum voltage, that is, 132 kV −20%, the winding current will be:

$$I_{hv} = \frac{30 \times 10^6}{\sqrt{3} \times 0.8 \times 132 \times 10^3} = 164\text{A}$$

And so the spill current now becomes:

$$I_{spill} = \frac{1575}{1600} - \frac{164}{\sqrt{3} \times 80} = 0.199\text{A}$$

It is clear from these figures that the presence of tap changers will lead to a further desensitization of the relay beyond the problems caused by CT ratio mismatch.

Insensitive relay settings and slow operating times can be overcome by using a biased differential relay. The application of biased protection and typical characteristics is shown in Figures 6.6 and 6.7, respectively. It is seen that the bias winding in the relay prevents operation under high through-current conditions.

This feature enables the overall protection to remain stable for external faults and increases its ability to withstand CT mismatch. Thus, the protection becomes more sensitive and can operate in a shorter time than the unbiased differential protection in an induction disc relay, giving a percentage characteristic.

6.4.5 Magnetic Inrush

When energized, a transformer can exhibit a phenomenon known as magnetic inrush. Figure 6.8 depicts typical line currents on an energized transformer winding, which show this effect. It is observed that inrush currents of up to six to eight times the rated transformer current can transiently occur. Inrush currents are caused by magnetic conditions within the core upon energization. The magnitude of inrush currents is determined by the point on the wave of the applied voltage at the point of switching and the state of the core flux prior to energization.

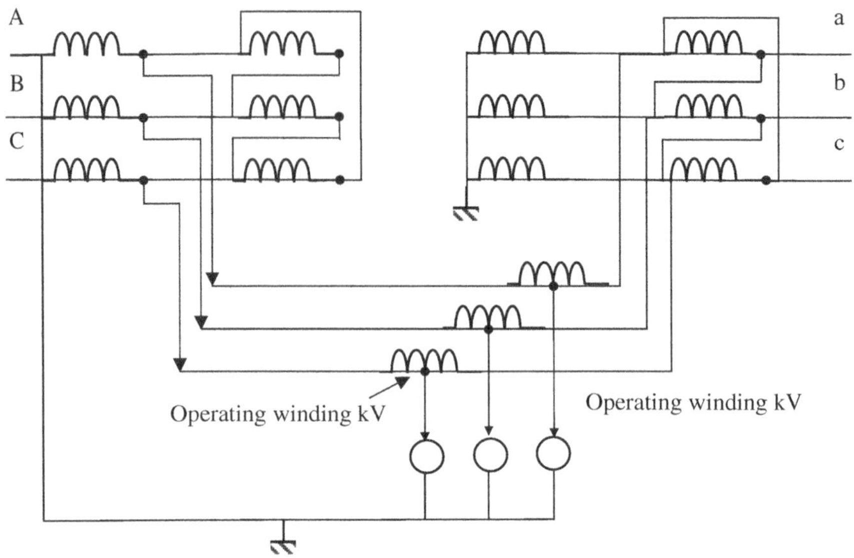

Figure 6.6 Biased differential scheme for delta/star transformer.

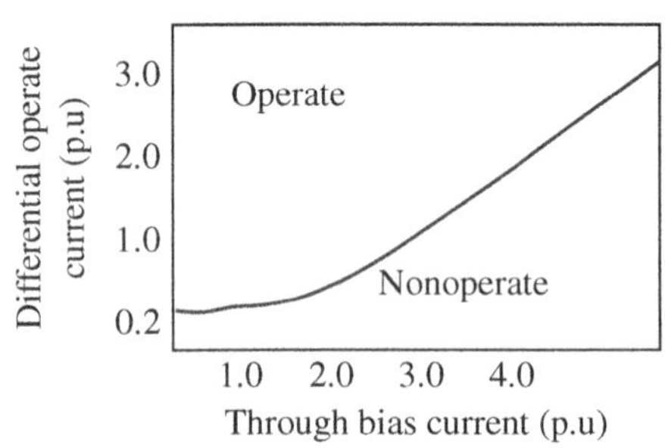

Figure 6.7 Typical biased characteristic.

Exercise 6.4 In Figure 6.9, can you sketch the shape of the steady-state flux in the core of the transformer which is produced by the shown sinusoidal winding voltage? What feature of the flux produces a positive half-cycle of the voltage.

From Faraday's law, the winding e.m.f. is proportional to the rate of change of flux, that is,

$$e \propto \frac{d\emptyset}{dt} \tag{6.6}$$

Since the winding voltage, or e.m.f. is sinusoidal, it follows from the above equation that Ø must be ***cosinusoidal*** and will appear as given in Figure 6.10. This figure shows that the winding e.m.f. positive half-cycle is produced as a result of the flux changing.

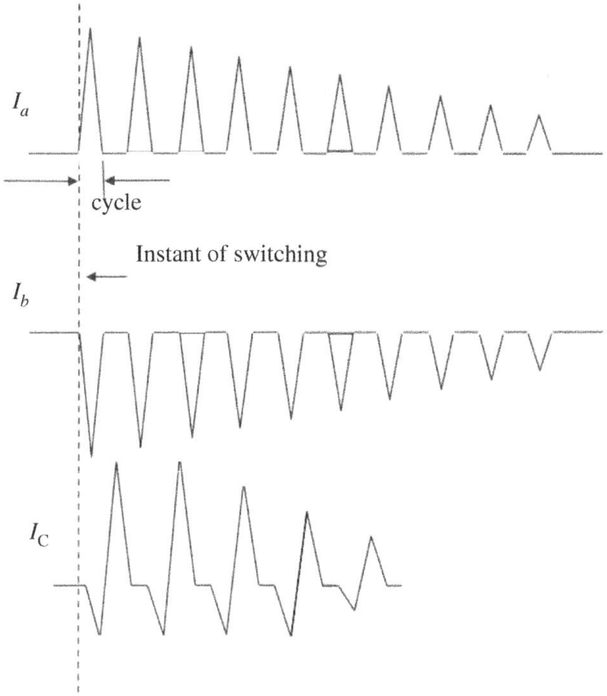

Figure 6.8 Typical magnetizing inrush currents for three-phase transformer.

Figure 6.9 Steady-state winding voltage.

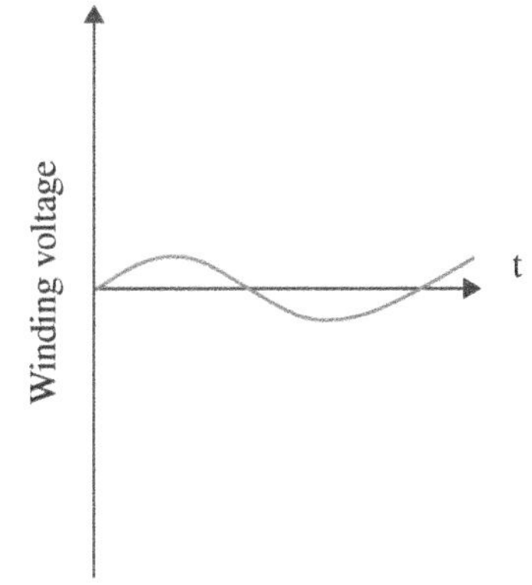

From its negative to positive peak, that is, a change of 2Ø, where Ø is the peak value of Ø. Note that the winding voltage is producing the flux rather than the other way around. In a transformer either explanation will be correct – it is a bilateral process – since the secondary voltage is produced as result of magnetic inrush,

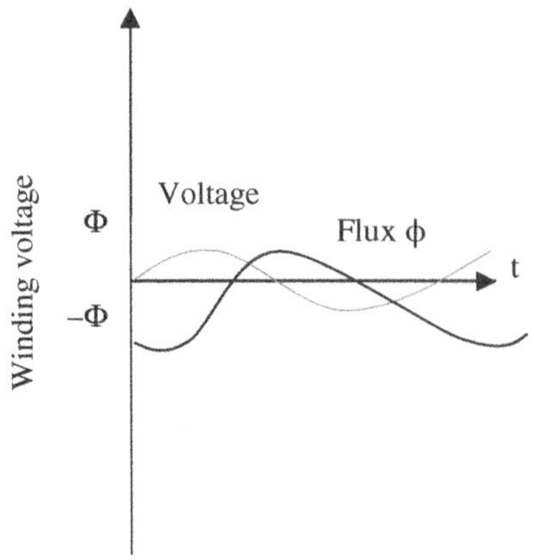

Figure 6.10 Steady-state winding voltage and flux.

which is only observed on the energized winding; it is the voltage causing the flux, although it is convenient, to aid understanding of this effect, and to think occasionally of the flux causing the voltage.

Exercise 6.5 If, in Figure 6.9, the voltage was not steady state, but applied at $t = 0$, how would the flux behave now to produce the first positive voltage half-cycle given that there is flux within the core prior to energization.

Simple answer to this question is that, again, the voltage positive half-cycle is caused by a sinusoidal change in the flux of 2Ø. Of course, in this situation, the flux must start from zero since this was the condition of the core prior to the voltage being applied. Hence, the flux curve will now look as shown in Figure 6.11, where it is seen to be almost fully offset for the first cycle. In order to produce such a large amount of flux, there must be a very large exciting current – the inrush current – and it is the nonlinearity of the core B-H loop which causes the initial currents, such as those shown in Figure 6.8, to distort to satisfy the flux conditions.

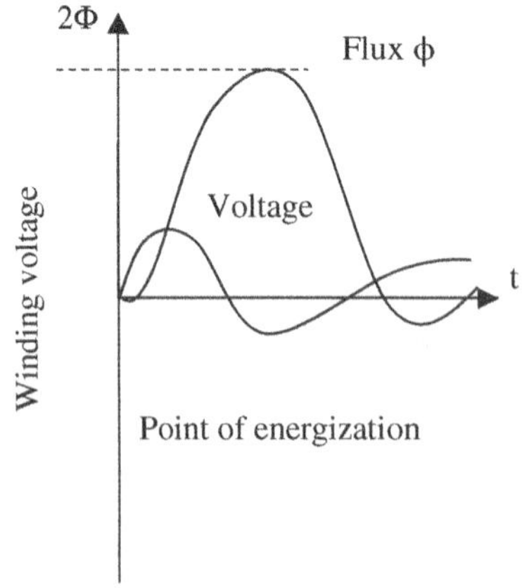

Figure 6.11 Transient winding voltage and flux.

Exercise 6.6 In the last exercise, how would be the flux behavior if the core was initially:

a) Magnetizing to +Ø
b) Magnetizing to −Ø

- If the core flux was initially +Ø, then the flux due to energization would be offset by Øand would thus begin at the end of 3Ø during the first half-cycle. The inrush current would thus be even larger and even more distorted.
- If the core flux was initially −Ø, then there would be no inrush current, and the flux would appear as given in Figure 6.9.

It is important to realize that inrush currents are largely unpredictable since they depend on the core flux at de-energization – remanent flux – and the point of the voltage wave at energization. On a three-phase transformer, they are unavoidable. In practice, the offset flux will decay away over time; the time constant for this increases with the size of the transformer and can be as large as 1 second for a large transformer. The occurrence of magnetic inrush can be detected for up to 30 minutes after energization on some transformers.

Since inrush currents only occur on the energized winding, they can appear to differential protection as an internal fault, and so this gives another constraint on the setting of differential protection. Although biased schemes may give good tolerance to through currents and CT mismatch, they need to be set high to avoid operation under inrush conditions unless other steps are taken to provide the protection immunity to this condition.

Various techniques exist for bias differential protection against the inrush current operation.

The simplest technique involves a time delay, which prevents any differential operation for up to 250 ms after energization. Refinement of this approach allows the timer to be overridden under heavy fault current conditions. The most commonly encountered bias technique uses the nonlinear feature of the inrush current, which has high second harmonic content. Thus, detection of the second harmonic can be used as a relay restraining quantity. A typical setting for the second harmonic would be to prevent relay operation if more than 15–20% of the second harmonic is detected.

6.5 Fuse Protection

For a smaller distribution transformer with no associated circuit breaker, the only option available for transformer protection is the fuse. Since fuses are basically overcurrent type devices, they have to withstand maximum load currents, short duration overload due to, for example, motor starting and magnetic inrush conditions described earlier. These constraints necessarily limit the sensitivity of the protection function. High rupturing capacity fuses are usually specified for transformers, yet, due to the slow operation of these fuses under three times rated current conditions, they are more likely to limit, rather than prevent transformer damage. How these slow operating times may aid coordination with protection devices on the secondary side of the transformer, since the fuse protection makes no allowance for heavy through currents.

6.6 Overcurrent Protection

Smaller transformers with a circuit breaker may use overcurrent relays as the primary protection, or alternatively overcurrent can be used as backup to differential protection. When used as primary protection, many of the previous remarks regarding fuse protection apply: the overcurrent relay must be set to avoid operating under short-duration overload and magnetic inrush (unless harmonic restraint is applied). Overcurrent protection will allow earth fault detection, although in view of the distribution of fault currents described in Section 6.1, sensitivity may be low. In addition, earth fault protection must be set to avoid operation due to single-phase loading of the transformer.

6.7 Restricted Earth Fault Protection

To overcome the problems of insensitive earth fault protection discussed in the previous subsection, particularly in the case of resistance earthing of the transformer star point, a unit type of earth fault protection referred to as restricted earth fault protection may be used.

Figure 6.12 shows the arrangement used to protect the star winding where the residual current from the three-line CTs is balanced against the current from a CT at the star point. This arrangement remains stable for all faults outside the protected zone and is affected by single-phase loading.

In addition, the relay will measure the total fault current and not just the contribution to the fault from the higher sensitivity. Restricted earth fault protection is also commonly applied to star winding with solid earthing, since substantial currents are measurable even for faults very close to the star point, and due to the high sensitivity of restricted earth fault protection, better overall winding protection may be achieved.

For protecting delta-connected windings, earth fault protection is inherently restricted since such windings will not allow the passage of zero sequence currents, and, again, highly sensitive relay settings may be used. Figure 6.12 shows the application of overcurrent and restricted earth fault transformer protection.

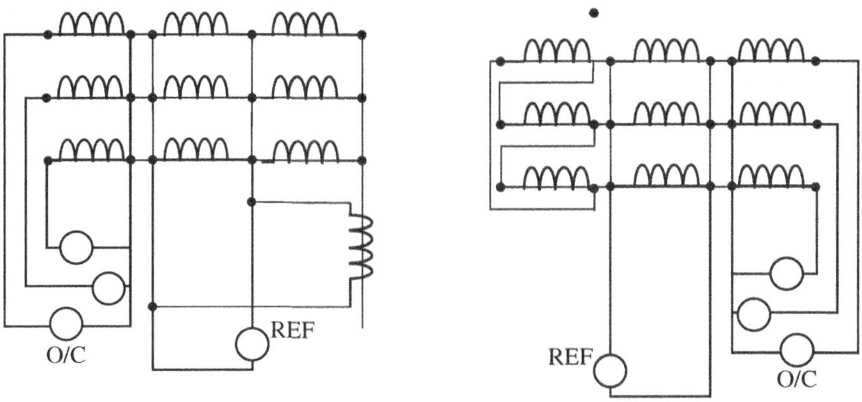

Figure 6.12 Overcurrent and restricted earth fault protection.

6.8 Buchholz Protection

The generation of gas within the transformer tank accompanies all faults; the volume of gas evolved will increase with the severity of the fault. The Buchholz relay detects gas evolution and applies to transformers with a separate oil conservator, typically greater than 1000 kVA. Figure 6.13 shows the basic outline of the Buchholz relay.

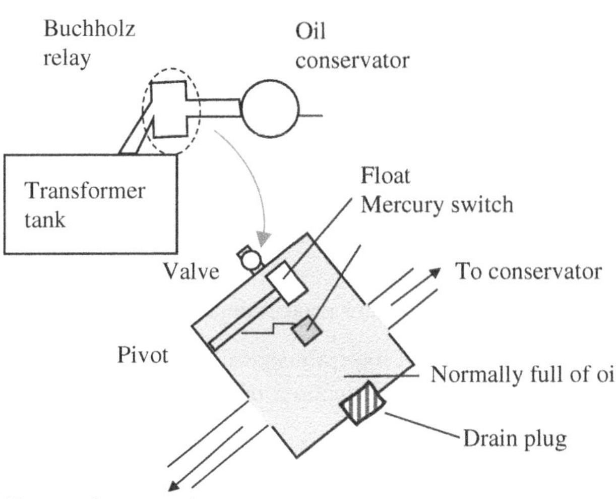

Figure 6.13 Buchholz relay.

If a gas bubble is produced in the transformer winding, the bubble will rise and become caught in the body of the Buchholz relay, allowing the pivoted float to fall and close the mercury switch. It is possible to mount two separate floats within the relay to give a graded indication of gas formation. Any air introduced to the oil system from another source, for example, the forced cooling system, can be periodically vented to the atmosphere through the valve at the upper point of the relay casing.

6.9 Winding Temperature Protection

Larger transformers with forced cooling will be provided with some form of winding temperature protection to prevent winding overheating due to either internal faults or failure of the cooling system. The sensing device detects internal faults or failures of the cooling system. The sensing device is installed at the top of the transformer tank. In addition to the temperature sensor, a heating element, powered from a line current CT, is placed in close proximity. The heater produces a local temperature rise proportional to loading, which, in conjunction with the oil temperature, allows the sensor to measure the average winding temperature.

The heater also takes into account the loading history of the transformer since it is arranged to have a thermal time constant similar to that of the transformer. Oil temperature of 95 °C or less are considered to be within the normal working range. Any elevation up to 10 °C above this temperature can be permitted for short periods, although long-term temperatures of this magnitude will cause damage to the windings. Winding temperature protection equipment typically sounds an alarm if the measured temperature is above 110 °C and will cause a trip if the temperature exceeds 125 °C.

6.10 Summary

The main factors to be considered for transformer protection are as follows:

- **Is A Circuit Breaker Fitted?:** if not, then fuse protection is the only option.
- **Size Transformer:** will dictate choice of protection: overcurrent/earth fault, biased differential.
- **Earthing Arrangement of Star Winding:** this will influence sensitivity of earth fault protection. Restricted earth fault protection will give the best sensitivity but needs extra CT.

- **CT Ratio:** may not be able to obtain an exact match, which will influence sensitivity of differential protection – especially if unbiased.
- **Tap Changing:** similar constraint to CT mismatch.
- **Sensitivity to External Faults:** heavy through currents may affect sensitivity of all types of protection except for biased differential.
- **Magnetic Inrush:** will affect all types of protection. The only exception is biased differential if some form of inrush restraint is applied, for example, second harmonic restraint.
- **Overheating/Core Fault Protection:** may need to fit over-heating protection and/or Buchholz relay, depending on the size of transformer.

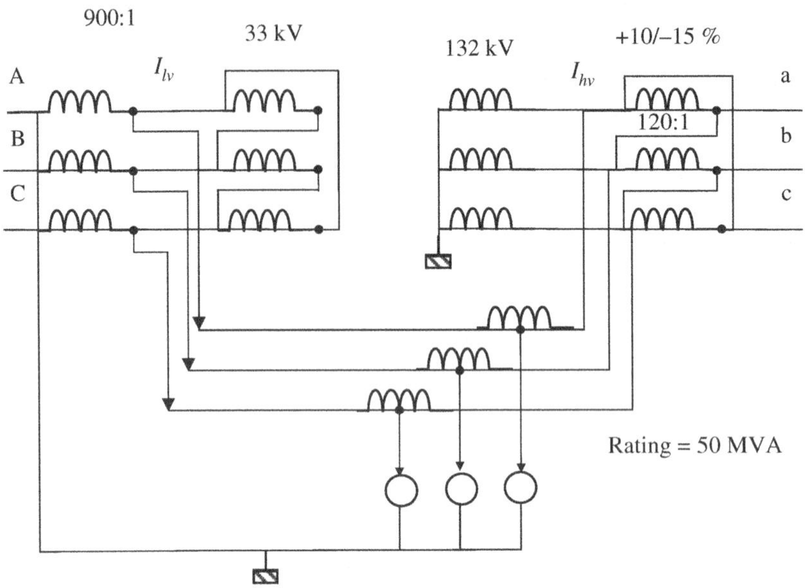

Figure 6.14 Calculating the suitable value for the relay sensitivity.

6.11 Self-Assessment Questions

Self-Assessment Question 6.1

Studying the unbiased differential scheme shown in Figure 6.14, calculate suitable value for the relay sensitivity, that is, minimum operating current. Explain your reasoning and comment on the answer.

Bibliography

Da Lei, Yunlei Yang, Mingjian Hu "Solutions for the Distributed Photovoltaic Access Distribution Network Relay Protection" published in 5th Asia Conference on Power and Electrical Engineering (ACPEE) in 2020.

L. Ndahepele, S. Chowdhurym "Impact of Distributed Generation on Traditional Protection in Distribution and Transmission Systems: A Review" published in IEEE PES/IAS PowerAfrica in 2020.

Cleberton Reiz, Jonatas Boas Leit "Impact Analysis of Distributed Generation on Protection Devices Coordination in Power Distribution Systems" published in IEEE PES Innovative Smart Grid Technologies Conference - Latin America (ISGT Latin America) in 2021.

Kun Yang, Zhen Xiang, Ruixiong Yang "Decentralized Protection Coordination Based on multi-agent system for Active Distribution Networks" Published in 7th Asia Conference on Power and Electrical Engineering (ACPEE) in 2022.

Renan H. Furlan, Carlos H. Beuter, Rodrigo P. Bataglioli "Improvement of overcurrent protection considering distribution systems with distributed generation" published in 18th International Conference on Harmonics and Quality of Power (ICHQP) in 2018.

Ping Zeng, Zhaoxin Du, Desheng Zhou, Xiaofeng Zou, Qian Feng "A Study on Adaptive Protection System of Distribution Network Based on Passive Protection Terminal of Ring Network Cabinet" Published in IEEE 6th Conference on Energy Internet and Energy System Integration (EI2) in 2022.

Xiaohui Song, Yu Zhang, Song Zhang, Shizhan Song, Jing Ma, Weibo Zhang "Active distribution network protection mode based on coordination of distributed and centralized protection" published in China International Electrical and Energy Conference (CIEEC). 2017.

Yuwei Xiong, Wei Chen "Positive sequence current differential protection based on least square method for distribution network with distribution generation" published in IEEE 4th International Electrical and Energy Conference (CIEEC). 2021.

Zong Guo, Jiarui Li, Hongbo Lian, Shihua Zhao, Yang Chen, Renjie Dai, Xiuneng Xu, Yufeng Gu, Yangyang Liu "Optimization Method for Protection of Distribution Network with MW-PV Based on Multi-level Cooperative Protection Strategy" published in IEEE Sustainable Power and Energy Conference (iSPEC). 2019.

Jamil Abdo, Tariq Masood "New Technique to develop Intelligent, highly reliable & Efficient Unmanned Power Generation Modular" published in 22nd International Conference on Renewable Energies and Power Quality (ICREPQ'24) Bilbao (Spain), 26–28 June 2024. https://www.icrepq.com/icrepq24/392-24-abdo.pdf.

Jamil Abdo, Tariq Masood "Operational Reliability- Production Best Practice Pragmatic Case Studies & Improvement" published in Proceedings of the 9th North American Conference on Industrial Engineering and Operations Management Washington D.C., USA, 4–6, 20 June 2024.

Jamil Abdo, Tariq Masood "Techno Economics Analysis to Reshape Global Clean and Green Energy" published in 22nd International Conference on Renewable Energies and Power Quality (ICREPQ'24) Bilbao (Spain), 26–28 June 2024. https://www.icrepq.com/icrepq24/415-24-abdo.pdf.

Jane Courtney, Aaron McDonnell "Impact on Distribution System Protection with the Integration of EG on the Distribution Network" published in 54th International Universities Power Engineering Conference (UPEC). 2019.

M. Saeed Ullah, Tariq Masood, Umair Zahid, H. Abuzar Ahsan "Electrochemical Oxidation of Dissolved organic Dye in Aqueous Solution" published in IMASCON-2019.

Nasser Alemadi, Tariq Masood, R Aggarwal, Suhail A Qureshi, D Kothari "Novel selection approach of STATCOM device Tailored and Optimized Engineering Configuration to Implement in between Qatar and Bahrain" published in International Journal of Automation and Power Engineering. 2014 10.14355/ijape.2014.0302.01.

Qin Lijun, Wang Ying, Zheng Shenrui, Guo Qing " Situation of study on distribution network protection containing distributed generation and application scheme of intelligent protection" published in 4th International Conference on Electric Utility Deregulation and Restructuring and Power Technologies (DRPT) 2011.

Rani Kumari, Bhukya K. Naick " Protection Coordination Impact on Reliability of Distribution Systems With Distributed Energy Resources" published in Third International Conference on Power, Control and Computing Technologies (ICPC2T). 2024.

Tariq Masood, Jamil Abdo "Maintenance Strategy Production Best Practices – Pragmatic Case Studies and Operations Improvement" published in Proceedings of the 9th North American Conference on Industrial Engineering and Operations Management Washington D.C., USA, 4–6, 20 June 2024.

Tariq Masood, Jamil Abdo "New Technique to Enhance Hybrid Renewable Energy Resources' Operational Flexibility and Reliability" published in 22nd International Conference on Renewable Energies and Power Quality (ICREPQ'24) Bilbao (Spain), 26–28 June 2024. https://www.icrepq.com/icrepq24/396-24-masood.pdf.

Tariq Masood, Jamil Abdo "Review: Uninterrupted Liquid Hydrogen (E-Methanol) Production framework to deliver Clean & Green Energy" Published 22nd International Conference on Renewable Energies and Power Quality (ICREPQ'24) Bilbao (Spain), 26–28 June 2024. https://www.icrepq.com/icrepq24/414-24-masood.pdf.

Tariq Masood, Abdel Aty Edris, Suhail Aftab Qureshi, Muhammad Tajammal, Murtaza Hashmi DP Kothari, Samer Kareem "SSSC tailored to optimization performance in Between United Arab Emirates and Oman on the GCC Power Grid" published in the 13th IET International conference on AC and DC power Transmission Conference on 14–16 February, 2017. http://ieeexplore.ieee.org/document/7934934/.

Tariq Masood, Abdel Aty Edris, Suhail Aftab Qureshi, Muhammad Tajammal, Murtaza Hashmi DP Kothari, Samer Kareem "Smart Grid Operations and Control Challenge by Implementing STATCOM Tailored to Optimization Performance at the GCC Power Grid" published in the 13th IET International conference on AC and DC power Transmission Conference on 14–16 February, 2017. http://ieeexplore.ieee.org/document/7934988/.

Tariq Masood, Muhammad Tajammal, Samer Kareem, Suhail Aftab Qureshi, Abdul Jabbar Khan, DP Kothari, Nair Jameel "Unified power Flow Controller Modeling and Analysis Technique on the GCC Power Grid Operations and Control" published in IEEE Energy Con, June 7–9 Limassol Cyprus 2018. https://ieeexplore.ieee.org/document/8398834/.

Tariq Masood, Muhammad Tajammal, Samer Kareem, Suhail Aftab Qureshi, Murtaza Hashmi, Nasir Jameel "Unified power Flow Controller (UPFC) Modeling and Analysis Technique by using Matlab and EMTP" published in (ICREQP,18 Salamanca Spain 21–23 March,2018). http://www.icrepq.com/icrepq18/250-18-masood.pdf.

Tariq Masood, Muhammad Tajammal, Samer Kareem, Suhail Aftab Qureshi, Murtaza Hashmi Naser Jameel "STATCOM Controller Performance Enhancement and Optimization Technique" published in (ICREQP,18 Salamanca Spain 21–23 March,2018). http://www.icrepq.com/icrepq18/249-18-masood.pdf.

Tariq Masood, Muhammad Tajammal, Samer Kareem, Suhail Aftab Qureshi, Murtaza Hashmi DP Kothari, "Smart Grid Operational functions and Control Challenges by Implementing SSSC Tailored to Optimize performance in between Qatar and KSA on the GCC Electrical-power grid" published in the ASTESJ Journal Advances in Science, Technology and Engineering Systems Journal Vol. 2, No. 1, 30.09.2017 2017. https://astesj.com/special-issue-4/.

Tariq Masood, N Al-Emadi, SA Qureshi, YY Almulla, AJ Khan, SK Shah "FACTS Control Devices (Statcom, SSSC and UPFC) Re-Configuration Techniques By PSIM/MATLAB" Published in: International Conference on Electrical Engineering 2007. https://ieeexplore.ieee.org/abstract/document/4287295.

Tariq Masood, Raj K Aggarwal, SA Qureshi, RAJ Khan "STATCOM and SVC control operations and optimization during network fault conditions" 2010. https://ieeexplore.ieee.org/document/5636868.

Tariq Masood, Raj K Aggarwal, Suhail Aftab Qureshi, RAJ Khan "STATCOM model against SVC control model performance analyses technique by MATLAB" published in International Conference on Renewable Energies and Power Quality (ICREPQ'10) 2010. https://icrepq.com/icrepq'10/655-Masood.pdf.

Tariq Masood, R.K Aggarwal, S.A Qureshi, R.A.J Khan "Statcom control reconfiguration technique for steady state and dynamic performance optimization during network fault conditions" published in International Conference on Renewable Energies and Power Quality (ICREPQ'10) Granada (Spain) 2010. https://www.icrepq.com/icrepq'10/656-Masood.pdf.

Tariq Masood, Samer Karim Shah, Muhammad Tajammal, "Unified power flow controller modeling and analysis technique on the GCC power grid" published in IEEE International Energy Conference (ENERGYCON) 2018. https://ieeexplore.ieee.org/abstract/document/8398834.

Tariq Masood, Suhail A Qureshi, RK Aggarwal, DP Kothari "Multiple FACTS devices integration through distributed control system and its application on the GCC interconnection" published in 7th IEEE GCC Conference and Exhibition (GCC) 2013. https://ieeexplore.ieee.org/abstract/document/6705750.

7

Busbar Protection

7.1 Objectives

When you have completed this section, you will be able to:

- List the main types of faults affecting the busbar
- Describe the different types of busbar protection available
- Determine some of the typical constraints and application considerations for different types of busbar protection

7.2 Introduction

Historically, busbars were not provided with specific protection schemes due to low fault levels and the low incidence of faults. Furthermore, it was feared that incorrect operations of busbar protection could lead to large-scale outages. Hence, the busbar relied on slow-operating backup protection.

As power system interconnectivity is increased, the effects of the few busbar faults that did occur were sufficiently severe to warrant the introduction of fast operating busbar protection schemes in large substations. The schemes in use today are predominantly frame-earth and differential protection, although phase comparison and directional locking protection have been used in the past.

The types of busbar faults encountered are dependent on busbar type and configuration. Phase segregated busbars, where each phase is separately contained within an isolation structure such as metal-clad switchgear, are likely to experience only earth faults. However, the possibility of accidental energization of the busbar with earth clamps attached precludes the sole use of earth fault protection since the latter fault will only be cleared with phase fault-sensitive protection. Open terminal busbars can experience all fault types, and so earth and phase fault protection is clearly required. The only possible exceptions are for distribution

Protection of Electrical Power Distribution Systems: Smart grid, Microgrid, AI, and Cyber Security, First Edition. Tariq Masood, Jamil Abdo, and Atif Iqbal.

busbars below 33 kV, where economic restraints may dictate a simpler form of protection.

There are many causes for busbar faults, including flashover, equipment failure, accidental contact, broken insulators, wind-driven objects, and small animals. The three biggest causes of faults are flashover, breaker failure, and insulation failure. The most frequently occurring types of busbar fault are, in rank order, single phase-to-earth, three phase-to-earth, and double phase-to-earth.

Before describing the various types of schemes, it is useful to consider initially the desirable characteristics of busbar protection. Busbar protection is used to protect against all faults occurring within the busbar zone of a busbar substation and should ideally initiate tripping of all circuit breakers required to isolate a fault. The main requirements are:

- Short operating time in order to minimize damage.
- High dependability – should always operate for internal faults. This is especially important given that busbar faults are relatively rare.
- High security – should remain stable for external faults. Again, this is very important because external faults are far more common.
- To give good discrimination. Many busbars are sectioned, and so it is clearly advantageous to restrict tripping to only the affected zone.

7.3 Busbar Fault Clearing Using Backup Protection

Backup line protection, such as overcurrent, earth fault, or distance protection, can be used as an indiscriminative form of busbar protection. For 11 kV and below, the costs of extra busbar protection can rarely be justified, and thus, time-delayed overcurrent or earth fault relays with a typical operating time in the region of 0.4 s will be used. For high-voltage busbars, distance protection A is likely to be used, and so the busbar will be included in zone 2; hence, fault clearance will be nondiscriminative and time delayed. For low-voltage busbars such as these, high discrimination protection will necessitate the fitting of frame-earth-type protection. For higher-voltage distribution, busbars with differential protection can be considered.

7.4 Frame-Earth Protection

7.4.1 Basic System

The simplest form of busbar protection, frame-earth protection, can be applied to busbars having metal-clad switchgear. Figure 7.1 illustrates the basic idea and depicts a four-breaker switchboard consisting of one incoming circuit and three outgoing circuits. For clarity, the three-phase circuits are represented as a one-line

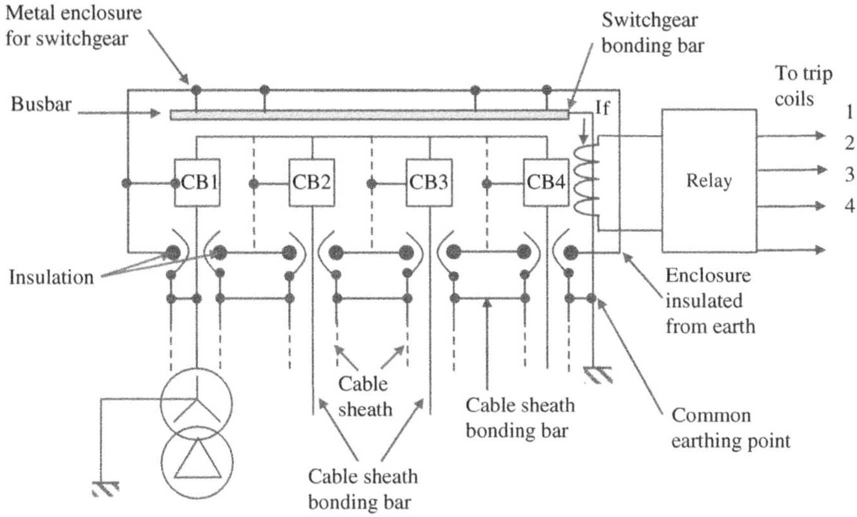

Figure 7.1 Basic frame earth scheme.

diagram. The metal switchgear enclosure is insulated from the earth by placing the unit on a concrete plinth, ensuring that the retaining bolts do not come into contact with any steel reinforcing bars. An insulation of a least 10 Ω between the enclosure and the earthing point is required. Within the enclosure, there are two separate earthing conductors – the switchgear bonding bar, used to connect the enclosure and breaker frames to earth, and the cable sheath bonding bar, used to connect the earth sheath of the incoming and outgoing circuit cables to earth. Note that depending on the earthing system employed, cable sheath earths may or may not be used. The two earthing bars are kept separate by ensuring that cable sheaths do not come into contact with the switchgear enclosure. Both the earthing bars are connected to a common earthing point. The current transformer (CT) of the frame-earth relay is connected to the switchgear bonding bar.

Exercise 7.1 Why do you think that the switchgear and cable earths are kept separate?

To answer the question, we need to recall that the third desirable characteristic of busbar protection should remain stable for external faults. By segregating the earths, the occurrence of a cable fault will not cause the busbar protection to operate since the fault current will not pass through the switchgear bonding bar or the relay CT. Of course, we would expect some form of protection for the cable; however, this is not the function of busbar protection. If a cable fault did occur on one of the outgoing circuits, then only that circuit breaker should be tripped. In contrast, a genuine busbar fault would necessitate all the circuit breakers being tripped, and so we see why discrimination is important to busbar protection. The frame-earth relay used in Figure 7.1 will be an instantaneously operating overcurrent relay. In this application, the relay is shown directly tripping the four circuit breakers. However, in more complex schemes, it is common to use an intermediate tripping relay to perform this task.

7.4.2 Frame-Earth with Check Relay

The disadvantages of the scheme of Figure 7.1 are twofold:

- First, insulation faults in the secondary wiring within the enclosure can cause current to flow through the switchgear bonding bar, with a consequent risk of operating the protection.
- Second, mechanical shock to the frame-earth relay can cause inadvertent tripping.

To overcome these disadvantages, it is common to use a relay to check the action of the frame-earth relay; the breaker will only be tripped if both relays indicate a fault.

Exercise 7.2 Referring to Figure 7.1 and given that the check relay is also an instantaneous overcurrent relay, where do you think the CT of this relay could be situated in order to overcome both of the of disadvantages stated above?

The check relay must also detect the presence of earth fault current. However, to eliminate disadvantage 1, the check relay must measure its current from a point outside the switchgear enclosure. The obvious choice in this case will be to position the CT on the neutral conductor of the transformer feeding the switchboard. This is shown in Figure 7.2 together with the associated d.c. tripping circuit since the frame-earth and check relays now operate tripping relays to initiate tripping of the circuit breakers.

7.4.3 Frame-Earth Protection for Sectioned Busbar

For a busbar containing a bus section circuit breaker, the frame-earth system can be arranged to trip only the side of the busbar containing the fault, as well as the bus section breaker. Figure 7.3 illustrates this arrangement for a sectioned busbar with two incoming circuits. A check relay R4 is provided, which is operated from the neutral current of either of supply transformer. The busbar is electrically isolated into three zones: the left-hand zone, the right-hand zone, and the zone containing the section breaker. The tripping relays, R5, R6, R7, and R8, are arranged so that a fault in the left-hand zone will leave the right-hand zone connected and vice versa. Thus, discrimination is ensured.

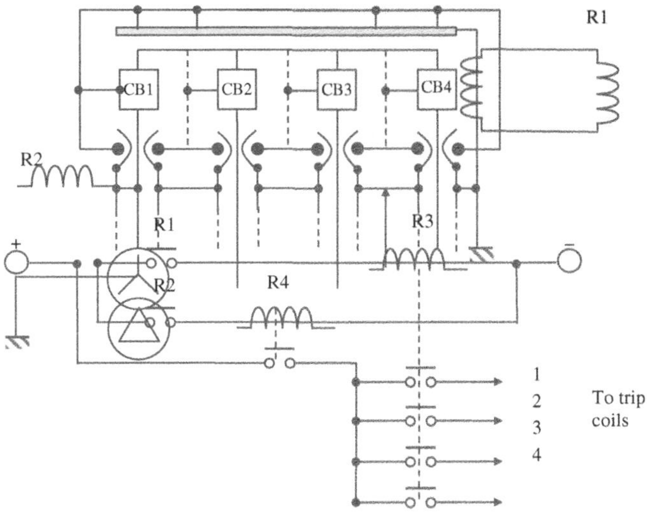

Figure 7.2 Frame-earth protection with neutral check.

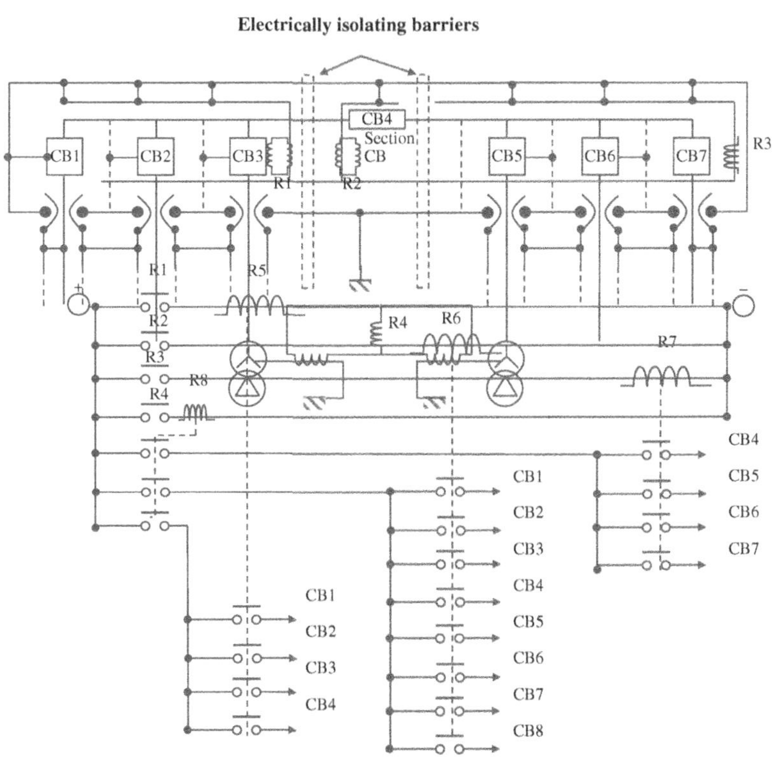

Figure 7.3 Frame-earth protection for a sectioned busbar.

7.4.4 Frame Earth Protection for Double Busbar Switchboard

Owing to the difficulty of insulating between the main and reserve busbars in a double bus switchboard, an arrangement similar to the single, sectioned busbar protection scheme is used, with the exception that all feeders connected to the reserve busbar will be tripped irrespective of the zone of the busbar fault.

7.5 Differential Protection

7.5.1 Basic Application

Circulating current differential system, as discussed in section five, may be applied to the protection of the busbar. Unlike the application of differential to line protection, a busbar is usually a multiterminal zone and so there will be many parallel connected current transformers. A typical application is shown in Figure 7.4.

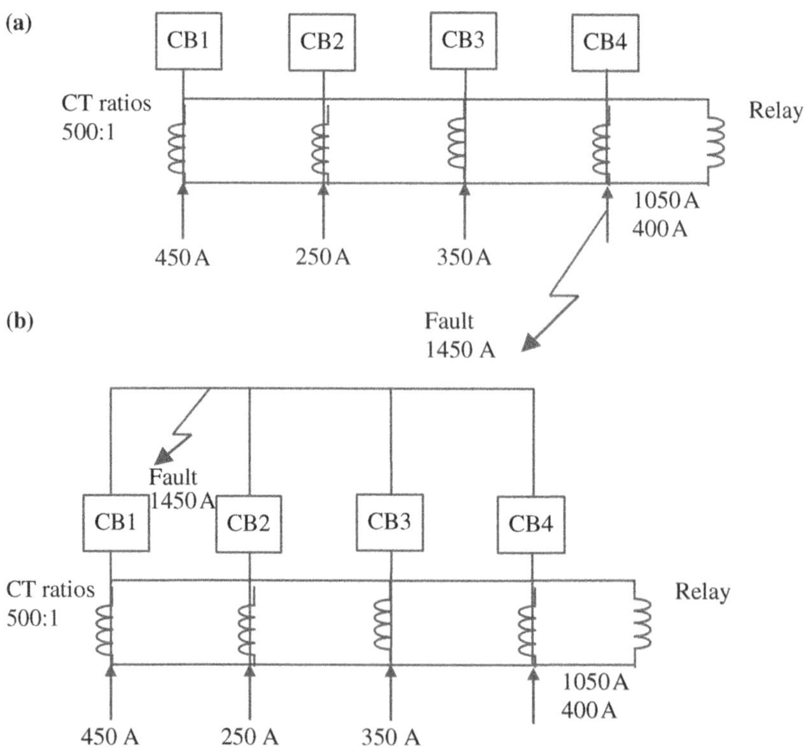

Figure 7.4 Differential busbar protection.

Exercise 7.3 Calculate the current through the relay for the external fault shown in Figure 7.4a?

Since the CT ratio is 500:1, the current through CB1, I_1, is given by:

$$I_1 = \frac{450}{500} = 0.9\ \text{A} \tag{7.1}$$

Similar

$$I_2 = \frac{250}{500} = 0.5\ \text{A} \tag{7.2}$$

$$I_3 = \frac{350}{500} = 0.7\ \text{A} \tag{7.3}$$

and

$$I_4 = \frac{-1050}{500} = -2.1\ \text{A} \tag{7.4}$$

Using the convention that current into the substation is positive.

From the connection of the CTs, the current through the relay, I_R, is given by:

$$I_R = I_1 + I_2 + I_3 + I_4 = 0.9 + 0.5 + 0.7 - 2.1 = 0 \tag{7.5}$$

Clearly, the relay will not operate for external faults.

Exercise 7.4 Calculate the current through the relay for the internal fault shown in Figure 7.4b?

By inspection, the relay current can be written as:

$$I_R = \frac{450}{500} + \frac{250}{500} + \frac{350}{500} + \frac{450}{500} = 2.9\ \text{A} \tag{7.6}$$

And hence the relay will operate for internal faults.

For schemes requiring only earth fault protection, the arrangement of Figure 7.5a, which shows a three-phase representation of a simple two-terminal zone, will suffice.

Note that it is also possible to provide earth fault protection by using a single-core balance CT, which encompasses all three phases on each terminal. Similarly, the arrangement of Figure 7.5b shows how both earth and phase faults may be accommodated. A scheme covering both earth and phase faults will be more expensive owing to the cost of extra wiring and relays. For lower-voltage substations, the cost of the scheme can be a limiting factor. For the sake of clarity, future figures in this section will not show three-phase detail; however, the basic connection of Figure 7.5a, b can easily be extended to multiterminal applications.

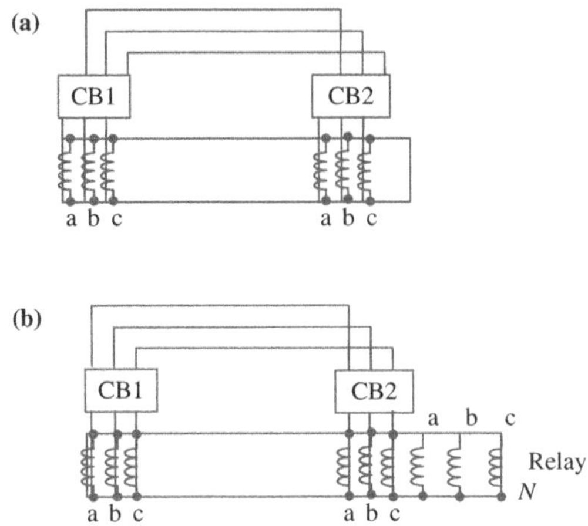

Figure 7.5 Differential scheme arrangement for earth fault protection.

7.5.2 Differential Protection for Sectioned Substation

Figure 7.6 shows the application of differential protection to a sectioned busbar. In common with frame-earth protection, the differential scheme will allow the unfaulted section of the busbar to operate by opening the section circuit breaker.

Exercise 7.5 Draw the d.c. tripping circuit for the differential scheme of Figure 7.6, given that relay 1 operates tripping relay 4, and relay 2 operates tripping relay 5?

The d.c. tripping circuit is shown in Figure 7.7.

As in the frame-earth protection case, it is common to provide a check relay to ensure dependable operation of the protection. The sectioned busbar differential scheme can be checked by a nondiscriminating differential scheme, relay 3 in Figure 7.8. The overall scheme would still allow for two zones of operation;

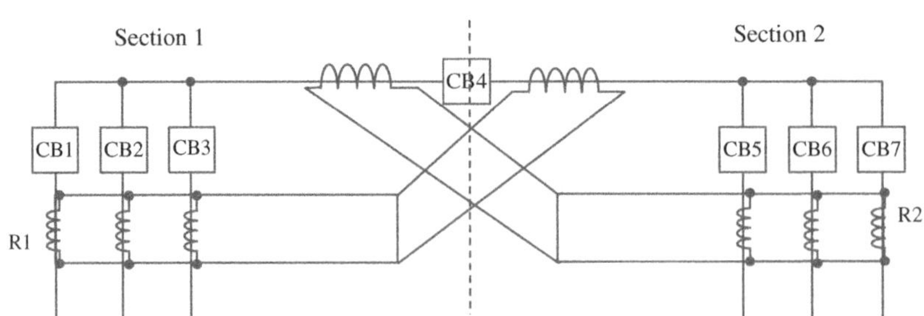

Figure 7.6 Differential protection for sectioned busbar.

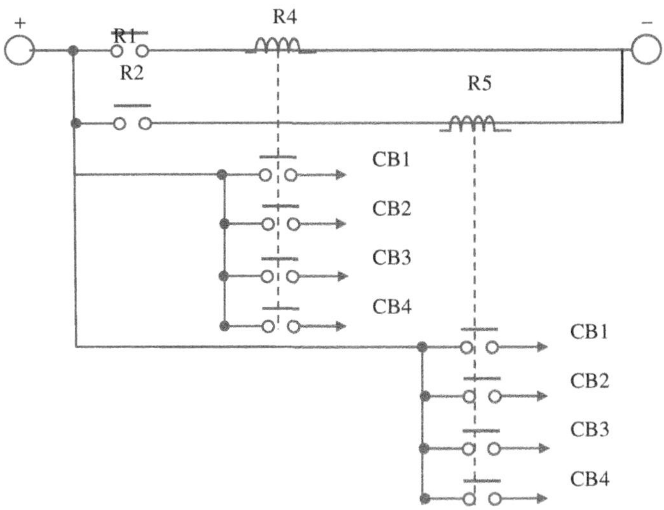

Figure 7.7 d.c. tripping circuit for differential scheme of Figure 7.6.

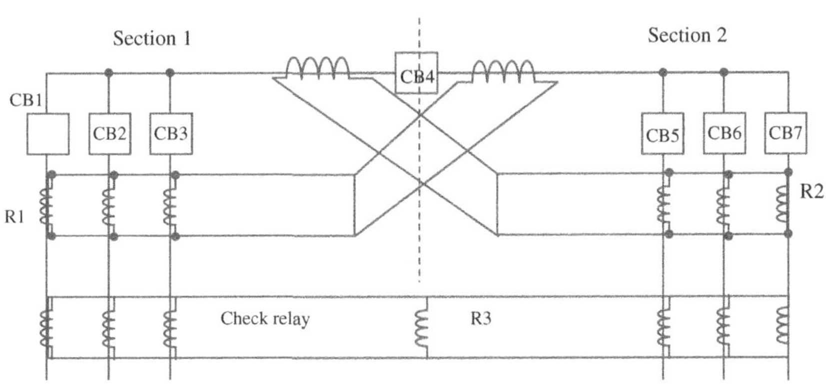

Figure 7.8 Differential protection for sectioned busbar with check relay.

however, the check relay feature would prevent operating contingency conditions, such as the discriminating scheme developing an open circuit CT.

7.5.3 Differential Protection for Double Busbar Substation

Compared with frame-earth protection, differential protection applied to a double bus substation gives the advantage of allowing the busbar to be protected in separate zones. A typical example of such a scheme is shown in Figure 7.9, although it is stressed that this is a somewhat simplified view since the scheme only

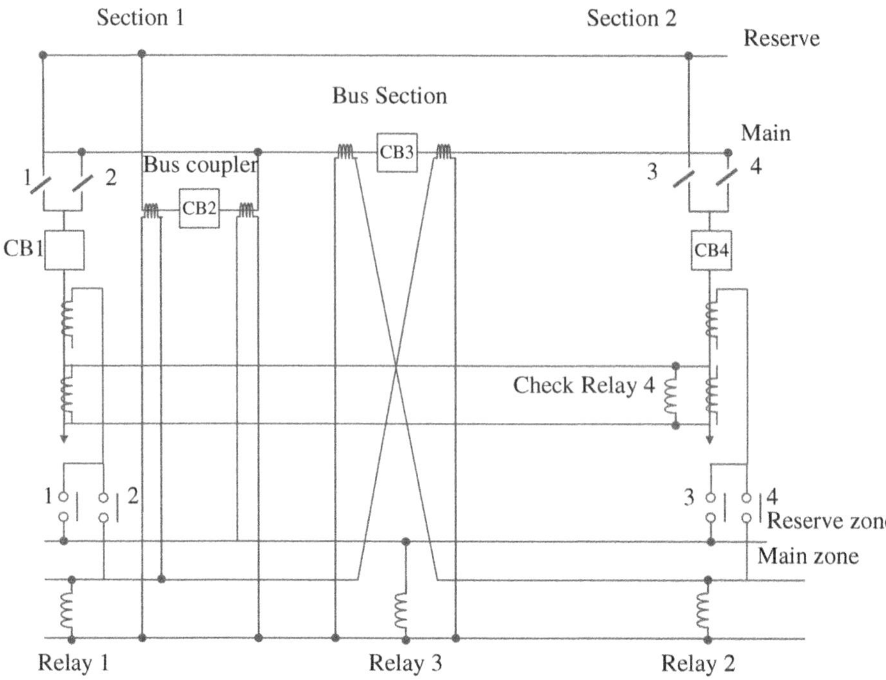

Figure 7.9 Differential protection scheme for double busbar switchboard.

encompasses two feeders, whereas double bus stations usually have more feeders, and the reserve bus is shown without isolators. Nevertheless, a clear understanding of the techniques described herein will allow you to understand more elaborate differential schemes. The substation consists of two busbars, the main and the reserve. The main busbar is sectioned into two sections: CB3, the bus section circuit breaker, and the busbars are connected together via CB2, the bus coupler.

Excluding the check relay (relay 4), there are three relays, one for each of the reserve busbars, the main zone 1, and the main zone 2. Owing to the necessity for the zone of a differential scheme to encompass all relevant circuit breakers, the CTs monitoring the feeders connected to CB1 and CB4 have to be switched into the main or reserve relay circuits, depending on the position of the isolators 1 and 2, or 3 and 4. This is achieved by auxiliary switches 1, 2, 3, and 4 which are usually automatically switched in sympathy with the isolators 2 is closed, thus connecting the feeder to the main busbar, then the zone is delimited by the CT downstream of CB1, the CT to the left of CB2, and the CT to the right of CB3, as shown in Figure 7.9.

Again, the check relay is nondiscriminative and will operate for all faults on the busbars. The check relay scheme is specifically designed to have no auxiliary switches and can thus provide backup if such switches fail the discriminating scheme.

7.5.4 High Impedance Relays for Differential Busbar Schemes

So far, we have not discussed the nature of the relay used in busbar differential schemes. Differential schemes applied to line protection usually consist of current sensing relays with some form of bias. Such relays are usually of the low impedance type. The bias is necessary to compensate for line charging and CT mismatches. Since there is relatively little capacitance in a busbar, there is no need to provide compensation.

Hence, provided that high relay settings are used to compensate for CT mismatch, there is no need to bias the busbar differential scheme. The main difficulty in setting busbar differential relays is to ensure that CT saturation does not lead to malfunctioning because this condition can affect both discriminative and check relay operation alike. CT saturation is a particular problem in busbar protection because of the multiterminal nature of the protected plant. For example, in Figure 7.4b, the internal busbar fault is seen to be fed from four sources, each of differing degrees of CT saturation.

A more onerous situation is shown in Figure 7.4a, where the external fault causes a very large current flow through the feeder attached to CB4. If the associated CT were to saturate, then a spill current would flow through the relay, causing a risk of incorrect operation. In early busbar differential schemes, this effect was not understood and consequently led to the use of biased differential schemes. However, it is possible to alleviate the effects of CT saturation by using *high impedance* relays, where a *stabilizing resistor* is inserted in series with the relay.

7.5.5 Stabilizing Resistance

To illustrate the effect of CT saturation and the calculation of the series resistance value, the example arrangement of Figure 7.10 will be used. This shows a simple two-terminal busbar scheme with the CTs represented by simplified equivalent circuits consisting of a "prefect CT" and secondary winding resistance R_{CT}.

Note that the magnetizing and core loss shunt branch elements are ignored for this example. The resistance of the wiring between the CTs and the relay is represented by RW1 and RW2. The relay is shown to have a coil resistance of RR and a stabilizing resistance of RS.

Under the fault-current conditions, the e.m.f. is from CT2.

Thus, a circulating current flows through the wiring, but not the relay, across which there is no e.m.f. under CT saturation conditions, the e.m.f. in the saturated

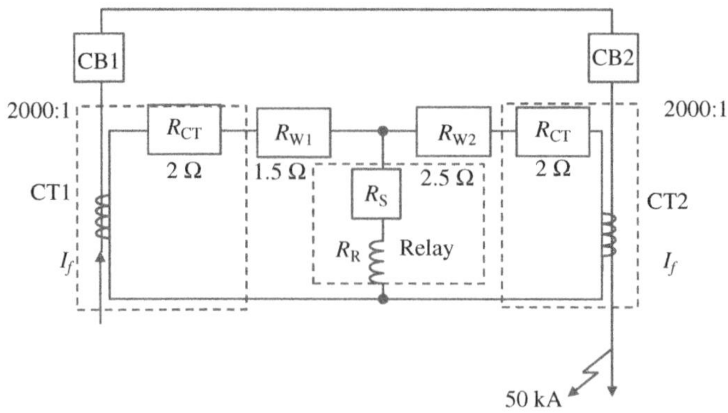

Figure 7.10 Equivalent circuit of high impedance differential scheme.

CT collapses to zero, thus imposing an e.m.f. across the relay. For a high impedance type relay, the setting current will be small compared to the circulating current. Thus, the relay e.m.f., V_s, will be the voltage drop across the combined resistance of the CT winding and connecting leads.

$$V_s = \frac{I_F}{N}(R_W + R_{CT}) \tag{7.7}$$

where N is the unsaturated CT turns ratio; I_F is the primary fault current; the total wiring resistance between the saturated CT and the relay i.e. if CT1 saturates. Clearly the relay must not operate in this situation, thus, stability condition is

$$I_s(R_R + R_S) \geq V_S \tag{7.8}$$

where I_S is the normal relay setting current. This expression allows the stabilizing resistance to be calculated.

Exercise 7.6 By considering the saturation of each CT of Figure 7.10, calculate the minimum value of the stabilizing resistor RS given that the relay coil resistance is 2 Ω and its setting is 0.5 A?

If CT1 is saturated, then the stability voltage is given by:

$$V_S = \frac{5000}{2000}(1.5 + 2) = 87.5 \text{ V} \tag{7.9}$$

whereas the saturation of CVT2 will give

$$V_S = \frac{5000}{2000}(2.5 + 2) = 112.5 \text{ V} \tag{7.10}$$

Clearly the saturation of CT2 is the more onerous condition since it leads to a high relay voltage. Rearranging the stability conditions, we obtain:

$$R_S \geq \frac{V_S}{I_S} - R_S \tag{7.11}$$

$$R_S \geq \frac{112.5}{0.5} - 2 \tag{7.12}$$

$$R_S \geq 223\ \Omega$$

Note that the relay coil resistance will change with the current setting. In this example, the fault current was specified. To ensure relay stability under all fault conditions, this calculation must be performed using the maximum prospective fault current. In the situation where the busbar fault level is lower, the switchgear rating is used, and it is prudent to take the maximum current based on the switchgear rating to allow for future capacity increases.

7.5.6 Relay Setting Calculation

In order to calculate the stabilizing resistance value, it is necessary to know the relay setting. To establish this, the effective relay setting IR may be calculated from:

$$I_R = I_S + nI_{eS}$$

where I_{eS} is the CT excitation current, and n is the number of parallel connected CTs. Multiplying I_R by the turn ratio, N, will give the primary effective setting which should be no greater than 30% of the minimum prospective fault current. The minimum fault current should be assessed bearing in mind the type of fault considered, that is, earth faults will give lower fault currents than the phase-to-phase faults, which will give lower fault currents than three-phase faults. Finally, we must ensure that the CTs will give sufficient output to operate the relay. This is achieved by making the CT knee-point voltage at least twice the relay operating voltage, that is, setting multiplied by the sum of stabilizing and coil resistance.

Exercise 7.7 A 33 kV single busbar switchgear has a total of 10 circuits and switchgear rated at 500 MVA. The system is earthed via a resistor which limits the fault current to 1000 A. The CTs are 500:1 and have an excitation current of 0.01 A. Calculate the relay setting current?

An earthing resistor is likely to affect earth faults only. Thus, the minimum prospective fault current can be taken to be 1000 V since any other fault type will give a higher current. This corresponds to a minimum secondary fault current

$\frac{1000}{500} = 2$ A. Hence, the effective setting should be $0.3 \times 0.2 = 0.6$ A. Since there are 10 circuits, we can assume that there are 10 parallel CTs. Hence, using the above equation:

$$0.6 = I_S + 10 \times 0.01$$

which gives

$$I_S = 0.5 \text{ A}$$

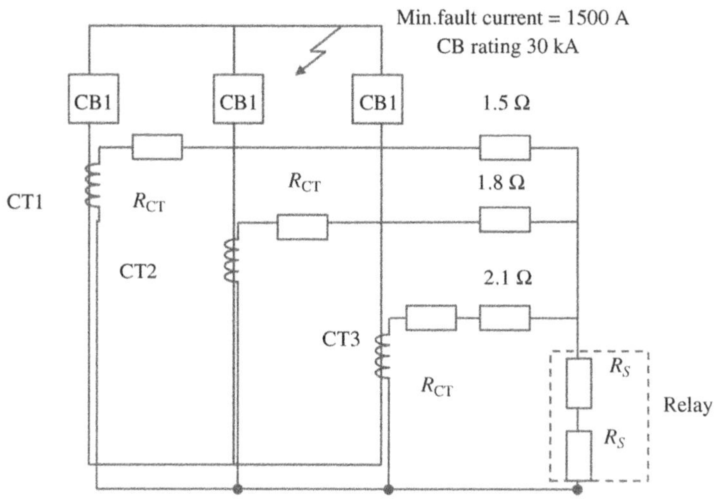

Figure 7.11 Equivalent circuit of high impedance differential scheme.

Table 7.1 Current versus coil resistance.

Setting current (A)	Coil resistance (Ω)
0.4	2.5
0.6	1.67
0.8	1.25

7.6 Self-Assessment Questions

Self-Assessment Question 7.1a

Figure 7.11 shows a high impedance differential busbar protection scheme. Using the data available, calculate a suitable value for the stabilizing resistance R_S (Table 7.1).

Bibliography

Jamil Abdo, Tariq Masood "New Technique to Develop Intelligent, Highly Reliable & Efficient Unmanned Power Generation Modular" published in 22nd International Conference on Renewable Energies and Power Quality (ICREPQ'24) Bilbao (Spain), 26–28 June 2024. https://www.icrepq.com/icrepq24/392-24-abdo.pdf.

Jamil Abdo, Tariq Masood "Operational Reliability – Production Best Practice Pragmatic Case Studies & Improvement" published in Proceedings of the 9th North American Conference on Industrial Engineering and Operations Management Washington D.C., USA, 4–6 June 2024.

Jamil Abdo, Tariq Masood "Techno Economics Analysis to Reshape Global Clean and Green Energy" published in 22nd International Conference on Renewable Energies and Power Quality (ICREPQ'24) Bilbao (Spain), 26–28 June 2024. https://www.icrepq.com/icrepq24/415-24-abdo.pdf.

Nasser Alemadi, Tariq Masood, R Aggarwal, Suhail A Qureshi, D Kothari "Novel Selection Approach of STATCOM Device Tailored and Optimized Engineering Configuration to Implement in Between Qatar and Bahrain" published in International Journal of Automation and Power Engineering 2014. 10.14355/ijape.2014.0302.01.

Jane Courtney, Aaron McDonnell "Impact on Distribution System Protection with the Integration of EG on the Distribution Network" published in 54th International Universities Power Engineering Conference (UPEC) in 2019.

Yu Deng, Dingkang Liang, Jiahao Wang, Xinyi Song, Zehua Li, Zhetong Hong "Reliability Evaluation of Distribution System Based on Dynamic Bayesian Network" published in 3rd Power System and Green Energy Conference (PSGEC) in 2023.

Zong Guo, Jiarui Li, Hongbo Lian, Shihua Zhao, Yang Chen, Renjie Dai, Xiuneng Xu, Yufeng Gu, Yangyang Liu "Optimization Method for Protection of Distribution Network with MW-PV Based on Multi-level Cooperative Protection Strategy" published in 2019 IEEE Sustainable Power and Energy Conference (iSPEC) 2019.

Hana A Halim, Melaty Amirruddin, Nur S Noorpi, Nurhakimah M Mukhtar "An Improved Protection Scheme for Smart Distribution Grid" published in 2013 1st International Conference on Artificial Intelligence, Modelling and Simulation 2013.

Seongil Kim, Byungchul Sung, Jeong-Hyeok Heo, Duck-Su Lee, "Protection Against Transient Overvoltages Caused by Intermittent Earth Faults in Unearthed Marine Power Distribution Network" published in IEEE International Conference on Power Systems Technology (POWERCON) in 2022.

Rani Kumari, Bhukya K Naick "Protection Coordination Impact on Reliability of Distribution Systems With Distributed Energy Resources" published in 2024 Third International Conference on Power, Control and Computing Technologies (ICPC2T) 2024.

Qin Lijun, Wang Ying, Zheng Shenrui, Guo Qing "Situation of Study on Distribution Network Protection Containing Distributed Generation and Application Scheme of Intelligent Protection" published in 2011 4th International Conference on Electric Utility Deregulation and Restructuring and Power Technologies (DRPT) 2011.

Tariq Masood, Jamil Abdo "Maintenance Strategy Production Best Practices – Pragmatic Case Studies and Operations Improvement" published in Proceedings of the 9th North American Conference on Industrial Engineering and Operations Management Washington D.C., USA, 4–6 June 2024.

Tariq Masood, Jamil Abdo "New Technique to Enhance Hybrid Renewable Energy Resources' Operational Flexibility and Reliability" published in 22nd International Conference on Renewable Energies and Power Quality (ICREPQ'24) Bilbao (Spain), 26–28 June 2024. https://www.icrepq.com/icrepq24/396-24-masood.pdf.

Tariq Masood, Jamil Abdo "Review: Uninterrupted Liquid Hydrogen (E-Methanol) Production Framework to Deliver Clean & Green Energy" published in 22nd International Conference on Renewable Energies and Power Quality (ICREPQ'24) Bilbao (Spain), 26–28 June 2024. https://www.icrepq.com/icrepq24/414-24-masood.pdf.

Tariq Masood, N Al-Emadi, SA Qureshi, YY Almulla, AJ Khan, SK Shah "FACTS Control Devices (Statcom, SSSC and UPFC) Re-Configuration Techniques By PSIM/MATLAB" published in 2007 International Conference on Electrical Engineering 2007. http://ieeexplore.ieee.org/abstract/document/4287295.

Tariq Masood, Raj K Aggarwal, Suhail Aftab Qureshi, RAJ Khan "STATCOM Model Against SVC Control Model Performance Analyses Technique by MATLAB" published in International Conference on Renewable Energies and Power Quality (ICREPQ'10) 2010. https://icrepq.com/icrepq'10/655-Masood.pdf.

Tariq Masood, Raj K Aggarwal, SA Qureshi, RAJ Khan "STATCOM and SVC Control Operations and Optimization During Network Fault Conditions" published in 2010 IEEE International Symposium on Industrial Electronics 2010. https://ieeexplore.ieee.org/document/5636868.

Tariq Masood, Suhail A Qureshi, RK Aggarwal, DP Kothari "Multiple FACTS Devices Integration Through Distributed Control System and Its Application on the GCC Interconnection" published in 2013 7th IEEE GCC Conference and Exhibition (GCC) 2013. https://ieeexplore.ieee.org/abstract/document/6705750.

Tariq Masood, Abdel Aty Edris, Suhail Aftab Qureshi, Muhammad Tajammal, Murtaza Hashmi, DP Kothari, Samer Kareem "SSSC Tailored to Optimization Performance in Between United Arab Emirates and Oman on the GCC Power Grid" published in the 13th IET International Conference on AC and DC Power Transmission Conference on 14–16 February 2017. http://ieeexplore.ieee.org/document/7934934.

Tariq Masood, Abdel Aty Edris, Suhail Aftab Qureshi, Muhammad Tajammal, Murtaza Hashmi, DP Kothari, Samer Kareem "Smart Grid Operations and Control Challenge by Implementing STATCOM Tailored to Optimization Performance at the GCC Power Grid" published in the 13th IET International conference on AC and DC power Transmission Conference on 14–16 February 2017. http://ieeexplore.ieee.org/document/7934988.

Tariq Masood, Muhammad Tajammal, Samer Kareem, Suhail Aftab Qureshi, Murtaza Hashmi, DP Kothari "Smart Grid Operational Functions and Control Challenges by Implementing SSSC Tailored to Optimize Performance in Between Qatar and KSA on the GCC Electrical-Power Grid" published in the ASTESJ Journal Advances in Science, Technology and Engineering Systems Journal Vol. 2, No. 1, 2017. https://astesj.com/special-issue-4.

Tariq Masood, Muhammad Tajammal, Samer Kareem, Suhail Aftab Qureshi, Abdul Jabbar Khan, DP Kothari, Nair Jameel "Unified Power Flow Controller Modeling and Analysis Technique on the GCC Power Grid Operations and Control" published in 2018 IEEE International Energy Conference, Limassol Cyprus, 7–9 June 2018. https://ieeexplore.ieee.org/document/8398834.

Tariq Masood, Muhammad Tajammal, Samer Kareem, Suhail Aftab Qureshi, Murtaza Hashmi, Nasir Jameel "Unified Power Flow Controller (UPFC) Modeling and Analysis Technique by Using Matlab and EMTP" published in International Conference on Renewable Energies and Power Quality (ICREQP'18) Salamanca (Spain) 21–23 March 2018. http://www.icrepq.com/icrepq18/250-18-masood.pdf.

Tariq Masood, Muhammad Tajammal, Samer Kareem, Suhail Aftab Qureshi, Murtaza Hashmi, Naser Jameel "STATCOM Controller Performance Enhancement and Optimization Technique" published in International Conference on Renewable Energies and Power Quality (ICREQP'18) Salamanca (Spain) 21–23 March 2018. http://www.icrepq.com/icrepq18/249-18-masood.pdf.

Tariq Masood, Samer Karim Shah, Muhammad Tajammal, "Unified Power Flow Controller Modeling and Analysis Technique on the GCC Power Grid" published in 2018 IEEE International Energy Conference (ENERGYCON) 2018. https://ieeexplore.ieee.org/abstract/document/8398834.

L Ndahepele, S Chowdhurym "Impact of Distributed Generation on Traditional Protection in Distribution and Transmission Systems: A Review" published in IEEE PES/IAS PowerAfrica in 2020.

Vassilis C Nikolaidis, Evangelos Papanikolaou, Anastasia S Safigianni "A Communication-Assisted Overcurrent Protection Scheme for Radial Distribution Systems With Distributed Generation" IEEE Transactions on Smart Grid Vol. 7, No. 1, 2015.

Luyao Pei, Cheng Xu, Jinsong Zhang, Bo Wang, Yining Ma "Artificial Intelligence Algorithm in Intelligent Power Supply and Distribution Network Design System" published in IEEE 3rd International Conference on Electronic Technology, Communication and Information (ICETCI) in 2023.

Cleberton Reiz, Jonatas Boas Leit "Impact Analysis of Distributed Generation on Protection Devices Coordination in Power Distribution Systems" published in 2021 IEEE PES Innovative Smart Grid Technologies Conference – Latin America (ISGT Latin America) in 2021.

M Saeed Ullah, Tariq Masood, Umair Zahid, H Abuzar Ahsan "Electrochemical Oxidation of Dissolved Organic Dye in Aqueous Solution" published in IMASCON-2019.

Tariq Masood, RK Aggarwal, SA Qureshi, RAJ Khan "Statcom Control Reconfiguration Technique for Steady State and Dynamic Performance Optimization During Network Fault Conditions" published in International Conference on Renewable Energies and Power Quality (ICREPQ'10) Granada (Spain) 2010. https://www.icrepq.com/icrepq'10/656-Masood.pdf.

Yuwei Xiong, Wei Chen "Positive Sequence Current Differential Protection Based on Least Square Method for Distribution Network with Distribution Generation" published in 2021 IEEE 4th International Electrical and Energy Conference (CIEEC) 2021.

Lü Zhining, Ning Baifeng, Luo Weifeng, Liu Wei "Distribution Networks Ground Fault Protection Method Based on Fuzzy Measure Fusion Criterion" published in IEEE 8th International Conference on Advanced Power System Automation and Protection (APAP) in 2019.

8

Motor Protection

8.1 Objectives

When you have completed this section, you will be able to:

- Define the basic motor characteristic involved in protection
- Define the motor protection requirements
- Describe the basic principles involved in motor protection against winding faults, overload, and phase unbalance

8.2 Introduction

A wide range of motors and motor characteristics exists because of the numerous functions for which they are used, and all motors need protection. In this respect, it is important to consider carefully the characteristics of the motor when applying protection. For example, the starting currents and starting times must be known when applying overload protection, and the thermal withstand under balanced and unbalanced conditions must be clearly defined. The cost and extent of the protection system must be weighed against the potential hazards. The size of the motor and the type of service will also influence the type of protection required. This book is concerned with looking at motor protection requirements and considers, in some detail, certain of the more important potential hazard such as winding faults (phase-phase and phase-ground), thermal damage from overload (continuous and intermittent), and abnormal conditions such as unbalanced operation and reversed phase against which the motors must be protected.

Protection of Electrical Power Distribution Systems: Smart grid, Microgrid, AI, and Cyber Security,
First Edition. Tariq Masood, Jamil Abdo, and Atif Iqbal.

8.3 Motor Characteristics Involved in Protection

The primary motor characteristics available and involved in protection are:

- Starting current curve
- Thermal capability curve, which should include the permissible locked rotor thermal limit

These characteristics are normally available from the motor manufacturers and are fundamental for the application of protection. Typical curves are illustrated in Figure 8.1.

The starting current curve is normally at rated voltage. The thermal limit curve represents three limits blended into one:

- The higher current portion which indicates the permissible locked rotor time, which is the time the rotor can remain stalled after motor energization before thermal damage ensues within the motor. In very large motors, this thermal limit can be reached in less than the motor's starting time, so that these motors must start rotating instantly to avoid this damage.
- The accelerating thermal limit curve from locked rotor current to the motor breakdown torque current, which is around 75% speed.
- The operating thermal limit curve which represents the motor overload capacity.

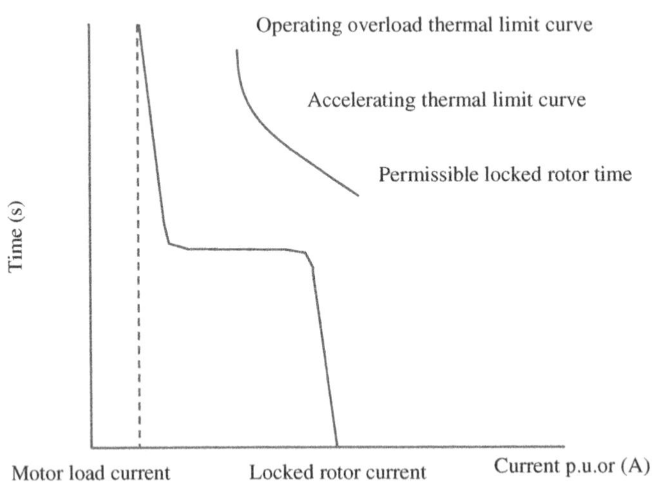

Figure 8.1 Typical induction motor characteristics.

8.4 Stator Winding Protection

Faults that occur within the motor windings are mainly earth faults caused by a breakdown in the winding insulation. This type of fault can be detected daily by means of a single instantaneous overcurrent relay, usually with a setting of approximately 20% of the motor full load current, connected in the residual section of the three current transformers (CTs) as shown in Figure 8.2. Care must be taken to ensure that the relay does not operate from spill current due to saturation of one or more CTs during initial peaks of the starting current, which can be as high as two and a half times the steady-state-rated value, as this may cause relay operation. To achieve stability under these conditions, it is normal to increase the minimum operating voltage of the relay by inserting a stabilizing resistor in series with it.

Exercise 8.1 How can saturation cause operations of the earth fault relay under normal motor starting conditions?

Motor starting current can have peaks of large magnitude, which in turn can saturate one or other of the CTs. The latter results in a system imbalance with a residual current flowing through the earth fault relay.

With regard to phase–phase, such failing is relatively rare because of the greater insulation levels between the phase windings. However, they do occur and must therefore be protected against. The phase fault current at the terminals of a motor is usually considerably larger than any normal current, such as motor starting current. A high setting instantaneous overcurrent phase relay thus provides a fast, reliable, inexpensive, and simple protection (this is also shown in Figure 8.2) when the starting current approaches the fault current, however, differential relaying becomes necessary, but this is normally provided on large, important motors only.

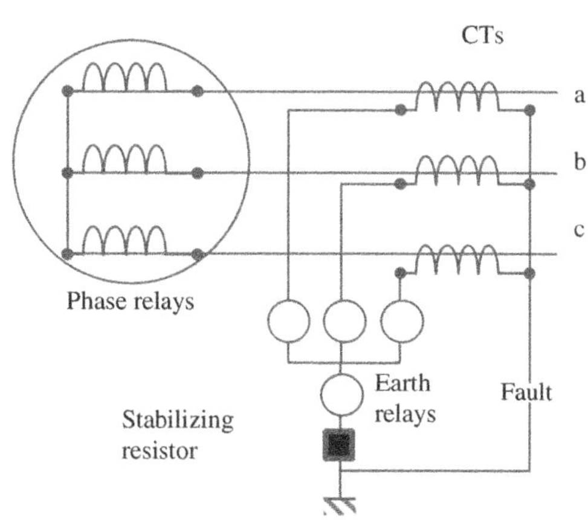

Figure 8.2 Phase and earth fault protection.

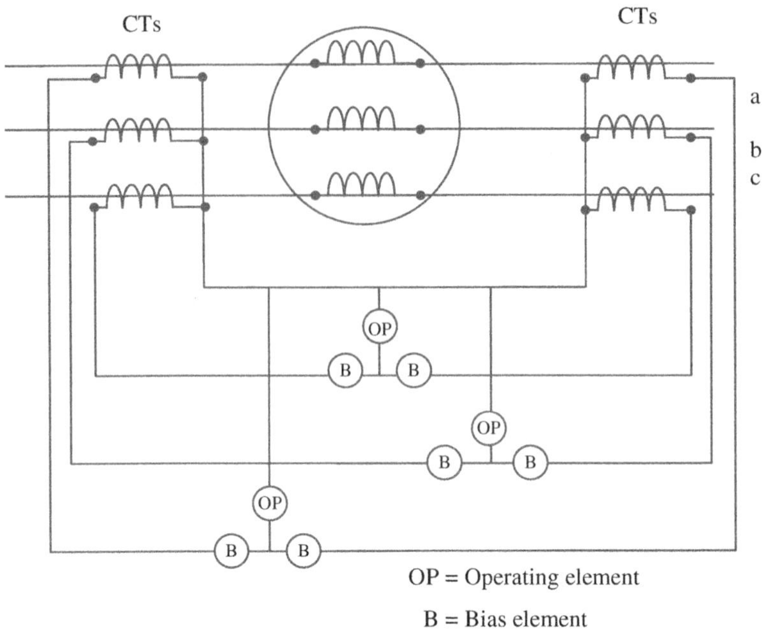

Figure 8.3 Differential protection for motors.

Figure 8.3 shows a conventional current differential protection scheme as applied to the protection of motor stator windings.

8.5 Overload Protection

Motor overloads are caused by mechanical loads. The wide diversity of motor duties and motor designs makes it difficult to obtain their heating curves under loads. Furthermore, these curves are an approximate average of an imprecise thermal zone, where varying degrees of damage or shortened insulation life may occur. It is thus difficult for any protection relay design to approximate these variable curves adequately over the range from light sustained overloads to severe fluctuating overloads.

Thermal overload involves the application of relays to match the thermal curve closely (shown in Figure 8.1). The relays should operate just before the limits are reached (or exceeded), and the most commonly used relay is the replica type, in which the thermal overload relay has a bimetal spiral in proximity to a heater, which acts as the contact activating device and which is supplied by a CT measuring current. The bimetal spiral, when heated, produces a torque to move the relay contact toward a fixed contact. The deflection is proportional to current squared and, thus, a motor operating at full load would move the contact three-quarters of the way toward a fixed contact set at say 120%. Hence, the operating time would only be a quarter of the time required for relay operation of the motor when it is lightly loaded.

However, the thermal time constant of the bimetal will respond in seconds while the motor temperature changes will take minutes or even hours. Because the bimetal is heated indirectly, that is, by heat radiated from the heaters, as opposed to direct heating by passing current through it, there is a time delay in its response.

This means that the moving contact will continue to move toward tripping even after the motor-starting current has disappeared, and this is called relay *overrun*. In practice, thermal relays have a typical 45% overrun at six times rated motor current. Thus, to avoid operation, the relay operating time must be at least twice the motor starting time. When using a static relay, however, the percentage overrun is small, so that the relay can be set to a lower operating time.

Exercise 8.2 What do you understand by the term relay overrun?

This is a problem with thermal relays, which comprise a bimetal strip, heated by radiation from a heater activated by motor current. The thermal response time of the bimetal strip is much faster than the temperature change time of the motor. Hence, the moving contact of the relay (in which moving time is proportional to temperature change) continues to move toward tripping even when the motor starting current has disappeared, and this is called relay overrun).

8.6 Phase Unbalance Protection

The voltage supplied to a three-phase induction motor can be unbalanced for a variety of reasons: single-phase loads, unbalance in feeders, or blown fuses in a power plant or power factor correction plant. In addition, the accidental opening of one phase load in the supply to the motor running, supplied by only two phases. An unbalance in the voltage (and therefore current) results in a negative-phase sequence (nps) component of current, and this is produced by the supplied voltage. This counter-rotating field will cause induced currents in the rotor at almost twice the normal system frequency, resulting in overheating of the motor rotor and possible damage. A voltage imbalance of 3.5% can produce approximately a 25% increase in motor temperature. Moreover, the 100 Hz current in the rotor also increases rotor resistance (due to skin effects), thus compounding the heating effect. Several protection techniques are available for unbalance detection, and they are as follows.

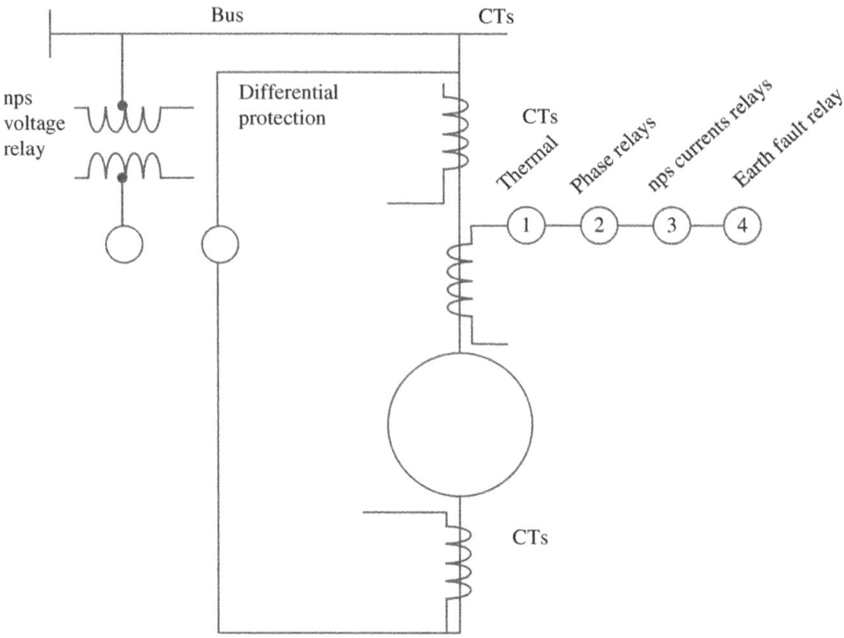

Figure 8.4 Typical motor protection arrangement.

- Magnitude difference between the three-phase currents
- The presence of nps current
- The presence of nps voltage is illustrated in Figure 8.4

The current balance type compares the phase current magnitude and operates when one phase current is significantly different in magnitude from the other phase currents. This is very effective protection for individual motor feeders, to detect open circuits or unbalance in that circuit. One relay should be applied to each load or feeder.

Another type responds to the nps current instantaneously. These relays are, however, not widely applied for motor protection. The nps voltage is employed for detecting phase imbalance or phase reversal in the supply or source circuits. Operating sensitivity is about 0.05 p.u. of the nps voltage. As shown in Figure 8.4, the relay is connected through a VT. Normally, sufficient nps voltage is available for an open phase in the source.

When the phases are reversed (and this can sometimes happen at the time of motor starting), the 1 p.u. positive phase sequence voltage becomes 1 p.u. nps voltage, and the nps relay responds positively to phase reversals.

8.7 Self-Assessment Questions

Self-Assessment Question 8.1

a) Why is it important to know the motor operating characteristics in motor protection design work?
b) What is the significance of nps current in the rotor of a motor?
c) The rotor heating characteristics of 415 V, 75 kW motor due to nps current are given approximately as:

$$I_2^2\, t = 2 \times 10^4 \tag{8.1}$$

where t is the time (s).

If nps current of 30% full load current flows in the motor due to phase imbalance, estimate the maximum time before which the imbalance must be detected in order to prevent rotor damage.

Bibliography

V. S. Sazanov, A. I. Kovalenko, A. A. Voloshin "The Development of System for Automatic Adaptive Change of Relay Protection Settings in Distribution Networks" published in 5th International Youth Scientific and Technical Conference on Relay Protection and Automation (RPA) 2022.

Lü Zhining, Ning Baifeng, Luo Weifeng, Liu Wei "Distribution Networks Ground Fault Protection Method Based on Fuzzy Measure Fusion Criterion" published in IEEE 8th International Conference on Advanced Power System Automation and Protection (APAP) in 2019.

Haiping Shi, Changjiang Zhao, Yuanliang Jin "Current Limiting Protection in Spacecraft Power System" published in 3rd International Conference on Electric Power Equipment – Switching Technology (ICEPE-ST) in 2015.

Haowei Lu, Hussain Biyawerwala, Hasnain Thakrawala "Polarized Distribution Protection Coordination Strategy Under the Impact from Various Distributed Energy Resources (DER) Generation Points" published in IEEE/PES Transmission and Distribution Conference and Exposition (T&D) in 2022.

Kun Liu, Xin Cui, Bin Wang, Xinzhou Dong "Study on High Impedance Grounding Fault Protection for Personal Safety Guard in Distribution Grid" published in IEEE 8th International Conference on Advanced Power System Automation and Protection (APAP) in 2019.

J. S. Kim, J. S. Song, C. H. Kim, N. H. Cho "A Study on the Protection Coordination Correction of Distribution System in Korea using EFI" published in IEEE 8th

International Conference on Advanced Power System Automation and Protection (APAP) in 2019.

Kun Yang, Zhen Xiang, Ruixiong Yang, Lin Jiang "Decentralized Protection Coordination Based on Multi-agent System for Active Distribution Networks" published in 7th Asia Conference on Power and Electrical Engineering (ACPEE) 2022.

Yu Zou, Minchuan Liao, Ruiqing Lian "Discussion on Lightning Protection of Distribution Network and its Differentiation" published in International Conference on Intelligent Computing, Automation and Systems (ICICAS) 2020.

A.A. Voloshin, E.A. Voloshin, A.I. Kovalenko, S.A. Danilov, D.A. Degtyarev; V.S. Sazanov "System Of Automatic Change of Relay Protection Operation Parameters in Distribution Networks" published in 4th International Youth Scientific and Technical Conference on Relay 2021.

Feng Deng, Xiang Ji, Guo-Fang Huang "A Centralized Distribution Protection System" published in China International Conference on Electricity Distribution 2012.

Bruno Fazio, Sanja Bajic, Serge St-Antoine "Relay Protection Coordination Analysis Using Fault Current Measurements" published in IEEE Power & Energy Society General Meeting (PESGM) 2022.

Dong Chen, Bing Hu, Mingjun Xue, Yinian Qi, Changsong "Research and Application of the System of Relay Protection and Self-Healing for Distribution Network with Mixed Use of Circuit Breaker and Load Switch" published in International Conference on Power System Technology (PowerCon) 2023.

Hana A. Halim, Melaty Amirruddin, Nur S. Noorpi, Nurhakimah M. Mukhtar "An Improved Protection Scheme for Smart Distribution Grid" published in 1st International Conference on Artificial Intelligence, Modelling and Simulation 2013.

Jamil Abdo, Tariq Masood "New Technique to Develop Intelligent, Highly Reliable & Efficient Unmanned Power Generation Modular" published in 22nd International Conference on Renewable Energies and Power Quality (ICREPQ'24) Bilbao (Spain), 26–28th June 2024. https://www.icrepq.com/icrepq24/392-24-abdo.pdf.

Jamil Abdo, Tariq Masood "Operational Reliability- Production Best Practice Pragmatic Case Studies & Improvement" published in Proceedings of the 9th North American Conference on Industrial Engineering and Operations Management Washington D.C., USA, 4–6, June 2024.

Tariq Masood, Jamil Abdo "Maintenance Strategy Production Best Practices – Pragmatic Case Studies and Operations Improvement" published in Proceedings of the 9th North American Conference on Industrial Engineering and Operations Management Washington D.C., USA, 4–6, June 2024.

Jamil Abdo, Tariq Masood "Techno Economics Analysis to Reshape Global Clean and Green Energy" 22nd International Conference on Renewable Energies and Power

Quality (ICREPQ'24) Bilbao (Spain), 26–28 June 2024. https://www.icrepq.com/icrepq24/415-24-abdo.pdf.

M. Saeed Ullah, Tariq Masood, Umair Zahid, H. Abuzar Ahsan "Electrochemical Oxidation of Dissolved Organic Dye in Aqueous Solution" published in IMASCON-2019.

Nasser Alemadi, Tariq Masood, R Aggarwal, Suhail A Qureshi, D Kothari "Novel Selection Approach of STATCOM Device Tailored and Optimized Engineering Configuration to Implement in Between Qatar and Bahrain" published in International Journal of Automation and Power Engineering 2014 10.14355/ijape.2014.0302.01.

Tariq Masood, Jamil Abdo "New Technique to Enhance Hybrid Renewable Energy Resources' Operational Flexibility and Reliability" published in 22nd International Conference on Renewable Energies and Power Quality (ICREPQ'24) Bilbao (Spain), 26–28 June 2024. https://www.icrepq.com/icrepq24/396-24-masood.pdf.

Tariq Masood, Jamil Abdo "Review: Uninterrupted Liquid Hydrogen (E-Methanol) Production Framework to Deliver Clean & Green Energy" Published in 22nd International Conference on Renewable Energies and Power Quality (ICREPQ'24) Bilbao (Spain), 26–28th June 2024. https://www.icrepq.com/icrepq24/414-24-masood.pdf.

Tariq Masood, Abdel Aty Edris, Suhail Aftab Qureshi, Muhammad Tajammal, Murtaza Hashmi DP Kothari, Samer Kareem "SSSC Tailored to Optimization Performance in Between United Arab Emirates and Oman on the GCC Power Grid" published in the 13th IET International conference on AC and DC power Transmission Conference on 14–16 February, 2017. http://ieeexplore.ieee.org/document/7934934/.

Tariq Masood, Abdel Aty Edris, Suhail Aftab Qureshi, Muhammad Tajammal, Murtaza Hashmi DP Kothari, Samer Kareem "Smart Grid Operations and Control Challenge by Implementing STATCOM Tailored to Optimization Performance at the GCC Power Grid" published in the 13th IET International conference on AC and DC power Transmission Conference on 14–16 February, 2017. http://ieeexplore.ieee.org/document/7934988/.

Tariq Masood, Muhammad Tajammal, Samer Kareem, Suhail Aftab Qureshi, Abdul Jabbar Khan, DP Kothari, Nair Jameel "Unified power Flow Controller Modeling and Analysis Technique on the GCC Power Grid Operations and Control" published in IEEE Energy Con, June 7–9 Limassol Cyprus 2018. https://ieeexplore.ieee.org/document/8398834/.

Tariq Masood, Muhammad Tajammal, Samer Kareem, Suhail Aftab Qureshi, Murtaza Hashmi, Nasir Jameel "Unified power Flow Controller (UPFC) Modeling and Analysis Technique by using Matlab and EMTP" published in (ICREQP,18 Salamanca Spain 21–23 March,2018). http://www.icrepq.com/icrepq18/250-18-masood.pdf.

Tariq Masood, Muhammad Tajammal, Samer Kareem, Suhail Aftab Qureshi, Murtaza Hashmi Naser Jameel "STATCOM Controller Performance Enhancement and Optimization Technique" published in (ICREQP,18 Salamanca Spain 21–23 March, 2018). http://www.icrepq.com/icrepq18/249-18-masood.pdf.

Tariq Masood, Muhammad Tajammal, Samer Kareem, Suhail Aftab Qureshi, Murtaza Hashmi DP Kothari "Smart Grid Operational functions and Control Challenges by Implementing SSSC Tailored to Optimize performance in between Qatar and KSA on the GCC Electrical-power grid" published in the ASTESJ Journal Advances in Science, Technology and Engineering Systems Journal Vol. 2, No. 1, 30.09.2017 2017. https://www.astesj.com/?sdm_process_download=1&download_id=2963.

Tariq Masood, Raj K Aggarwal, Suhail Aftab Qureshi, RAJ Khan "STATCOM Model Against SVC Control Model Performance Analyses Technique by MATLAB" published in International Conference on Renewable Energies and Power Quality (ICREPQ'10) 2010. https://icrepq.com/icrepq'10/655-Masood.pdf.

Tariq Masood, N Al-Emadi, SA Qureshi, YY Almulla, AJ Khan, SK Shah "FACTS Control Devices (Statcom, SSSC and UPFC) Re-Configuration Techniques by PSIM/MATLAB" Published in: International Conference on Electrical Engineering 2007. http://ieeexplore.ieee.org/abstract/document/4287295.

Tariq Masood, Raj K Aggarwal, SA Qureshi, RAJ Khan "STATCOM and SVC Control Operations and Optimization During Network Fault Conditions" 2010. https://ieeexplore.ieee.org/document/5636868.

Tariq Masood, RK Aggarwal, SA Qureshi, RAJ Khan "Statcom Control Reconfiguration Technique for Steady State and Dynamic Performance Optimization During Network Fault Conditions" published in International Conference on Renewable Energies and Power Quality (ICREPQ'10) Granada (Spain) 2010. https://www.icrepq.com/icrepq'10/656-Masood.pdf.

Tariq Masood, Samer Karim Shah, Muhammad Tajammal, "Unified Power Flow Controller Modeling and Analysis Technique on the GCC Power Grid" published in IEEE International Energy Conference (ENERGYCON) 2018. https://ieeexplore.ieee.org/abstract/document/8398834.

Tariq Masood, Suhail A Qureshi, RK Aggarwal, DP Kothari "Multiple FACTS Devices Integration Through Distributed Control System and its Application on the GCC Interconnection" published in 7th IEEE GCC Conference and Exhibition (GCC) 2013. https://ieeexplore.ieee.org/abstract/document/6705750.

9

Embedded Generation Protection Systems

9.1 Objectives

At the end of this section, you will have had an introduction to the implications of introducing embedded generation into a utility supply network and the techniques used to protect the embedded generator and the utility intertie. You will be able to

- Appreciate the implications of introducing embedded generation on safety, quality of supply, short-circuit protection, and stability
- Translate these into the types of protection required to minimize safety hazards
- Translate these into the types of protection required to minimize failures in providing the required quality of supply
- Translate these into the types of protection used to protect the utility from short-circuit faults fed by the embedded generator
- Translate these into the types of protection used to protect the embedded generator from short-circuit faults
- Translate these into the associated protection scheme required to prevent abnormal operation of the embedded generator and its interconnection to the utility network

9.2 Introduction

9.2.1 Embedded Generation

Since the mid-1970s, there has been a growing interest in using small and medium-sized generation units operating in parallel with the local utility's electricity supply. Several names have been used to describe these generators, including

Protection of Electrical Power Distribution Systems: Smart grid, Microgrid, AI, and Cyber Security,
First Edition. Tariq Masood, Jamil Abdo, and Atif Iqbal.

- private generators
- independent power producers
- nonutility generators (NUGs)
- distributed storage and generation units (DSGUs)
- embedded generation

The interest in embedded generation was supported by government action, which gave legal rights of private generation to be connected to the utility system. In the United Kingdom, the 1983 Energy Act gave independent generators the right to generate and to expect their local utility to purchase any electricity offered. These rights were reinforced in the 1989 Electricity Act, which also gives the newly privatized electricity distribution companies the right to generate up to 15% of their needs. Since this Act, the electricity companies have become very active in embedded generation.

9.2.2 Types of Generation

The types of embedded generation vary considerably and include the following:

- Small-scale combined heat and power units
  (10 KVA-200 kVA)
- Large-scale industrial combined heat and power units
  (1 MVA-50 MVA)
- Waste energy recovery and energy from waste units
  (1 MVA-10 MVA)
- Wind farms
  (500 kVA-10 MVA)
- Small power station
  (10 MVA-400 MVA)
- Voltage control station
  (50 MVA-200 MVA)
- Standby generation
  (300 kVA-10 MVA)

The majority of embedded generation consists of a rotating generator driven by a prime mover, which uses an internal combustion engine. The generator could be a:

- self-excited asynchronous generator
- mains-excited asynchronous generator

- power factor-corrected asynchronous generator
- synchronous generator

Asynchronous generators are used at the lower power levels, and synchronous generators at the medium and high-power levels. In general, small generators rated at less than 100 kVA are single-phase units and are connected to the low-voltage (LV) system. Large generators of up to 500 kVA may still be connected to the LV system, but will invariably be three-phase units. Medium-sized generation, with ratings ranging from 250 kVA to 10 MVA, will generally be three-phase units connected to the high-voltage (HV) distribution system. Large, embedded generations, with a rating of 25 MVA and above, will be directly connected to the sub-transmission network.

Exercise 9.1 Briefly explain why the presence of an embedded generator creates problems for the safety and control of a distribution system.

The difficulties with safety and the control of a distribution system introduced by the presence of embedded generation arise from the ability of the embedded generator to provide power to the system, which is outside the direct control of the utility control center.

Safety problems arise partly from the possibility that a part of the utility network may be kept energized following isolation from the grid of the main source of power. Under these conditions, utility staff may wrongly assume that the system is de-energized and hence put themselves at risk. Alternatively, the earth connection may be lost when the connection to the grid of mains is lost, and hence the neutral voltage may become significant with respect to ground potential, placing both utility staff and the general public at risk.

Control problems also stem from the embedded generator's ability to supply power. Reclosing the connection between the utility system and that part of the distribution system left connected to the embedded generator is significantly more complicated than if the system were de-energized. Out-of-step reclosure between the two networks could damage any rotating machinery connected to the utility network, including the embedded generator.

Other difficulties arise from the choice of settings for the protection equipment, the quality of supply provided to customers adjacent to the embedded generator, and the increased sophistication of the protection and control schemes. The growth in embedded generation over recent years has created the need to resolve these problems.

9.3 Problem Introduced by Connecting Embedded Generating

The problems introduced by connecting small and medium-sized generation to a utility network can be divided into three headings: safety, quality of supply, and the ability to cause damage or instability.

9.3.1 Safety

Safety problems arise from the embedded generator's ability to supply power to the network when the utility's primary source of power has been disconnected, as well as from the earthing arrangement used in distribution networks.

Traditionally, distribution networks were designed to obtain their power from main generating stations or grid supply points. The power flow was radially from the point of supply to the loads, and there were only a few complications due to the interconnection of the system. Loss of the main supply resulted in a nonenergized and hence inherently safe system. Restoration of supply was generally straightforward, in that breakers would connect an energized system to an unenergized system.

In general, there have been few changes to the distribution system to facilitate the introduction of embedded generation. However, with the embedded generator connected, loss of the main source no longer guarantees that the system is de-energized, and that part connected to the embedded generator may become a self-sufficient power island, posing a safety hazard for utility personnel. Restoration of supply also becomes complicated, in that the load side of circuit breakers can no longer be assumed to be unenergized, and there is a possibility of reconnecting the main source of generation to the embedded generator while they are running out of synchronization.

Specialized protection is therefore required to detect when a part of the system containing embedded generation becomes disconnected from the main source of supply and operates as a power island. As soon as this condition is detected, the protection should trip the intertie breaker. This type of protection was not required in a traditional power system, and, therefore, new relaying techniques have had to be developed. It is referred to by several names, including "loss of mains protection," "loss of grid protection," or "islanding protection."

The second safety problem involves the earthing of an embedded generator and the possibility that the embedded generator will become isolated from earth if the connection to the main source of utility generation is lost.

In the United Kingdom, it is normal practice to use single-point earthing on the HV system by using a delta-star transformer in the distribution system and earthing the transformer's star point. Therefore, when an embedded generator is

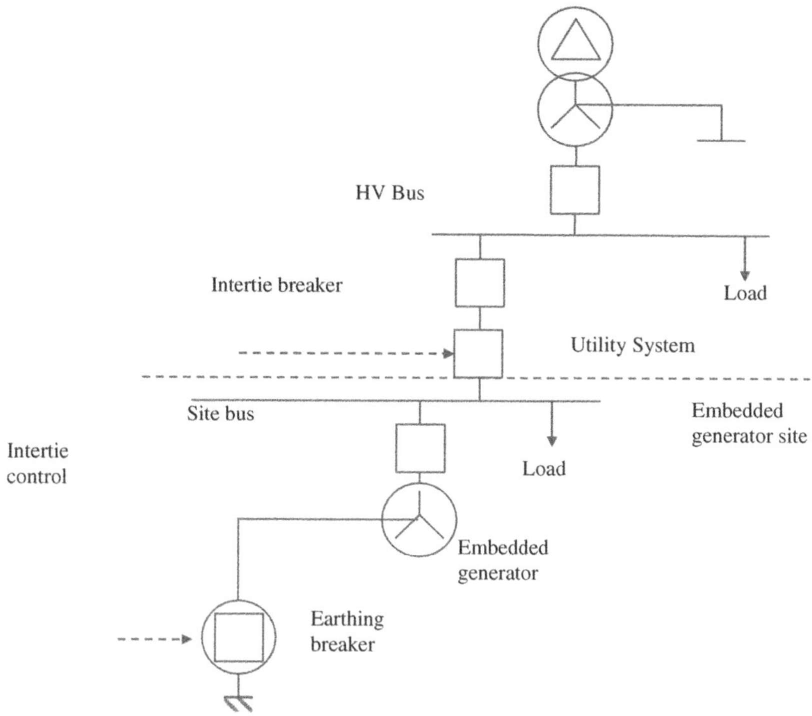

Figure 9.1 Earthing system of HV HV-connected embedded generator.

operated in parallel with the utility supply, its star point is isolated from the earth. When it is operated independently from the utility, the generator's star point is earthed either solidly or using an earthing impedance designed to limit fault currents.

A simple technique to provide this facility is to link the operation of the earthing breaker to the intertie breaker such that whenever the intertie breaker is closed, thereby connecting the embedded generator to the utility, the earthing breaker is open. Conversely, whenever the intertie breaker is open and the embedded generator is operating isolated from the utility supply, the embedded breaker is closed and the generator and the system it is feeding are properly earthed as illustrated in Figure 9.1.

9.3.2 Quality of Supply

In general, the utilities are required to ensure that the presence of the embedded generator will not detract from the quality of supply that they provide to all of their customers. In the United Kingdom, power providers are required by legislation to supply electricity within prescribed tolerances of voltage and frequency. Should the supply move outside these limits, the supply utility will be held responsible.

Before allowing an embedded generator to be connected to their system, the supply utilities insist that steps are taken to ensure that an embedded generator will not detract from their ability to meet these supply requirements. Protection relays are included to monitor the supply's voltage and frequency, and should their trip levels be exceeded, the connection between the embedded generator and the utility system is automatically broken by opening the intertie breaker.

This protection scheme involves four protection functions, namely, under- and overvoltage relaying and under- and over-frequency relaying. For an embedded generator operating in parallel with a UK/USA electricity board system, supplies must remain within the statutory voltage limits of $+10\%$ and -6% and a frequency limit of $\pm 1\%$.

In the near future, the quality of supply will include standards for the harmonic content of voltage supplied, and to ensure that an embedded generator does not contribute excessive harmonic content, additional relaying measuring harmonic content may be required.

9.3.3 Short-Circuit and Damage Prevention Protection

The protection systems associated with an embedded generator are required to ensure that:

- The embedded generator is protected from external faults.
- The embedded generator is protected from internal faults.
- The utility is protected from abnormal operating conditions resulting from the presence of the embedded generator.
- Automatic synchronism supervision is provided to ensure that the connection between the utility and the embedded generator is executed under conditions that will not result in a disturbance to the utility supply or damage to the embedded generator.

The protection scheme must therefore include:

1) Short-circuit protection for external faults
2) Short-circuit protection for internal faults
3) Short-circuit protection for earth faults
4) Neutral displacement protection
5) Phase unbalance protection
6) Field failure protection
7) Pole slipping protection
8) Reverse power protection
9) Synchronization control of the connection circuit breaker

10) Under- and overvoltage protection
11) Under- and over-frequency protection
12) Loss of grid protection
13) Protection supervision

A typical protection scheme for an embedded generator operating in parallel with the utility system is shown in Figure 9.2

9.3.4 Protection Guidelines

Several large utilities have produced guidelines defining the type of protection scheme that they require before they will agree to an embedded generator operating in parallel with their network[1,2]. The UK Electricity Board requirements are specified in the Electricity Council Engineering Recommendation G59[1].

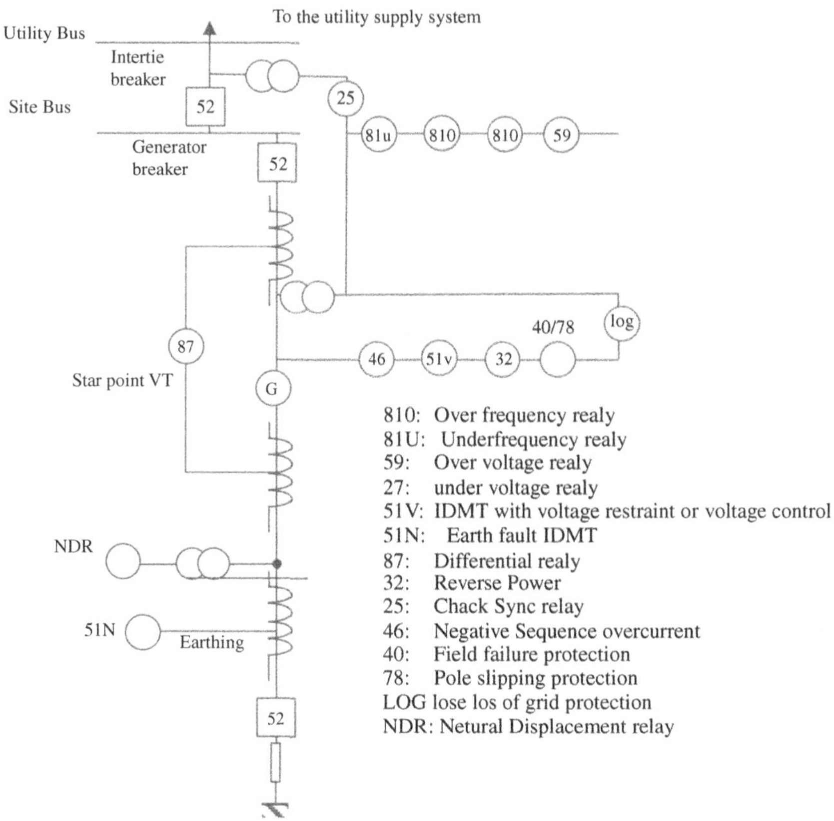

Figure 9.2 General protection scheme for an embedded generator.

The individual generator is responsible for providing and maintaining the required protection and control equipment associated with its plant.

Before agreeing to allow an embedded generator to operate in parallel with their system, most utilities require that the protection scheme ensure that:

- The embedded generator cannot be connected to their system unless all phases of their supply are energized and operating within agreed limits.
- The embedded generator is disconnected from the utility when an abnormality occurs that results in an unacceptable deviation in the voltage or frequency at the point of supply.
- The embedded generator is disconnected from the network when the utility supply is lost.
- The embedded generator cannot be connected to the utility system, or if already connected, then it is disconnected, whenever there are failures in the protection systems or their supplies.

9.4 Utility Protection Requirements for an Embedded Generator

9.4.1 Short-Circuit Protection for External Faults

The simplest protection against external short-circuit faults is provided by three-phase inverse definite minimum time (IDMT) delayed overcurrent relaying monitoring the output current from the generator. The relay's natural timing characteristic will provide an inherent tolerance to switching transients and magnetizing currents. This relay is shown as a (51 V) unit in Figure 9.2.

The complication for these types of faults when fed from an embedded generator is the generator's inability to maintain its output voltage during a low impedance fault and therefore its inability to maintain the level of current expected at the normal supply voltage. To increase the relay's sensitivity during voltage collapse or depression, it is normal practice to use a voltage-controlled IDMT relay. These relays have two operating curves, one of which is used when the terminal instantaneous undervoltage element switches the relay characteristic from the high setting, corresponding to normal supply voltage, to the sensitivity setting, corresponding to depressed voltage.

9.4.2 Short-Circuit Protection for Internal Faults

The protection used for the short-circuit protection for external faults also provides the first line of defense against internal faults on the generator and its associated cabling. For an internal fault when the embedded generator is operating in parallel with the utility supply, the voltage is supported by the utility generation, and it is unlikely to be depressed by the same amount as when supported by the embedded generator alone. Therefore, the (51 V) relay will normally operate on its standard characteristic curve rather than its sensitivity curve[3]. The generator's stator windings are also protected by phase-by-phase current differential relaying. This unit protection scheme is equally effective when the generator is running in parallel with the utility supply as when the generator is operating alone.

An inherent limitation of this type of differential relaying is that it does not provide 100% protection for the stator winding. The need for an operating current above the relay settings prohibited trip operation for faults close to the neutral point. In Figure 9.2, this relay (87) is designed.

When the embedded generator is operating independently of the utility supply and the earthing breaker is closed, additional earth fault protection is provided by an IDMT relay monitoring the current in the earthing circuit. If the site load is designed to be balanced, this relay can use a sensitivity setting. In Figure 9.2, the relay is designated (51 N).

Exercise 9.2 What action or actions should occur following the tripping of the short-circuit protection for an internal fault on an embedded generator?

Following the tripping of the short-circuit protection of an internal fault on an embedded generator, the circuit breaker connecting the embedded generator to the utility system should be opened, and then the prime mover should be shut down. An internal fault in the embedded generator will continue to do damage until all sources of power are disconnected. The generator from the utility removes one source of power, but the generator action, powered by the prime mover, will continue as a power source until the prime mover stops.

9.4.3 Neutral Displacement Protection

The neutral displacement protection monitors the voltage between the generator star point and earth. It is only required when the embedded generator is connected to a single-point earthed system, the earthing breaker is open, and the embedded generator is connected to the utility supply by the intertie breaker. Under these conditions, the system provides the earthing point for the embedded generator, the local site load, and the network to which it is connected.

Should the utility earth be lost, and an unbalance exist in the grounding impedance, the star point will adopt a voltage other than earth potential. If this voltage exceeds the setting on the neutral displacement relay, it will trip the intertie breaker and close the earthing breaker. Since the relay may also be tripped by a short-circuit fault even after the system is earthed, its operation is time delayed.

Exercise 9.3 Will the neutral displacement protection always respond to a loss of grid connection?

Unfortunately, the phase displacement protection will not always respond to a loss of grid connection. An islanded power system will operate with the neutral point at or near zero volts when the system is balanced. Also, during unbalanced operating conditions with the grid connection in place, the star point on an embedded generator connected to a system which is remotely earthed will deviate from zero volts.

9.4.4 Phase Unbalance Protection

An unbalanced operation will cause heating within the generator and subsequently damage the machine. The presence of negative sequence current provides an immediate indication of unbalance and provides a measurement for relaying; therefore, a relay designed to trip for a negative sequence overcurrent will protect the generator from this potentially damaging situation. This relay is designated as type (46) and its connection in the protection scheme is shown in Figure 9.2.

Negative sequence component produces a reaction field which rotates counter to the dc field system and flux, which cuts the rotor at twice the rotational velocity. This induces double-frequency currents in the field system and in the rotor body. The resulting eddy currents can be very large and cause severe heating of the rotor.

The negative sequence component can be detected by a filter circuit, and the relay set to operate when the negative sequence current power exceeds that power equivalent to the continuous thermal withstand of the machine. The relay's operating times are dependent on the inverse of the negative sequence current flowing in the generator and can vary between 0.3 and 100 seconds.

9.4.5 Field Failure and Pole-Slipping Protection

Field failure protection and pole-slipping protection are required for a synchronous generator. They are fundamentally related and can be detected using similar types of relays, or relays that include both functions. Field failure protection is designed as type (40) and pole-slipping protection type (78).

The excitation of a synchronous generator is supplied by an auxiliary supply, and it can fail for a variety of reasons. When there is a failure of the excitation and field failure, the generator is unable to deliver its output power into the system and it accelerates to a higher speed. Unable to operate as a synchronous machine, it operates as an induction generator and produces about half its rated power. As a result, a high proportion of the input power is dissipated in the generator as an induction generator, which is excited from the system in the form of reactive power. A synchronous generator can withstand a short loss of excitation, but if it persists for more than 10 seconds, mechanical damage will result. The field failure will immediately compromise the quality of supply provided to adjacent consumers since the voltage could be reduced or even collapse, and the local system may become unstable.

Pole slipping can result when there is an imbalance between the mechanical power input into a synchronous generator and the electrical power that it provides to the system. When the mechanical power input exceeds the system's ability to absorb electrical energy, or through reduced excitation, the generator is accelerated and loses synchronization with the utility network. When the machine loses synchronism, it passes through a motoring mode until it is able to deliver electrical power again. Repeated pole-slipping will damage the machine and cause significant power disturbances, which will compromise the quality of supply provided to adjacent consumers.

The traditional techniques for detecting field failure and pole slipping are based on monitoring the impedance of the generator as seen from its terminals.

A typical field failure may use an offset-mho characteristic lying in the negative reactance area and centered on the reactive axis as shown in Figure 9.3. Under normal operation, the impedance of the generator will lie in the first quadrant with the value represented by N. Following field failure, the machine will absorb reactive power, and hence its impedance will be dominated by a negative reactance. Time-delayed tripping ensures that the protection will not trip for synchronizing surges and transient conditions. This time delay is chosen to match the ability of the machine to withstand field failure, and it is generally greater than 0.5 seconds.

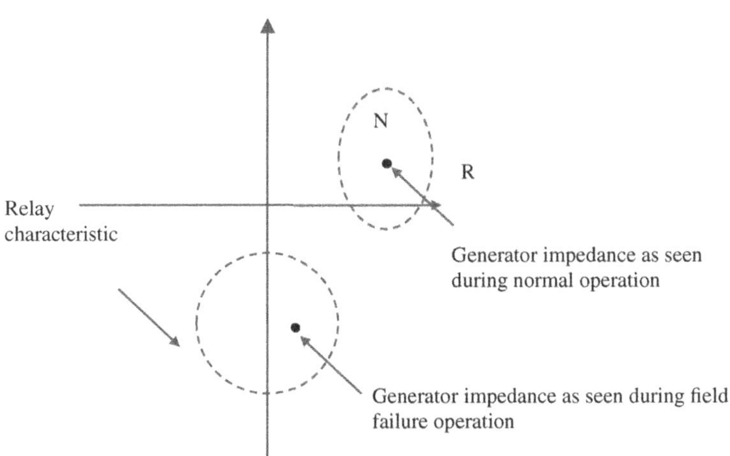

Figure 9.3 Relay characteristic for detecting field failure.

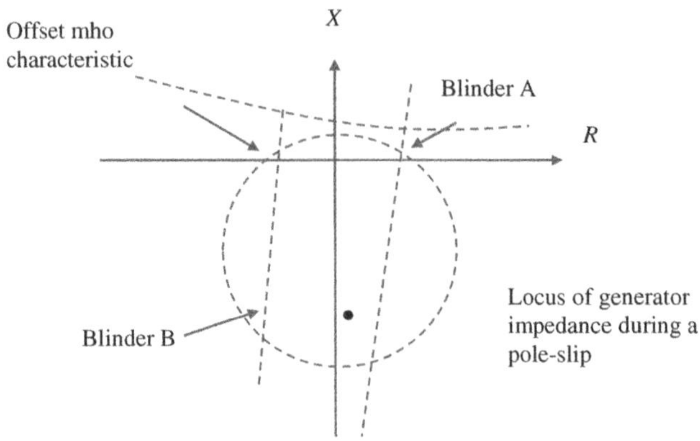

Figure 9.4 Relay characteristic for detecting pole slipping.

Traditional pole-slipping protection monitors the path of the apparent impedance of the generator and its rate of change. The simplest scheme uses a single mho characteristic and trips whenever the generator's impedance enters the characteristic. Unfortunately, this approach may malfunction for recovery swings. A more secure scheme uses a combination of offset mho and two blinder characteristics, as shown in Figure 9.4. During the pole slip, the impedance seen at the generator's terminal moves across the impedance plane and enters the offset mho and then crosses the two blinder characteristics in the appropriate order, indicating that a pole-slip is occurring, which leads to a relay trip. The operating time of the pole-slipping protection is not time delayed.

9.4.6 Reverse Power Protection

Reverse power protection protects the generator from prime mover failure and can assist in the instability problem. The relay monitors the direction of power flow at the generator terminals and can disconnect the machine in the event of system failure. Inverse time or instantaneous relays can be used depending on the machine and its prime mover. Reverse power protection is designated as (32).

For small generators, the machine can be assumed to be operating under balanced conditions and so a single-phase relay will suffice. These relays monitor the phase angle between the line current and phase voltage and trip when cos Ø becomes negative. For this application, the relays have to be sensitive to reverse real power flow and have to operate at low power factors. A typical setting would be equivalent to a 5% power factor, and the relay will be time delayed to prevent maloperation for transient phenomena. The time delay setting will generally be greater than 0.5 seconds and may be as high as 10 seconds.

9.4.7 Synchronization Control of the Connection Circuit Braker

The arbitrary connection of a generator to a source of alternating current can cause severe damage to that generator and disturb the source of supply provided to adjacent customers. Automatic check synchronization is therefore required on a circuit breaker that is used to connect the generator to the utility network to ensure that the two systems are synchronized at the moment that they are connected together.

The embedded generator's control system is used to adjust its output voltage and frequency prior to connecting it to the supply network. The check synchronization protection monitors the voltage difference between the two sides of the circuit breaker before it closes and the phase displacement between the two systems together with the frequency of the two supplies. Although the instruction to close the breaker can be given either manually or automatically, this protection only allows the breaker to close when the voltage difference and phase displacement are within the specified limits, and the two systems are operating at the same frequency.

The check synchronizing relays are designated as type (25) and connected across the circuit breaker that is used to connect the embedded generator to the utility supply. In the situation illustrated in Figure 9.2 where the site bus may be energized from either the utility supply or the embedded generator or both, the check synchronization relay will be connected across both breakers and live line/dead bus, deadline/live bus supervision may allow the closing of either breaker to supply the local load.

9.4.8 Under-/Overvoltage and Under-/Over-frequency Protection

Under-/overvoltage and under-/over-frequency relaying provides an immediate check that the system's voltage and frequency are held within statutory limits. Under abnormal conditions, for whatever cause, should these limits be broken, the intertie is disconnected, and the embedded generator is prevented from supplying adjacent utility consumers. These relays are set to respond in typically 0.5 seconds.

The frequency relays are designed as type (81), the undervoltage relay as type (27), and the overvoltage relays as type (59). They may be connected to monitor the utility supply or the generator output terminals as shown in Figure 9.5.

9.4.9 Loss of Grid Protection

Loss of grid protections is required to protect against accidental isolation of the two systems, in which a portion of the utility's load becomes isolated from the source

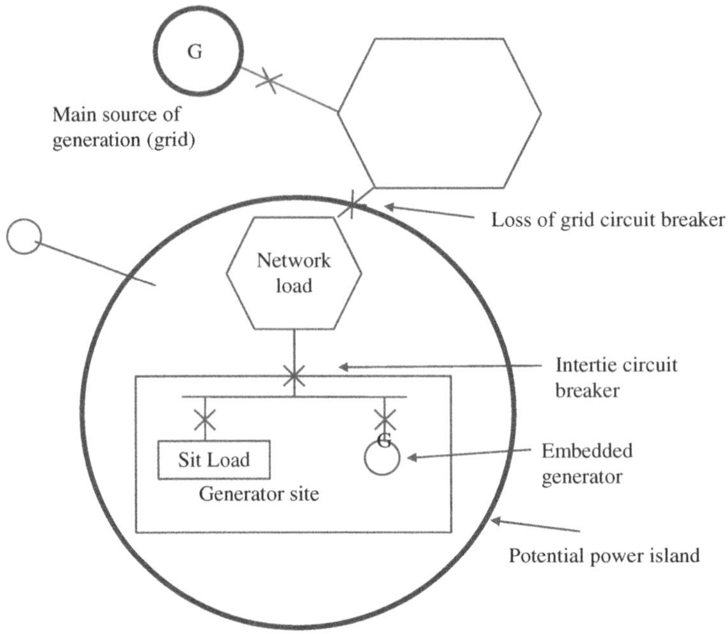

Figure 9.5 A utility network including embedded generation illustrating a loss of grid.

of supply but remains connected to the embedded generators. Under these conditions, the isolated system may continue to operate independently of the utility and form a power island.

The situation is illustrated in Figure 9.5. Noting that the grid breaker can be a breaker in the utility network, which provides a link between the embedded generator and the utility's main source of power. As yet, loss of grid protection has not been assigned an ANSI Safety Standard type number.

Island operation is not a problem provided that the voltage and frequency remain within statutory limits. The main issue is the possibility of restoration of the utility's supply out of synchronism with the embedded generator due either to a manual or automatic reclosure of the utility circuit breaker. Under these conditions, the generator should be disconnected from the network in less than 500 ms.

In cases where the embedded generator is intended only to provide power to the site load and not to export power to the utility system, loss of grid can be detected using a reverse power relay on the intertie between the embedded generator's site load and the utility.

For small generators, typically less than 200 kVA, which can export power to the utility, loss of grid generally results in overloading of the generator, and both the output voltage and frequency fall. This causes the undervoltage and under-frequency protection to operate.

Large generators fitted with high-speed automatic voltage regulators may well be able to match the subsequently island load, and there is a possibility that the voltage and frequency will be maintained within preset limits. Under these conditions, loss of grid becomes more difficult to detect, so a sophisticated protection is required.

Several protection schemes have been developed to detect a loss of grid. These can be divided into two fundamental groups.

The first uses active techniques, which directly interact with the operations of the power system, and the second uses passive techniques, which monitor the power system behavior. The only commercially available active loss of grid protection technique is the reactive export error detector[4]. Passive techniques include under-/overvoltage under-/over-frequency, rate of change of frequency, and phase displacement monitoring. We have installed low- and high-voltage power labs at Frostburg State University, concentrating on an alternative technique that monitors fluctuation in the power output of the embedded generator to determine whether or not the connection to the utility's primary source of power is intact.

Exercise 9.4 Briefly explain why the possibility of restoration of the utility's supply of synchronism with the embedded generator is a major concern following as loss of grid connection?

Out of set restoration of the utility supply to a system containing an embedded generator could cause damage to the generator as to is forced into synchronism with the utility system, and the associated current surges could cause dramatic voltage fluctuation in the circuits connecting the two systems together. This could stress the conductors and circuit breaker, and cause disturbance to other customers connected to the utility system. These disturbances would appear as voltage flicker, which could disturb a variety of customer loads.

Reactive Export Error Detector[4]

The reactive export error detector controls the embedded generator's excitation current so that it generates a level of reactive current in the utility to the generator site intertie, which cannot be supported unless the utility source is connected. The relay trips when the level of reactive current is not maintained at the required level for greater than the present period. This time delay prevents maloperation for short-lived supply fluctuation. An immediate advantage of this relay is that it can detect loss of grid where the disconnection does not change the generator load.

The time delay is typically set between 2 and 5 seconds, and the relay is therefore inherently slow. The approach is also limited when the power factor compensation capacitors remain connected to the island, and hence, the reactive currents can be maintained. The relay requires both voltage and current inputs, as well as connections to the voltage regulator.

This technique is one of the few techniques suitable for small power supply system where other relays fail to differentiate between loss of grid and abnormal system fluctuations. It is also used to provide backup protection to other faster schemes.

Rate of Change of Frequency[4]

This protection operates when the rate of change of frequency exceeds a present limit. The frequency is monitored either at the generator terminals or on the utility to customer intertie using a voltage input. The relay setting is chosen such that it will operate for the embedded generator load changes associated with the loss of grid, but it will not operate for fluctuations governed by the utility time constant. The setting is critical since too sensitive a setting could result in unnecessary tripping during system disturbances, and too high a setting could cause failure to detect loss of grid.

Under loss of grid, any changes in the load connected to the embedded generator will result in a df/dt governed by the inertia constant and rated capacity of the embedded generator. For small and medium-sized embedded generator, a trip setting Hz/s has been found to be optimal. The relay algorithm is designed to ignore slow changes in frequency, which could be caused by load changes on the utility network but to respond quickly to rapid changes in frequency when embedded generation becomes isolated. The rate of change of frequency relay can operate in the range 0.3–0.7 seconds, and trips in 80 ms with large load changes following dramatic disconnection. An inherent advantage of a rate of change of frequency relay is that when loss of grid does not result in a significant change in the generator loading and the relay initially fails to operate, it will trip when a subsequent load change produces the required df/dt.

Phase Displacement Monitor[5]

The phase displacement monitor operates when there are sudden changes in the phase displacement in the system voltage waveform. These phase displacements are a direct result of changes in the generator load and its operating frequency. The relay can respond in 50 ms at the instant of loss of grid, the generator's output power changes by more than 5% of its rating. As with the rate of change of frequency relay, the phase displacement monitor will also operate for any subsequent load change that produces the required change in phase displacement, should the loss of grid fail to cause tripping.

Generator Output Power Fluctuations Monitor[6]

This technique detects the loss of grid by monitoring the behavior of the power system to disturbances to the system and observing the difference in response between those experienced when the main utility supply is connected to the embedded generator and those when the embedded generator is operating in isolation. The algorithm uses changes in the power output from the embedded generator to provide an insight into the transfer function of the generating plant feeding the power system. Under normal conditions with the utility supply connected, the transfer function reflects the combination of the main source and the embedded generation. Following loss of grid, or while the embedded generator is operating in isolation, it reflects the characteristics of the embedded generator alone.

Referring to the simplified representation of a power system shown in Figure 9.2, the embedded generator and the main utility supply can be represented by idealized generators of capacity G_g and G_m with the inertia constant H_g and H_m, respectively. While the utility's source of supply is connected to the network, a change in the system load of ΔP_S produces a change in the distributed generator's loading of ΔP_g defined by:

$$\Delta P_g = \Delta P_S \frac{H_g G_g}{H_g G_g + H_m G_m} \tag{9.1}$$

But if the distributed generator operates independently from the utility supply, for example, following loss of grid, any change of system load will result in the distributed generator's load changing by ΔP_g defined by:

$$\Delta P_g = \Delta P_S \tag{9.2}$$

Accepting that the utility has a greater capacity than the embedded generator and that its inertia constant will also be greater, the difference in response to a disturbance provides an immediate basis for determining whether or not the utility supply is connected to the portion of the network containing the embedded generation.

An additional advantage of using the power as the relaying measurement is that the power measurement is readily obtained from the instantaneous measurement of the generator's terminal voltage and output currents. Providing that the samples are taken at the same instant, they also do not have to be synchronized to the power system frequency.

In a typical protection scheme associated with an embedded generator, these measurements are also required for other relaying functions required to protect the machine. This loss of grid protection is therefore suitable for inclusion in an integrated protection package for a small or medium-sized embedded generator without dramatically increasing the hardware requirements.

In addition to reliably detecting the occurrence of a loss grid, the protection algorithm must remain stable during conditions of extreme load unbalance, loss of phase, and during load flicker. These conditions have been accommodated by amplitude limiting measurement to changes in power output from the embedded generator, and integrating samples over a moving window period. The protection's trip criterion is defined by:

$$\left[\sum_{n=-t_x}^{n=0} fn(\Delta P_g)_n\right] > k_s \tag{9.3}$$

where fn is the amplitude limiting function, n is the sampling instant $(\Delta P_g)_n$, k_s is the trip setting, and t_x is the length of the sampling window.

Extensive simulation studies have been undertaken to study the effects of both loss of grid and other disturbances on a power system containing an embedded generator. These studies were conducted using both the alternative transient version of the EMPT program, ATP,[7] and programs developed in-house.

The amplitude limiting function is used to control the minimum operating time of the protection, and the length of the sampling window controls the nominal maximum operating time. For the initial cycle and a maximum of six cycles, simulation studies have shown that the algorithm can reliably trip for loss of grid conditions resulting in a load change at the generator's terminal of greater than 1% of the generator's rating, and that the operating times are as designed. An attractive feature of this approach is that it responds to out-of-synchronism reclosures of the main source of supply to the embedded generator and can therefore trip the intertie breaker to minimize subsequent damage. Further work on this technique is continuing, and the algorithm has been implemented in a laboratory microprocessor-based demonstration relay.

9.4.10 Protection Supervision

Some form of protection supervision is normally required with a protection scheme as involved as that used for embedded generation. If discrete relays are used, this may consist of ensuring that the adjacent protection schemes provide a level of backup protection and that the power supplies to the relaying panel are supervised.

Where modern digital protection is used and in particular microprocessor-based relaying, self-monitoring functions are generally available, which trips if there is an internal failure in the relay. In the case of failure of the protection system, the general agreement between the embedded generator and the utility requires the embedded generator to be disconnected from the utility system.

9.5 Self-Assessment Questions

Self-Assessment Question 9.1

Briefly explain why phase-by-phase current differential protection is unable to provide 100% protection for an embedded generator's stator winding and describe what action is required should a stator winding fault be detected. Some research (recommended) may be required to answer this question.

Self-Assessment Question 9.2

Briefly explain why loss of grid protection presents a major challenge to protection relay designers.

Bibliography

Leuven EMTP Center, 'Alternative transients program rule book' Belgium 1987. Wanjing Xiu, Yuan Liao Scientific Research Publishing.

Jamil Abdo, Tariq Masood "New Technique to develop Intelligent, highly reliable & Efficient Unmanned Power Generation Modular" published in 22nd International Conference on Renewable Energies and Power Quality (ICREPQ'24) Bilbao (Spain), 26th–28 June 2024. https://www.icrepq.com/icrepq24/392-24-abdo .pdf.

Haowei Lu, Hussain Biyawerwala, Hasnain Thakrawala "Polarized Distribution Protection Coordination Strategy Under the Impact from Various Distributed Energy Resources (DER) Generation Points" published in IEEE/PES Transmission and Distribution Conference and Exposition (T&D) in 2022.

Kun Liu, Xin Cui, Bin Wang, Xinzhou Dong "Study on High Impedance Grounding Fault Protection for Personal Safety Guard in Distribution Grid" published in IEEE 8th International Conference on Advanced Power System Automation and Protection (APAP) in 2019.

Qiushi Li, Xianghua Tang, Xiongjie Shi "Demonstration and Application of AC/DC Hybrid Power Supply System in Building" Published in 2nd IEEE Conference on Energy Internet and Energy System Integration (EI2) in 2018.

Xian Wang "Application of Equation of Light Shutting in Selective Protection Device" published in International Conference on Networking Systems of AI (INSAI) in 2021.

Hadi Zayandehroodi, Azah Mohamed, Hussain Shareef, Marjan Mohammadjafar, Masoud Farhoodnea "A Novel Protection Coordination Strategy using Back Tracking Algorithm for Distribution Systems with High Penetration of DG" published in IEEE International Power Engineering and Optimization Conference Melaka, Malaysia in 2012.

Seongil Kim, Byungchul Sung, Jeong-Hyeok Heo, Duck-Su Lee, "Protection Against Transient Overvoltages Caused by Intermittent Earth Faults in Unearthed Marine Power Distribution Network" published in IEEE International Conference on Power Systems Technology (POWERCON) in 2022.

Yu Deng, Dingkang Liang, Jiahao Wang, Xinyi Song, Zehua Li Zhetong Hong "Reliability Evaluation of Distribution System Based on Dynamic Bayesian Network" Published in 3rd Power System and Green Energy Conference (PSGEC) in 2023.

Jamil Abdo, Tariq Masood "Operational Reliability- Production Best Practice Pragmatic Case Studies & Improvement" published in Proceedings of the 9th North American Conference on Industrial Engineering and Operations Management Washington D.C., USA, 4–6 June 2024.

Tariq Masood, Jamil Abdo "Maintenance Strategy Production Best Practices – Pragmatic Case Studies and Operations Improvement" published in Proceedings of the 9th North American Conference on Industrial Engineering and Operations Management Washington D.C., USA, 4–6 June 2024.

Jamil Abdo, Tariq Masood "Techno Economics Analysis to Reshape Global Clean and Green Energy" published in 22nd International Conference on Renewable Energies and Power Quality (ICREPQ'24) Bilbao (Spain), 26–28 June 2024. https://www.icrepq.com/icrepq24/415-24-abdo.pdf.

M. Saeed Ullah, Tariq Masood, Umair Zahid, Hafix Abuzar Ahsan "Electrochemical Oxidation of Dissolved organic Dye in Aqueous Solution" published in IMASCON-2019.

Nasser Alemadi, Tariq Masood, R Aggarwal, Suhail A Qureshi, D Kothari "Novel Selection Approach of STATCOM Device Tailored and Optimized Engineering Configuration to Implement in Between Qatar and Bahrain" published in International of Journal Automation and Power Engineering 2014 10.14355/ijape.2014.0302.01.

Tariq Masood, Jamil Abdo "New Technique to Enhance Hybrid Renewable Energy Resources' Operational Flexibility and Reliability" published in 22nd International Conference on Renewable Energies and Power Quality (ICREPQ'24) Bilbao (Spain), 26–28 June 2024. https://www.icrepq.com/icrepq24/396-24-masood.pdf.

Tariq Masood, Jamil Abdo "Review: Uninterrupted Liquid Hydrogen (E-Methanol) Production framework to deliver Clean & Green Energy" Published in 22nd International Conference on Renewable Energies and Power Quality (ICREPQ'24) Bilbao (Spain), 26–28 June 2024. https://www.icrepq.com/icrepq24/414-24-masood.pdf.

Tariq Masood, Abdel Aty Edris, Suhail Aftab Qureshi, Muhammad Tajammal, Murtaza Hashmi DP Kothari, Samer Kareem "SSSC Tailored to Optimization

Performance in Between United Arab Emirates and Oman on the GCC Power Grid" published in the 13th IET International conference on AC and DC power Transmission Conference on 14–16 February, 2017. http://ieeexplore.ieee.org/document/7934934/.

Tariq Masood, Abdel Aty Edris, Suhail Aftab Qureshi, Muhammad Tajammal, Murtaza Hashmi DP Kothari, Samer Kareem "Smart Grid Operations and Control Challenge by Implementing STATCOM Tailored to Optimization Performance at the GCC Power Grid" published in the 13th IET International conference on AC and DC power Transmission Conference on 14–16 February, 2017. http://ieeexplore.ieee.org/document/7934988/.

Tariq Masood, Muhammad Tajammal, Samer Kareem, Suhail Aftab Qureshi, Abdul Jabbar Khan, DP Kothari, Nair Jameel "Unified power Flow Controller Modeling and Analysis Technique on the GCC Power Grid Operations and Control" published in IEEE Energy Con, 2018 7–9 June Limassol Cyprus 2018. https://ieeexplore.ieee.org/document/8398834/.

Tariq Masood, Muhammad Tajammal, Samer Kareem, Suhail Aftab Qureshi, Murtaza Hashmi, Nasir Jameel "Unified Power Flow Controller (UPFC) Modeling and Analysis Technique by using Matlab and EMTP" published in (ICREQP,18 Salamanca Spain 21–23 March,2018) http://www.icrepq.com/icrepq18/250-18-masood.pdf.

Tariq Masood, Muhammad Tajammal, Samer Kareem, Suhail Aftab Qureshi, Murtaza Hashmi Naser Jameel "STATCOM Controller Performance Enhancement and Optimization Technique" published in (ICREQP,18 Salamanca Spain 21–23 March,2018). http://www.icrepq.com/icrepq18/249-18-masood.pdf.

Tariq Masood, Muhammad Tajammal, Samer Kareem, Suhail Aftab Qureshi, Murtaza Hashmi DP Kothari "Smart Grid Operational functions and Control Challenges by Implementing SSSC Tailored to Optimize Performance in Between Qatar and KSA on the GCC Electrical-Power Grid" published in the ASTESJ Journal Advances in Science, Technology and Engineering Systems Journal Vol. 2, No. 1, 2017 30.09.2017. https://astesj.com/special-issue-4/.

Tariq Masood, Raj K Aggarwal, Suhail Aftab Qureshi, Raj Khan "STATCOM Model Against SVC Control Model Performance Analyses Technique by MATLAB" published in International Conference on Renewable Energies and Power Quality (ICREPQ'10) 2010. https://icrepq.com/icrepq'10/655-Masood.pdf.

Tariq Masood, N Al-Emadi, SA Qureshi, YY Almulla, AJ Khan, SK Shah "FACTS Control Devices (Statcom, SSSC and UPFC) Re-Configuration Techniques By PSIM/MATLAB" Published in 2007 International Conference on Electrical Engineering 2007 https://ieeexplore.ieee.org/abstract/document/4287295.

Tariq Masood, Raj K Aggarwal, SA Qureshi, RAJ Khan "STATCOM and SVC Control Operations and Optimization During Network Fault Conditions" 2010. https://ieeexplore.ieee.org/document/5636868.

Tariq Masood, RK Aggarwal, SA Qureshi, Raj Khan "Statcom Control Reconfiguration Technique for Steady State and Dynamic Performance Optimization During Network Fault Conditions" published in International Conference on Renewable Energies and Power Quality (ICREPQ'10) Granada (Spain) 2010. https://www.icrepq.com/icrepq'10/656-Masood.pdf.

Tariq Masood, Samer Karim Shah, Muhammad Tajammal, "Unified Power Flow Controller Modeling and Analysis Technique on the GCC Power Grid" published in 2018 IEEE International Energy Conference (ENERGYCON) 2018. https://ieeexplore.ieee.org/abstract/document/8398834.

Tariq Masood, Suhail A Qureshi, RK Aggarwal, DP Kothari "Multiple FACTS Devices Integration through Distributed Control System and its Application on the GCC Interconnection" published in 2013 7th IEEE GCC Conference and Exhibition (GCC) 2013. https://ieeexplore.ieee.org/abstract/document/6705750.

VS Sazanov, AI Kovalenko, AA Voloshin "The Development of System for Automatic Adaptive Change of Relay Protection Settings in Distribution Networks" published in 2022 5th International Youth Scientific and Technical Conference on Relay Protection and Automation (RPA) 2022.

AA Voloshin, EA Voloshin, AI Kovalenko, S.A. Danilov, DA Degtyarev, VS Sazanov "System Of Automatic Change Of Relay Protection Operation Parameters In Distribution Networks" published in 2021 4th International Youth Scientific and Technical Conference on Relay 2021.

Feng Deng, Xiang Ji, Guo-Fang Huang "A centralized Distribution Protection System" published in 2012 China International Conference on Electricity Distribution 2012.

Bruno Fazio, Sanja Bajic, Serge St-Antoine "Relay Protection Coordination Analysis using Fault Current Measurements" published in 2022 IEEE Power & Energy Society General Meeting (PESGM) 2022.

Dong Chen, Bing Hu, Mingjun Xue, Yinian Qi, Changsong Oin "Research and Application of the System of Relay Protection and Self-Healing for Distribution Network with Mixed Use of Circuit Breaker and Load Switch" published in 2023 International Conference on Power System Technology (PowerCon) 2023.

Zhichao Lin, Yingmin Wang, Riping Zhong, Busheng Luo, Qiujie Ma, Chuankun Yu "Coordinated Optimization Method of Distributed Protection for Distribution Network Based on Multi-agent" publish in 2023 8th Asia Conference on Power and Electrical Engineering (ACPEE) 2023.

Xiaohui Song, Yu Zhang, Song Zhang, Shizhan Song, Jing Ma, Weibo Zhang "Active Distribution Network Protection Mode Based on Coordination of Distributed and Centralized protection" published in 2017 China International Electrical and Energy Conference (CIEEC) 2017.

10

Autoreclose Relaying

10.1 Objectives

When you have completed this section, you will be able to:

- List the main causes of faults on distribution systems
- Appreciate the likelihood of faults being permanent
- Describe the operation of single-shot autoreclose relays
- Explain the operation of multi-shot autoreclose relays

10.2 Introduction

The objective of autoreclose relaying is to return the power system to its normal operating condition automatically as quickly as possible following a break in service caused by a short-circuit fault in that system. Autoreclosing can only be effective where there is a good probability that the fault is temporary and can be cleared by disconnecting it from the source of supply. In general, this only applies to overhead line feeders where the faults are transient or semipermanent and where they can be successfully burned away by the fault current. Therefore, autoreclose schemes are not used with underground cable feeders, since most short-circuit faults on these systems are permanent, being caused by insulation degradation resulting from ageing, overvoltage, overheating, mechanical damage, or some combination of these.

In an autoreclose scheme, the short circuit fault will be detected by the relevant protection relay, and either the operation of that protection or the opening of the associated circuit breaker will be used to initiate the autoreclose switching sequence. This will reclose the circuit breaker after a preset time delay and, in the majority of cases, will restore the power system to its healthy prefault configuration. If the reclosure is unsuccessful, then the protection will trip again and open the circuit breaker to isolate the fault.

Protection of Electrical Power Distribution Systems: Smart grid, Microgrid, AI, and Cyber Security, First Edition. Tariq Masood, Jamil Abdo, and Atif Iqbal.

10.2.1 The Main Cause of Faults on Overhead Distribution Lines

The most common cause of short faults on overhead distribution lines is a lightning strike close to the conductors. This type of fault accounts for more than 80% of all faults affecting overhead distribution lines. Indirect lightning strikes induce high voltages into the conductors, which in turn lead to insulation failure. If the fault is disconnected quickly, there is a high probability that there will be no permanent damage to the power system plant, and the line can be returned to service. This type of fault is referred to as a transient fault.

When lightning strikes directly, those conductors that are hit by the lightning generally contain sufficient concentrated energy to cause permanent damage.

After these faults, the power system equipment will be damaged, and this damage must be repaired before the line can be returned to service. Reclosing the circuit breaker onto these faults will reinstate the fault and cause the protection to trip again. Such faults are therefore defined as permanent faults. For overhead distribution lines, vegetation, such as trees and creepers, can also cause short-circuit faults. These can be associated with the support poles where creepers grow up the poles and across the insulators. Similar faults can be caused midspan, where timber can foul the conductors. For both these conditions, the short-circuit fault can be described as semipermanent since the fault current effectively burns the fault path away and removes the cause of the problem.

In tall grasslands, bush fires can cause transient faults by creating an atmosphere sufficiently ionized to cause arcing. A common feature of these types of faults is that they are relatively high impedance.

10.2.2 Transient, Semipermanent, and Permanent Faults

The majority, between 80 and 90%, of faults on overhead distribution lines are transient in nature and can be cleared by disconnecting the fault from the source of power. As soon as the circuit breaker or breaker has isolated the faulted plant from the source of power, the fault arc is extinguished, and when the dielectric strength of the insulation has been restored, the plant can be safely reconnected to the power system and thus re-energized.

The remaining 10–20% of faults are classified as permanent faults, and any attempt to reconnect them to the power system will result in fault current and the re-tripping of the protection system. In general, the damage caused by permanent faults must be repaired before the plant can be restored to normal service.

Semipermanent faults are generally associated with vegetation fouling the overhead lines and can be cleared by burning the fault path away with the fault current.

Naturally, any transient fault, which is not disconnected from the source of supply, will eventually cause permanent damage to the power system plant and hence develop into a permanent fault. The shorter the duration of the fault, the less likely that the fault will cause long-lasting damage, and it is therefore beneficial to minimize fault clearance times.

10.2.3 Fault Clearance

Tripping the circuit breaker or circuit breaker connecting the faulted plant to the source or source of supply does not instantaneously remove the fault from the power system. The circuit breaker requires time to separate its contacts mechanically and to rupture the arc formed by the opening contacts, finally breaking the current flow into the fault.

The process involved in rupturing the circuit breaker arc also leads to the rupturing of the fault arc. These two arcs are in series. This may require only a few half cycles, but it can take longer depending on the type of switchgear and the characteristics of the circuit involved.

Before the circuit breaker can be reclosed, thereby re-energizing the previously faulted plant, the dielectric strengths of the insulation must have recovered sufficiently to withstand the system voltages involved. If the dielectric strength has not recovered, the re-energization voltage will restrike the arc, and the fault will be re-established.

10.2.4 Circuit Breaker Operation

In addition to considering the state of the power system in assessing the criteria for successful autoreclosure, the switchgear must also be suitable for this duty. The autoreclose cycles impose severe stresses on the circuit breaker. Also, before reclosure can be contemplated, the circuit breaker must be ready to close its contacts and also trip, without delay, should the fault persist. In most cases, the times required to allow the circuit breaker to reset are less than that required for the de-ionization of the fault path.

Exercise 10.1 Explain why the operation of the circuit breaker reclosure contact should be delayed after the circuit breaker tripping contact has operated.

Before the circuit breaker can be reclosed, time must be given for the circuit breaker to complete its operation and reset following a trip. The circuit breaker has to be ready not only to reclose but also to trip again should the fault not be cleared.

Time must also be allowed for the ionization caused by the fault to disperse before the overhead distribution line can be re-energized so that the dielectric strength of the insulation can recover. In general, the time required for de-ionization of the fault path is greater than that required for the circuit breaker to reset.

10.3 Autoreclosing Relays

Although included in the general definition of protection relays, autoreclosing relays do not detect fault conditions nor cause circuit breaker tripping, but are involved instead in control sequence associated with circuit breaker reclosing. Autoreclosing relays are switching sequence controllers and can be implemented using electromechanical, static, programmable logic controller, and microprocess technology. They do, though, have to operate in schemes including and alongside protection relays; they share the same harsh electrical environment as the protection relays.

10.3.1 The Single-Shot Autoreclosing Relay Sequence

The single-shot reclosing relay's switching sequence can be triggered either by the operation of the protection trip relay's output or by an auxiliary contact on the circuit breaker. Although triggering from the protection relay is the more common, using an auxiliary contact on the circuit breaker has some advantages over using the protection trip output. The circuit breaker contact operates when the breaker's main contacts open and not when the instruction to trip is given. Assuming that the sequence is triggered by the protection relay, the switching sequence begins with a time delay to allow the circuit breaker to break the fault current and deionize the fault path. This time delay between the autoreclose scheme being initiated and the operation of the relay to close the circuit breaker is defined as the dead time. After the dead time, the autoreclose relay initiates circuit breaker closure. When the circuit breaker's contacts have had time to close, power is restored to the circuit. For a transient fault, the fault will have been cleared by the first trip, the power will be restored to the power supply circuit, and the autoreclose can be considered to be successful. But if the fault was permanent, the reclosing of the circuit breaker would have re-energized the fault. The protection relay would trip again, opening the circuit and the circuit breaker, and disconnecting the faulted plant. The autoreclose scheme's output contact, which initiates circuit breaker reclosure, also initiates the scheme's resetting timer, the reclaim timer. If the reclosure is successful and the protection does not trip during the reclaim time, the autoreclose scheme is reset, awaiting the next fault condition. The reclaim time is chosen to be sufficiently long to ensure that the system has been safely restored, but short enough to ensure that a future independent fault is not mistaken for a continuation of the original fault. This is particularly important in areas that are prone to lightning strikes and where storms could produce a series of independent faults. These switching operations for the single-shot autoreclose scheme are

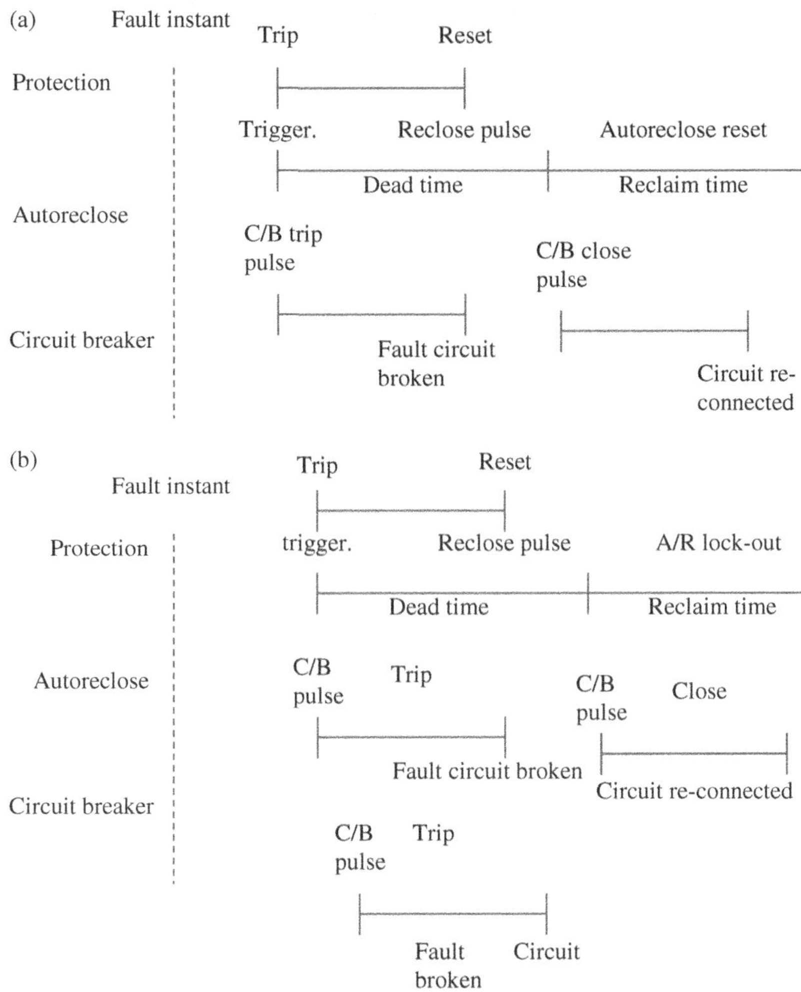

Figure 10.1 Autoclose switching sequence for transient and permanent faults. (a) Reclosure of the transient fault. (b) Reclosure of permanent fault.

illustrated in Figure 10.1, which shows the timing sequence for both transient and permanent faults.

A typical schematic for a single-shot auto reclose scheme is shown in Figure 10.2. The autoreclose scheme could be triggered by an inverse definite minimum time delay (IDMT) type of overcurrent relay, an instantaneous

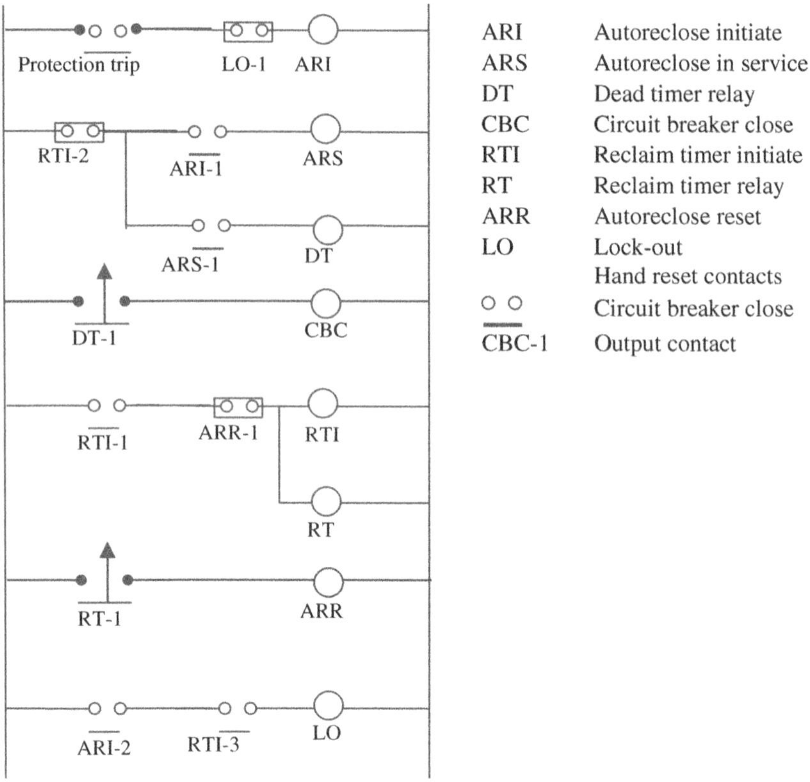

Figure 10.2 Typical schematic for a single-shot autoreclose scheme.

overcurrent relay, a differential protection, or any other type of protection used to protect the feeder circuit.

10.3.2 Choice of Dead Time

The choice of dead time setting depends on the power system configuration to ensure that the fault has been cleared from all sources of power, the ability of the circuit breaker to prepare for reclosure and possible subsequent tripping, for the fault path to de-ionize and recover its dielectric strengths so as to support the energizing voltage and type of load that the circuit is supplying.

When the distribution line is fed from two or more ends, consideration must be given to the operation of the protection and autoreclose schemes at all of the feeder's terminals. It can generally be assumed that relaying schemes at the different terminals are not synchronized, and that a communication system does not exist for this duty. The fault path will be maintained until all sources of power are disconnected from the faulted feeder, and delays between the tripping of the protection at the different ends of the feeder therefore require a longer dead time than with a radially fed system.

Before the autoreclose can take place, the circuit breaker must be ready to both reclose and trip, should the fault be permanent. This may involve recharging the closing and breaking mechanisms. In general, clearing the fault arc of a transient fault and restoring the dielectric strength, if the insulation only takes a few cycles of the power system waveform, and, therefore, is covered by the recovery time of the switchgear. Different loads can place severe demands on the choice of dead time. The least interruption to supply is provided by a short dead time, typically 0.5 seconds. Where such a short dead time may compromise or damage either the load or the system supplied by the feeder, longer times are used. The least interruption to supply is provided by a short dead time, typically 0.5 seconds. Where such a short dead time may compromise or damage either the load or the system supplied by the feeder, longer times are used.

Where the load contains synchronous machines, the dead time is chosen to be sufficiently long to allow them to be disconnected from the system before the reclosure. This is also applied to the system with embedded generation, where an out-of-step reclosure between the main supply and the "small" generator could cause power surges and bring about severe damage. For a system containing induction motors, the dead time is chosen to allow the motor residual voltage to fall to a third of the rated value.

A common concern with modern domestic supplies is the response of modern appliances to supply interruption. For example, television receivers can be damaged by short breaks in supply, and a delayed reclosure will dramatically reduce the possibility of damage.

Where the dead time is set to less than 1 second, the autoreclose scheme is defined as a high-speed scheme. Where the dead time is set to greater than 1 second, the scheme is defined as a delayed autoreclose scheme.

10.3.3 Multiple-Shot Autoreclose Schemes

With multiple-shot schemes, multiple dead times and reclosures are used before the scheme locks out. If any of the reclosures is successful in restoring the supply, the reclosing sequence is halted, and after the reclaim time, the system resets.

In general, the number of reclosures is limited to three or four before a lockout. Depending on the particular autoreclose scheme, different dead times can be used for each reclosure, or one dead time can be used for the first reclosure and a second for subsequent autoreclosures. The choice of dead time depends on the complexity of the system, the technology being used, and the price that can be justified.

One popular protection and autoreclose used widely on radially fed overhead distribution feeder circuits uses a relatively sensitive instantaneous overcurrent relay to provide the first trip and initiate the autoreclose sequence. This provides immediate tripping for faults in a large proportion of the protected feeder system, and this will operate before other local protected feeder systems and, as such, will operate before other local protection has had time to trip, be it an IDMT relay or a fuse. Where the first reclosure is successful, the power is restored to all the customers and the system returns to its prefault status.

If this first autoreclose is not successful and the fault is not cleared, then the second trip will be controlled by an IDMT overcurrent relay rather than the instantaneous relay. The settings for this IDMT relay are chosen to grade with the downstream protection and therefore allow time for the downstream relays or fuses to trip. If they clear the fault, this IDMT relay will not trip, and only those customers directly affected by the fault will lose supply. If the fault persists, eventually the IDMT relay will trip, thus opening the circuit breaker and initiating the second reclosure. Using a longer dead time than the first shot, the system is allowed longer to recover before reclosing in case the fault was a transient one.

Should the fault persist, the reclosure will re-energize the fault, and the IDMT will repeat its tripping sequence. The time required for the IDMT relay to trip allows some types of faults to burn away and be cleared by tripping the circuit again. Successful reclosure may still result.

Eventually, after several failed attempts, it is accepted that the probability of success is low, and the autoreclose scheme locks out and must be reset by an operator.

An example of the schematic diagram of a multi-shot autoreclose scheme based on relay logic is shown in Figure 10.3. Such a scheme could be directly implemented using electromechanical or electromechanical and static electronic technologies. A feature of this scheme is that it uses separate timers for the four dead times, which can therefore have individual settings.

With the introduction of micro-electronics into automatic sectionalizers, the switching sequence associated with multi-shot autoreclose schemes can be used to control their tripping. These devices can monitor the passage of fault current into the feeder they protect, and they can be programmed to trip after several failed reclosures have occurred in a set period, thus disconnecting the suspect feeder from the source of supply while the system is not energized.

Where the autoreclose relay is used with automatic sectionalizers, a single IDMT relay may be used to detect the faults rather than using a combination of instantaneous and IDMT.

Exercise 10.2 Describe the operation of the reclaim timer, RT, in the multi-shot autoreclose scheme shown in Figure 10.3?

The reclaim timer, RT, is activated after the completion of each autoreclose dead time, noting that this is the same time as the circuit breaker closing pulse is provided.

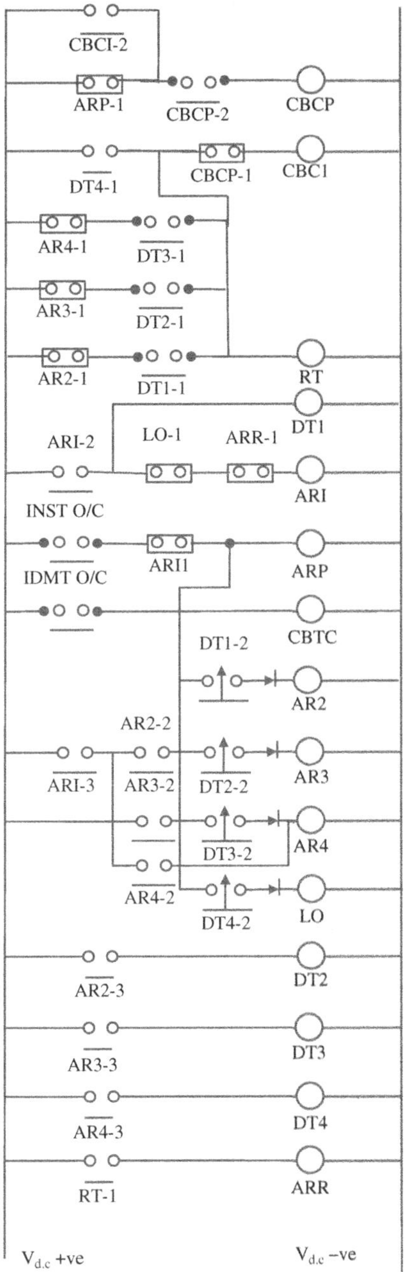

ARI	Autoreclose initiate
AR2	Autoreclose stage 2
AR3	Autoreclose stage 3
AR4	Autoreclose stage 4
ARP	Autoreclose prepare (delay on pickup)
ARS	Autoreclose in service
DT1	Dead timer relay 1
DT2	Dead timer relay 2
DT3	Dead timer relay 3
DT4	Dead timer relay 4
CBCI	Circuit breaker close
CBCP	Circuit breaker closing pulse timer
CBCP	Circuit breaker close
CBTC	Circuit breaker trip coil
RTI	Reclaim timer initiate
RT	Reclaim timer relay
ARR	Autoreclose reset
LO	Lock-out
	Hand reset contacts

Figure 10.3 Schematic for a multi-shot autoreclose scheme.

For the first autoreclose, the reclaim time RT is started by the closing of contact DT-1. If the reclose is successful, the timer RT operates and activates relay ARR, which resets the autoreclose system and enables it to respond to any future protection trips.

If the autoreclose is not successful and the next protection trip occurs within the reclaim time, the closing of the trip contact would cause relay AR2 to operate and open contact AR-1, thus de-energizing the reclaim timer RT and resetting it. The reclaim timer would then be activated after the second dead time, controlled by the dead time DT2. Again, if the autoreclose were successful, the reclaim timer RT would operate and reset the scheme using relay ARR.

If this autoreclose were also unsuccessful, a similar sequence to the above would occur, and if the autoreclose were successful, operation of the reclaim timer would reset the system. If this autoreclose were unsuccessful, the next protection trip would energize the lockout relay, LO via the then closed contact DT4-2. This would de-energize the autoreclose scheme and use contact LO-1 to prevent further operation until the scheme has been manually reset

10.3.4 Check Synchronism Relays

Check synchronism relays are increasingly being used to provide an automatic check on phase angle and voltage differences between two power systems that are to be connected by circuit breakers. This could be associated with a feeder connecting multiple sources of power together, or with a system that includes embedded generation. Check synchronism relays operate when the differences in phase angle and voltage are within the present limit and restrain whenever these differences are outside the set limits. The operating relay's output is used to permit the circuit breaker's closing operations, such that the reclosure only occurs when the two systems are synchronized, and therefore it will not cause a major disturbance to the system.

A typical operating characteristic for a check synchronizing relay is shown in Figure 10.4. The relay inputs VA are obtained from voltage transformers on either side of the circuit breaker. This figure shows the characteristics where the phase angle setting is ±20 and its voltage unbalance is set to ±10. By including a time setting to operate with the phase angle setting, the relay can restrain whenever the slip frequency between the two systems is above a set limit.

Whenever the system connected to one side of an open circuit breaker is de-energized, that is, dead, the closing of the circuit breaker can be executed at any time without the problem of causing a system disturbance. A live line/dead bus or live bus/deadline check relay will operate whenever the system on one side of the circuit breaker is de-energized. This type of relay uses voltage relays in a scheme similar to that shown in Figure 10.5, and the sensitivity of these relays is chosen such that any induced voltage from parallel lines will not cause malfunctioning.

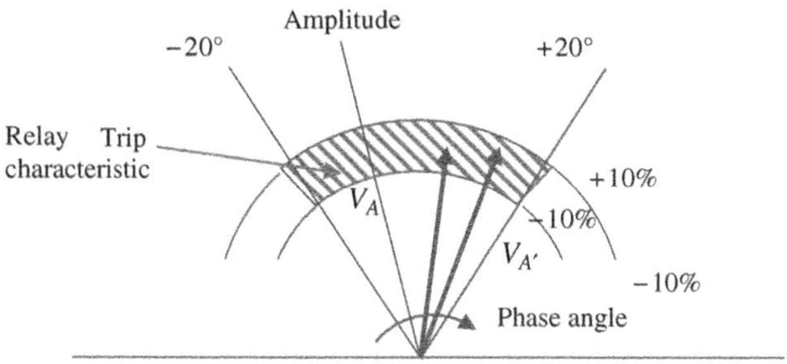

Figure 10.4 Typical operating characteristic for a check synchronization.

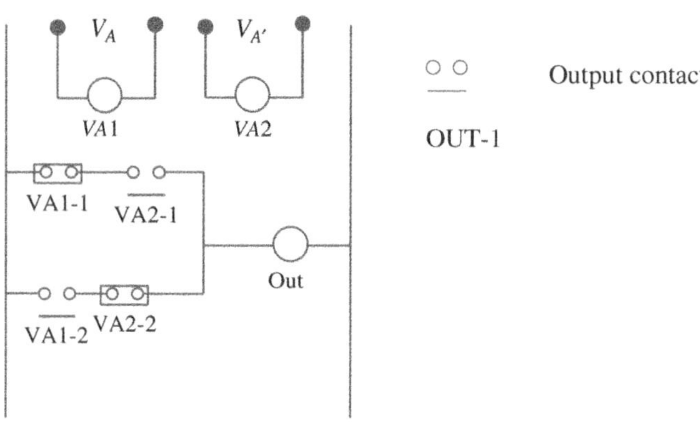

Figure 10.5 Live line/dead bus, live bus/deadline check supervision relay.

10.4 Self-Assessment Questions

Self-Assessment Question 10.1

Briefly describe the characteristics of the following types of faults and comment on the possible success of autoreclosing a circuit breaker following the protection trip:

a) Transient
b) Permanent
c) Semipermanent

Self-Assessment Question 10.2

Describe the duties of the lockout relay in an autoreclose scheme commenting on its operation in the single-shot and multi-shot autoreclose schemes shown in Figure 10.2 and 10.3, respectively.

Bibliography

Leuven EMTP Center, 'Alternative transients program rule book' Belgium July 1987. Wanjing Xiu, Yuan Liao Scientific Research Publishing.

Rani Kumari, Bhukya K. Naick "Protection Coordination Impact on Reliability of Distribution Systems With Distributed Energy Resources" published in Third International Conference on Power, Control and Computing Technologies (ICPC2T) 2024.

Jane Courtney, Aaron McDonnell "Impact on Distribution System Protection with the Integration of EG on the Distribution Network" published in 54th International Universities Power Engineering Conference (UPEC) 2019.

Hana A. Halim, Melaty Amirruddin, Nur S. Noorpi, Nurhakimah M. Mukhtar "An Improved Protection Scheme for Smart Distribution Grid" published in 1st International Conference on Artificial Intelligence, Modelling and Simulation 2013.

Qingfa Chai, Jiayin Jiang, Yi Tang, Anning Wang, Congcong Li "High Reliability Relay Protection Setting Scheme of Distribution Network" published in International Conference on Power System Technology (PowerCon) 2023.

Jamil Abdo, Tariq Masood "New Technique to Develop Intelligent, Highly Reliable & Efficient Unmanned Power Generation Modular" published in 22nd International Conference on Renewable Energies and Power Quality (ICREPQ'24) Bilbao (Spain), 26–28 June 2024. https://www.icrepq.com/icrepq24/392-24-abdo.pdf.

Qiushi Li, Xianghua Tang, Xiongjie Shi "Demonstration and Application of AC/DC Hybrid Power Supply System in Building" Published in 2nd IEEE Conference on Energy Internet and Energy System Integration (EI2) in 2018.

Xian Wang "Application of Equation of Light Shutting in Selective Protection Device" published in International Conference on Networking Systems of AI (INSAI) in 2021.

Arun G. Phadke, Peter Wall, Lei Ding, Vladimir Terzija "Improving the Performance of Power System Protection Using Wide Area Monitoring Systems" published in Journal of Modern Power Systems and Clean Energy 2016. 10.1007/s40565-016-0211-x.

P. Chandhra Sekhar, R. A. Deshpande, V. Sankar "Evaluation and Improvement of Reliability Indices of Electrical Power Distribution system" Published in National Power Systems Conference (NPSC) in 2016.

Jing Xiao, Ning Wu, Shuai Han, Xiaorui Wu, Wenlan Gong "Electric Vehicle Support Distribution Grid Emergency Power Protection Strategy after a Major Power Outage" Published in 3rd International Conference on Energy Engineering and Power Systems (EEPS) in 2023.

Liu Chenyi, Xie Lei, Hu Shuilian; Zhous Hengdong "Research of Comprehensive Application of Intelligent Low-Voltage Power Distribution Units in Improving Power Quality" published in China International Conference on Electricity Distribution (CICED) in 2018.

Jibran Ali, Federico Silvestro, Mahmood Jamil "Virtual Power Plant for Improving Power System Protection Issues: Solution to the Problem of Power System Reliability under Distributed Energy Resources" published in Australasian Universities Power Engineering Conference (AUPEC) in 2018.

Jamil Abdo, Tariq Masood "Operational Reliability- Production Best Practice Pragmatic Case Studies & Improvement" published in Proceedings of the 9th North American Conference on Industrial Engineering and Operations Management Washington D.C., USA, 4–6, June 2024.

Tariq Masood, Jamil Abdo "Maintenance Strategy Production Best Practices – Pragmatic Case Studies and Operations Improvement" published in Proceedings of the 9th North American Conference on Industrial Engineering and Operations Management Washington D.C., USA, 4–6, June 2024.

Jamil Abdo, Tariq Masood "Techno Economics Analysis to Reshape Global Clean and Green Energy" published in 22nd International Conference on Renewable Energies and Power Quality (ICREPQ'24) Bilbao (Spain), 26–28 June 2024. https://www.icrepq.com/icrepq24/415-24-abdo.pdf.

M. Saeed Ullah, Tariq Masood, Umair Zahid, H. Abuzar Ahsan "Electrochemical Oxidation of Dissolved organic Dye in Aqueous Solution" published in IMASCON-2019.

Nasser Alemadi, Tariq Masood, R Aggarwal, Suhail A Qureshi, D Kothari "Novel Selection Approach of STATCOM Device Tailored and Optimized Engineering Configuration to Implement in Between Qatar and Bahrain" published in International Journal of Automation and Power Engineering 2014. 10.14355/ijape.2014.0302.01.

Tariq Masood, Jamil Abdo "New Technique to Enhance Hybrid Renewable Energy Resources' Operational Flexibility and Reliability" published in 22nd International Conference on Renewable Energies and Power Quality (ICREPQ'24) Bilbao (Spain), 26–28 June 2024. https://www.icrepq.com/icrepq24/396-24-masood.pdf.

Tariq Masood, Jamil Abdo "Review: Uninterrupted Liquid Hydrogen (E-Methanol) Production framework to deliver Clean & Green Energy" published in 22nd International Conference on Renewable Energies and Power Quality (ICREPQ'24) Bilbao (Spain), 26–28th June 2024. https://www.icrepq.com/icrepq24/414-24-masood.pdf.

Tariq Masood, Abdel Aty Edris, Suhail Aftab Qureshi, Muhammad Tajammal, Murtaza Hashmi DP Kothari, Samer Kareem "SSSC tailored to optimization performance in Between United Arab Emirates and Oman on the GCC Power Grid" published in the 13th IET International conference on AC and DC power Transmission Conference on 14–16 February, 2017. http://ieeexplore.ieee.org/document/7934934/.

Tariq Masood, Abdel Aty Edris, Suhail Aftab Qureshi, Muhammad Tajammal, Murtaza Hashmi DP Kothari, Samer Kareem "Smart Grid Operations and Control Challenge by Implementing STATCOM Tailored to Optimization Performance at the GCC Power Grid" published in the 13th IET International conference on AC and DC power Transmission Conference on 14–16 February, 2017. http://ieeexplore.ieee.org/document/7934988/.

Tariq Masood, Muhammad Tajammal, Samer Kareem, Suhail Aftab Qureshi, Abdul Jabbar Khan, DP Kothari, Nair Jameel "Unified power Flow Controller Modeling and Analysis Technique on the GCC Power Grid Operations and Control" published in IEEE Energy Con, June 7–9 Limassol Cyprus 2018. https://ieeexplore.ieee.org/document/8398834/.

Tariq Masood, Muhammad Tajammal, Samer Kareem, Suhail Aftab Qureshi, Murtaza Hashmi, Nasir Jameel "Unified power Flow Controller (UPFC) Modeling and Analysis Technique by using Matlab and EMTP" published in (ICREQP,18 Salamanca Spain 21–23 March, 2018). http://www.icrepq.com/icrepq18/250-18-masood.pdf.

Tariq Masood, Muhammad Tajammal, Samer Kareem, Suhail Aftab Qureshi, Murtaza Hashmi Naser Jameel "STATCOM Controller Performance Enhancement and Optimization Technique" published in (ICREQP,18 Salamanca Spain 21–23 March, 2018). http://www.icrepq.com/icrepq18/249-18-masood.pdf.

Tariq Masood, Muhammad Tajammal, Samer Kareem, Suhail Aftab Qureshi, Murtaza Hashmi DP Kothari, "Smart Grid Operational functions and Control Challenges by Implementing SSSC Tailored to Optimize performance in between Qatar and KSA on the GCC Electrical-power grid" published in the ASTESJ Journal Advances in Science, Technology and Engineering Systems Journal Vol. 2, No. 1, 30.09.2017 2017. https://astesj.com/special-issue-4/.

Tariq Masood, Raj K Aggarwal, Suhail Aftab Qureshi, RAJ Khan "STATCOM Model Against SVC Control Model Performance Analyses Technique by MATLAB" published in International Conference on Renewable Energies and Power Quality (ICREPQ'10) 2010. https://icrepq.com/icrepq'10/655-Masood.pdf.

Tariq Masood, N Al-Emadi, SA Qureshi, YY Almulla, AJ Khan, SK Shah "FACTS Control Devices (Statcom, SSSC and UPFC) Re-Configuration Techniques By PSIM/MATLAB" published in International Conference on Electrical Engineering 2007. https://ieeexplore.ieee.org/abstract/document/4287295.

Tariq Masood, Raj K Aggarwal, SA Qureshi, RAJ Khan "STATCOM and SVC control operations and optimization during network fault conditions" 2010. https://ieeexplore.ieee.org/document/5636868.

Tariq Masood, RK Aggarwal, SA Qureshi, RAJ Khan "Statcom Control Reconfiguration Technique for Steady State and Dynamic Performance Optimization During Network Fault Conditions" published in International Conference on Renewable Energies and Power Quality (ICREPQ'10) Granada (Spain) 2010. https://www.icrepq.com/icrepq'10/656-Masood.pdf.

Tariq Masood, Samer Karim Shah, Muhammad Tajammal, "Unified Power Flow Controller Modeling and Analysis Technique on the GCC Power Grid" published in IEEE International Energy Conference (ENERGYCON) 2018. https://ieeexplore.ieee.org/abstract/document/8398834.

Tariq Masood, Suhail A Qureshi, RK Aggarwal, DP Kothari "Multiple FACTS Devices Integration Through Distributed Control system and its Application on the GCC Interconnection" published in 7th IEEE GCC Conference and Exhibition (GCC) 2013. https://ieeexplore.ieee.org/abstract/document/6705750.

11

Coordinated Protection and Control

11.1 Objectives

When you have completed this section, you will be able to:

- List the main protection, control, and monitoring functions that exist in a substation
- Explain why conventional protection and control schemes are difficult to test, maintain, and expand
- Show how coordination of protection and control gear can be achieved in the substation using computer-controlled equipment
- Describe the relevant digital technology used in coordinated substation, particularly in digital communications
- Describe the structure of a modern coordinated substation

11.2 Conventional Protection and Control Systems

We shall begin by considering the overall arrangement of protection and control systems that exist in a conventional substation. The main functions found within a substation are:

- protection of power system equipment
- control and interlocking of switchgear
- generation of alarms for system/local control
- event recording
- disturbance recording
- automatic reclosing/restoration
- synchronizing checks
- load shedding

A substation may have all or only some of these functions, depending on its size.

Protection of Electrical Power Distribution Systems: Smart grid, Microgrid, AI, and Cyber Security, First Edition. Tariq Masood, Jamil Abdo, and Atif Iqbal.

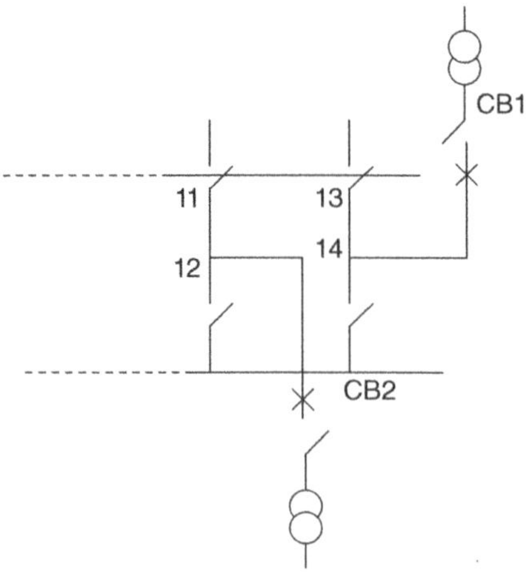

Figure 11.1 Section of a busbar.

Exercise 11.1 Can you summarize these functions into three main categories?

Essentially, all substation functions can be classified as protection, control, or monitoring functions. Some functions may inherit characteristics of two of these, for example, load shedding, or autoreclosures schemes could be construed as being either protection or control.

As an example of a conventional isolator interlocking function, consider Figure 11.1 which shows a section of a distribution substation busbar. The isolators 11, 12, 13, and 14 cannot be switched on load.

Exercise 11.2 Under what conditions (i.e. status of isolators and circuit breaker) can isolator 11 be opened?

Since isolator 11 cannot switch load current, the following switching conditions are allowed:

1) **Circuit Dead:** CB1 and CB2 open
2) **Circuit Live:** CB1, CB2, CB12, and CB14 closed, CB13 open

In order to ensure that 11 cannot be inadvertently operated, an interlocking control system is used. In a conventional substation, this will consist of a hard-wired logic arrangement using auxiliary relays and auxiliary contacts on the switchgear connected together by wiring. Since the components of the substation of Figure 11.1 are distributed around the site, this will lead to large amounts of wiring. As the complexity of the substation increases, so does the complexity of the interlocking control system, the number of auxiliary relays, and the amount of wiring needed. Additionally, there will be event recording

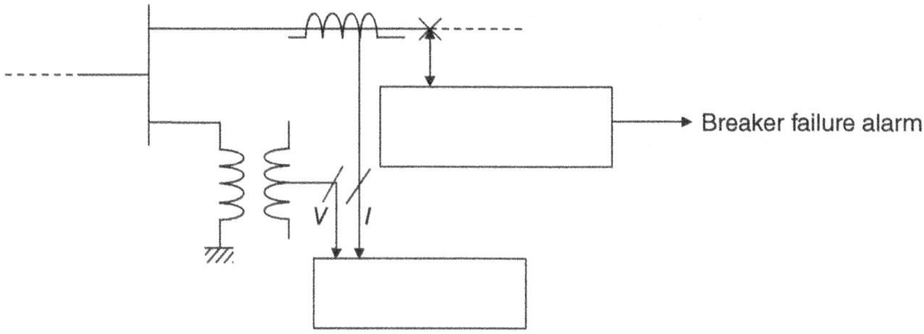

Figure 11.2 Connection between and distance relay.

equipment to log the operation of the switchgear, transducers to measure busbar voltage, current, frequency, and complex power, and the ability to communicate with the substation operator, either onsite or at a remote location.

Figure 11.2 shows the connection of a distance relay to a major distribution level overhead line. The current transformer is usually conveniently located on the circuit breaker bushings, whereas the voltage transformers are freestanding. Thus, wiring from two areas provides the relay with the input signal. It is common to back up the main distance protection with another type of secondary protection. To keep the high security of protection, the relay transducers will be duplicated, so even more wiring is needed. As a result of relay operation, the breaker control will initiate breaker operation. The breaker auxiliary contacts are monitored to ensure that the breaker operates as instructed, otherwise breaker fail alarm may be generated; again, this requires more wiring. In the case of a substation with many converging lines, it is common for all relays to be located in the same building, and so the wiring needed becomes complicated. Other functions which will be provided with the protection include disturbance allowing post-fault analysis of waveform, and a communication system to inform the operator of the operation of any protective gear.

Exercise 11.3 What operational problem do you expect with the conventional approach t_0 substation protection and control?

The main disadvantages of conventional protection and control systems are experienced during maintenance and when the systems need expansion. During maintenance and testing, equipment needs to be removed from service, which can necessitate changes to the hard wiring, thus complicating maintenance procedures. If the substation needs expansion, then nearly all the subsystems will need to be changed due to their "hard-wired" nature. Additionally, installation and commission are complicated since errors in the wiring can be difficult to trace.

11.3 Coordinated Protection and Control

In the previous section, we examined typical substation protection and control systems. In each system, we saw that various tasks need to be performed, such as operating switchgear, monitoring busbar voltage, current, frequency, and power levels, and reporting information. Although the result of performing these tasks is different in either case, the tasks are, nonetheless, similar. Substation coordination aims to combine the redundant tasks used in protection and control systems to reduce the overall complexity. This has distinct advantages for the design, operation, and maintenance of a substation. Closely allied to coordination and integration are sometimes referred to as the same idea, strictly speaking, integrating substation systems and protection functions. This approach has the disadvantage of reducing overall dependability since the failure of one component can endanger both functions. Coordination, however, leaves the system separate yet reduces redundancy.

Coordination of protection and control in the substation has largely arisen from recent developments in computer-based equipment such as digital protection relays, digital communications, and computer-based control and monitoring equipment. When computer-based equipment is extensively used in a substation, coordination becomes straightforward since digital information can be easily shared between different equipment and, most importantly, supplied to all areas of the substation without creating large looms of wiring. Figure 11.3 shows a conceptual view of a coordinated protection and control system using a hierarchical arrangement. In this arrangement, the main interface with the plant occurs at level 1, where data acquisition units record information and interact with circuit breakers. This information is then available at all levels through the hierarchy. At level 2, the relay computers are responsible for the relevant protection. Level

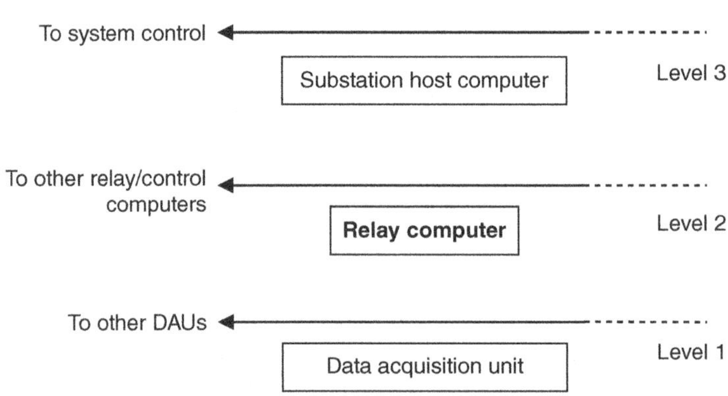

Figure 11.3 Concept of substation with coordinated protection and control.

3 consists of a host computer, which is responsible for all substations' control functions, monitoring functions, and communication with remote control centers.

The concept of a coordinated substation, as depicted in Figure 11.3, was originally developed in the 1970s. Since then, there has been a revolution in the development of computers. The modern trend in computing has moved away from using large, powerful, central computers to using many smaller computers. This is referred to as distributed computing and has arisen due to developments in both microprocessors and digital communications. Distributed computing will be apparent in the office, where typically each desk has its own personal computer connected to central resources, such as hard disks and printers, which can be shared by all in the office. The impact of distributed computing has resulted in the structure of a modern coordinated substation being less hierarchical than that shown in Figure 11.3, in particular, there is rarely an overall computer (level 3) in a modern coordinated substation. In order to understand this, we need to examine modern digital technology explained in this section before proceeding to the structure of the modern coordinated substation in Section 11.4.

11.4 Digital Technology

11.4.1 Digital Communications

Digital communication allows the transmission of information based on numerical values rather than as a continuous analogue signal. The two most common formats for sending information digitally are asynchronous and synchronous. Figure 11.4 shows the structure of a single number sent asynchronously. Notice how only two signal levels are used in this signal, which correspond to the binary digits, or "ones" and "zeros", used in digital devices. Asynchronous means that the data can be sent at any time to the recipient. For this reason, it is necessary to inform the recipient that a number is being sent by sending a start bit followed by seven bits representing the number. In this case, the binary number sent is 0100110. For example, if the number is encoded according to the American Standard Code for Information Interchange (ASCII), then it would refer to a printable character (the character "%"). Thus, the number may be part of a message reporting, for example, a breaker operation; conversely, the number may represent entirely numeric data, for example, a voltage level on a busbar. Following the seven bits of data, a parity bit is sent which acts as a simple error checker by informing the recipient whether there is an odd or even number of "ones" sent to the data. Finally, the stop bit informs the recipient that the number has been sent. Notice how the recipient knows if the data has been sent correctly by examining the start, stop, and parity bits. This ability to check data is one of the

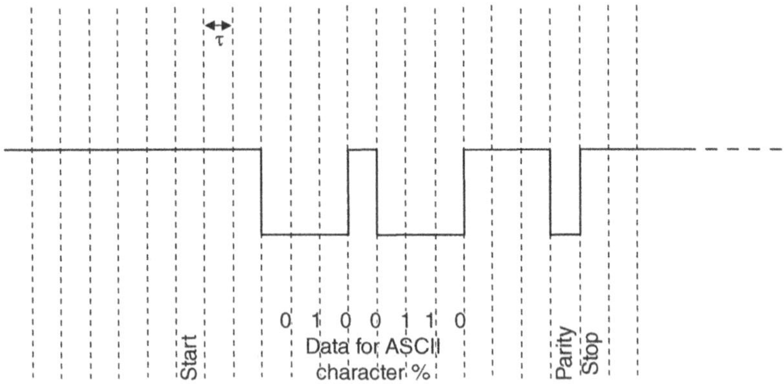

Figure 11.4 Data sent by asynchronous communication.

main advantages of sending data digitally. The time required to send one "bit," that is, a one or a zero, is shown in Figure 11.4 as τ. The reciprocal of this time is referred to as the baud rate – named in honor of its French inventor. This type of digital communication is classed as serial since the binary digits, or bits are sent serially, that is, one after the other. It is also possible to send data in parallel, for example, eight bits could be sent simultaneously by using eight parallel channels. Since parallel transmission requires, in general, dense cabling (minimum 12 cores per cable), it is usually used only for very short distance communication, for example, between a computer and printer.

Exercise 11.4 If asynchronous 7-bit data transmission is used to send ASCII characters, and $\tau = 417$ μs what is the baud rate? What is the character rate? How many ASCII characters are there?

The baud rate $= 1/\tau = 1/417$ μs $= 2400$ bits/s. Assuming that characters are transmitted sequentially without delay, then it will take 10 bits to transmit a character. Thus, the character rate is $2400/10 = 240$ characters/s. Using seven bits, it is possible to form 128 separate characters; however, not all of these are "printable" characters. Some of them are used for other tasks in a message, such as carriage return and ringing a bell; others are used to control the transmission. Note that in asynchronous transmission, before the transmission takes place, both the sender and recipient of the information have agreed on what the baud rate will be.

In synchronous transmission, again, the baud rate is known in advance, but the sender will always send the data at a specific time, that is, there can be no gaps between messages. If no useful information is being sent, then a special character is transmitted to represent this so that the recipient remains in synchronism with the sender. For this reason, there is no need to include stop and start bits in synchronous transmission, and hence data can be sent with higher efficiency than with asynchronous transmission. Also, error checking is usually performed by sending a separate number after, say, 209 numbers of data are sent, rather than using a single parity bit.

Whether asynchronous or synchronous transmission is used will depend on the application. If a human is interacting with a computer using a terminal, then asynchronous transmission is commonly used. If two computers are communicating with each other, then asynchronous transmission is more likely. If data is only being sent from the sender to the recipient, then we speak of simplex transmission. If there are two separate transmission routes allowing data to be sent in both directions simultaneously, then this is classed as duplex transmission. Two-way data may also be exchanged between two parties, but only using one transmission route, which can work in either direction; this is half-duplex transmission.

The information shown in Figure 11.4 could be directly sent along a wire, usually a twisted pair. This could be arranged by defining voltage levels to represent a "one" and 0 V could represent a "zero"; this is used in the transmission standard that you may have heard called RS232. For sending data over long distances, this approach is impractical since the sharp edges of the pulses become "rounded" after a certain distance and the data may become unreadable. For long-distance transfer, some form of *modulation* is required. This may take the form of transmitting one frequency over the wire to represent a "one" and another frequency to represent a "zero." The frequencies chosen will be suited to the wiring being used and will allow the information to be transmitted over long distances. This form of modulation – *frequency modulation* (FM) is widely used for data transmission.

11.4.2 Computer Networking

Exercise 11.5 In Figure 11.5, there are four computer, how could the computers be connected together using data transmission channel so that each computer can communicate with itself and its three neighbors?

Figure 11.5 The problem of interconnecting four computers.

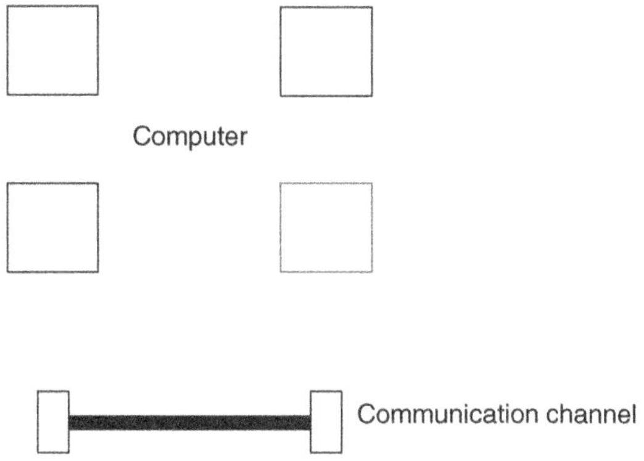

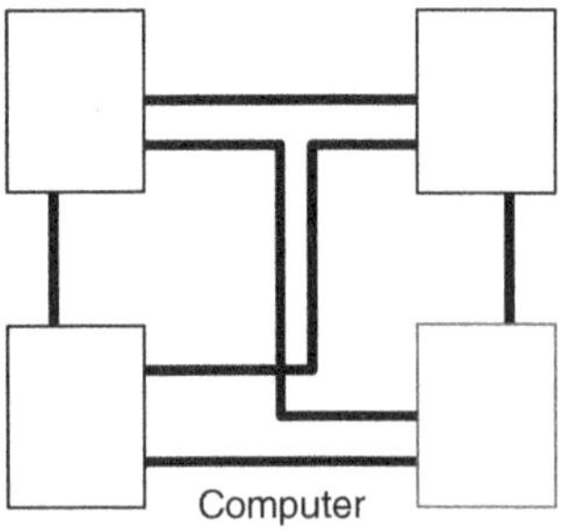

Figure 11.6 Full connected system.

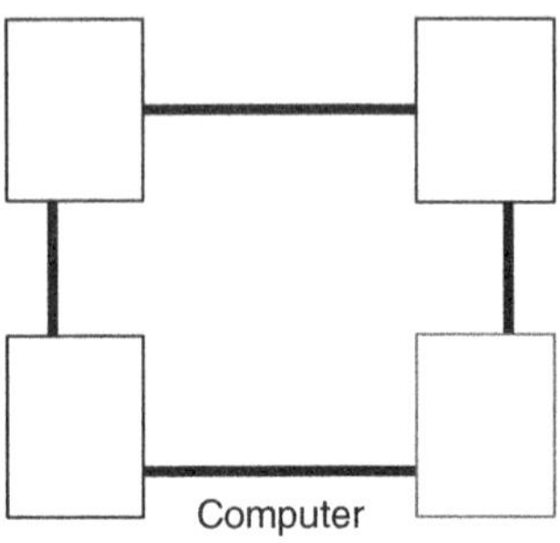

Figure 11.7 Daisy chain connected system.

One possible method of connecting the computers together is a network that uses dedicated links between each computer. This is referred to as a fully connected system, as shown in Figure 11.6. This would be rather extravagant since we would need a total of six separate channels. Also, each computer would need equipment for communicating with three other computers. Since a computer can only do one thing at a time, the transfer of data would only ever occur along one of its channels at a time.

Thus, this approach would be highly redundant in terms of the usage of communications equipment. Another approach, see Figure 11.7, would be to connect the computers together like a daisy chain. Now, each computer only has two communicating devices, and only four channels are used. However, if computers diagonally opposite each other need to communicate, then interposing computers will spend some of their time unproductively forwarding information.

The best solution to this problem is to give each computer one communicating device and connect them together with only one communicating device and with only one channel, in the form of a bus structure as shown in Figure 11.8. The advantage of this system is that only one communicating device is required per computer; the disadvantage is that any message sent on the bus will be received by all computers. This type of communication, where the transmission media is shared by all users, is referred to as a broadcast system.

Figure 11.8 Bus structure.

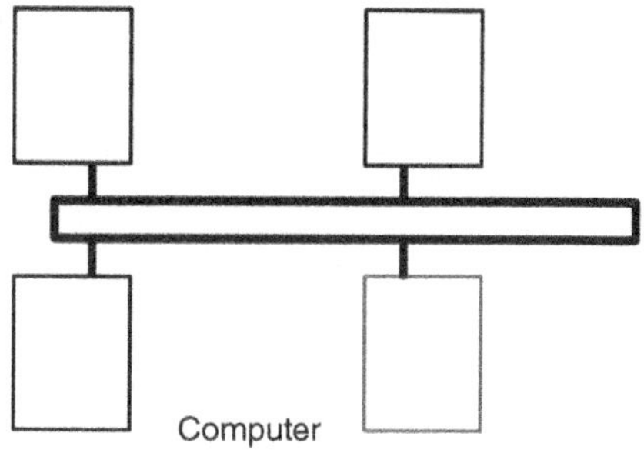

For computers to communicate together on a broadcast system, it is necessary to establish a *media access protocol*. This is a simple procedure for sending information on the shared media, which is agreed in advance by all the computers on the network. The first requirement of this procedure, which is found on all computer networks, is that a computer does not send information immediately and frequently on the network, but stores it up over a period of time and places it in *packet*. Thus, message, or *traffic* on the network, consists of bursts of concentrated information (packets) followed by silence. The second requirement establishes when a computer is allowed to send packet. There are two basic types of access protocols dealing with this aspect:

- In the first type, the bus structure is connected in the form of a ring; packets can only be sent if a computer has a special piece of data called a token. The token is sequentially passed around the ring. A computer sends its information when it possesses the token and sends the token to its nearest neighbor when the information is sent, or if it does not require access to the network. This is token ring network, which was originally developed by IBM.
- In the second type of network, information can be sent by any computer at any time. Of course, there is the possibility that two computers will send packets simultaneously, which is referred to as a collision. In this case, each computer waits for a reply to its message. If none is received, then the original message is resent after a random period of time. This type of network is technically referred to as a carrier-sensing multiple access with collision detection (CSMA/CD). The most common implementation of CSMA/CD is known under the name of its coaxial wire communication media Ethernet: Ethernet does not require a ring structure.

Both token ring and ethernet are extensively used for network computers in a local area thus giving the term local/area network (LAN). Communication of computers over a long distance, a wide area network (WAN), is also possible, but this is usually achieved by connecting a dedicated communication link between two LANs.

11.4.3 Communication Media for Data Communications

The development of different types of communication media is also very important. The simplest form of media is a twisted wire pair. Due to the high modulation frequencies used for computer communications, this medium is unlikely to be used for distances any longer than a meter or two, although metallic pilot wires have been used for many years in protection to convey analogue signals usually at power system frequency. Coaxial cable is common for distances up to several hundred meters and is generally used to form the backbones of LANs. Long-distance communication relies either on radio communications, such as direct line-of-sight microwave systems, or on fiber optics. Fiber optics is the most significant development in the communications industry in recent years due to its high bandwidth; the bandwidth of a single optical fiber can exceed the whole of the radio broadcast spectrum.

Exercise 11.6 Can you think of another advantage of optical fiber that makes this media especially useful in the substation environment?

A distinct advantage of optical fiber technology is its noise immunity. Since data is conveyed as pulses of light, there is no need for any metallic conductor in an optical fiber. Thus, fiber-optic data links are increasingly being used in substations due to their ability to function in an electromagnetically harsh environment. For example, the operation of any switchgear will cause electromagnetic radiation, which can prevent data communication from being correctly received when metallic media is used, but would not affect fiber optic data transmission. One limitation of optical fiber is that they cannot readily be used in the bus structure demanded by, say, Ethernet. Thus, their use in LANs is restricted to replacing coaxial cable in electromagnetically "noisy" area, or optical fiber can be made of either glass, which allows transmission of, currently, up to 150 km in a single stretch, or lower-cost plastic fibers which can only work on short distance of several hundred meters.

11.4.4 Digital Relays

Modern protection relays are based on microprocessor technology. Relays can be classified as digital, where microprocessors are used in the implementation, and numeric of the relaying decision is derived from a numerical formulation of the relaying process executed on a microprocessor. Digital relays offer many advantages compared to conventional electronic relays; these include:

- reduced cost
- reduced size

- self-checking
- predictable behavior
- increased reliability
- increased dependability and security

Digital relays also offer digital communication facilities. Many relays can be interfaced by connecting the relay to some form of serial communications link. This enables the settings of the relay to be changed remotely, that is, the engineer does not need to visit the substation. Additionally, unit type relays exchange information using digital communication techniques. Some relays allow direct fiber digital communication techniques. Some relays allow direct fiber optic inputs in the relay casing, thus reducing the probability of electromagnetic interference affecting relay operation.

11.5 Typical Structure of a Modern Coordinated Substation

Previous sections introduced the concept of a coordinated substation and discussed the impact of modern technology. We will now see the structure of a modern coordinated substation. In Section 11.2, we saw that coordination could be achieved by a hierarchy of computers. Modern computing has moved toward the distributed computing approach, and so, although many computers are used to provide the functions shown in Figure 11.3, the hierarchical structure is not apparent in a modern substation.

To afford coordination of functions, it is necessary to divide the substation into manageable sections that can be computer-controlled. For example, in the section of the substation shown in Figure 11.9, two separate areas called bays are shown. In general, each bay has its own local control, protection, and monitoring system. How big or how encompassing a bay will depend on the specific substation, but the minimum would be one busbar and one breaker.

Figure 11.10 shows the overall structure of the coordinated substation approach. At the bottom of Figure 11.10, we see the protection, control, and monitoring associated with one bay; each device is referred to as a terminal. Communications between the bay and the rest of the substation are conducted through the bay control terminal. All terminals in the bay communicate along an object bus which, due to the interconnections with, and proximity to, power system plant, is implemented using fiber-optical links for noise immunity. In the bay shown in Figure 11.10, we see that two-line protection relays, one autoreclose relay, a breaker terminal, and a monitoring terminal. The bay control terminal has an interface with the plant, and this can access busbar transducers as well as circuit breakers.

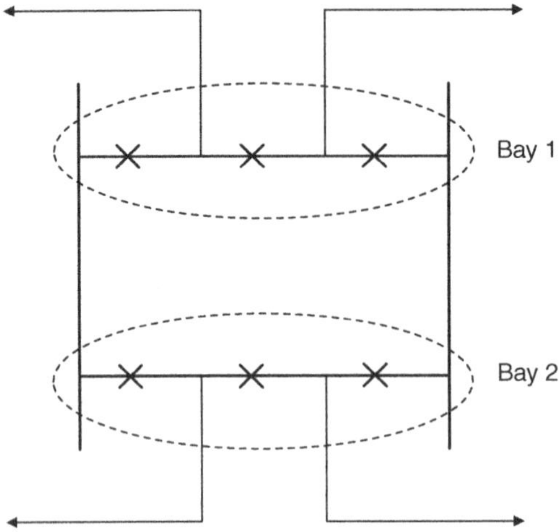

Figure 11.9 Bay structure of the substation.

Exercise 11.7 Is it necessary for other terminals within the bay to have a plant interface?

The answer to this question is both yes and no. All terminals in the bay have to access the bay control terminal's plant interface through the object bus. However, to ensure the dependability of the operation of all bay terminals, it is advantageous for other terminals also to have plant interfaces. Thus, if the bay control terminal's plant interface fails, then others can take its place. Exactly how many plant interfaces are used will depend on the power utility; however, it is likely to be more than just the bay control terminals interface.

The monitoring terminals allow recording of events and switching operations and communication with the substation monitoring system through a dedicated communication link. All bays are connected together throughout the substation using the station bus, which uses coaxial cable as the medium. The station bus allows an operator to control the substation via an interface on the station bus that includes a workstation and a printer for logging events. Since many substations are unmanned, remote operation of the substation is available by connecting the station bus to a remote computer via a device called a gateway, which gives access to a wide area network. Thus, the substation operator need not to be on site.

It will be apparent from Figure 11.10 that the protection, monitoring, and control substation functions are present in the coordinated and control substation. However, the amount of wiring needed has been reduced since all communication now takes place along shared media. The coordinated substation thus overcomes the disadvantages experienced with the conventional substation in that it can be easily maintained, tested, and expanded.

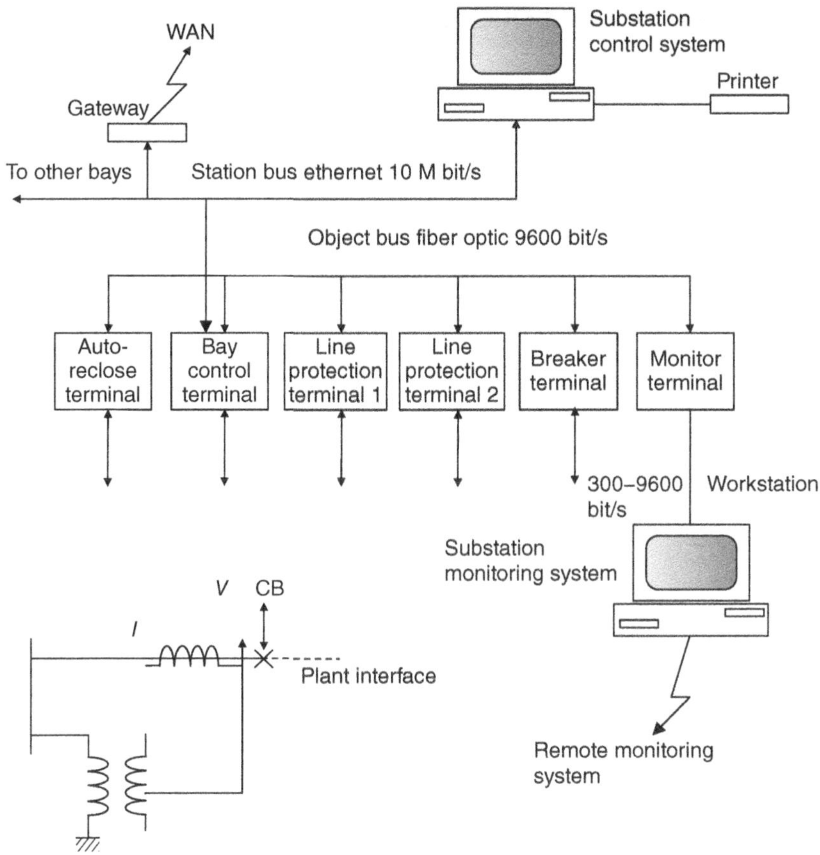

Figure 11.10 Bay structured coordinated protection and control systems of a substation.

11.6 Self-Assessment Questions

Self-Assessment Question 11.1

What are the functions found in a substation? Why are conventional substation systems difficult to maintain, test, and expand? How are these problems overcome in a modern coordinated substation?

Self-Assessment Question 11.2

Describe how data is communicated

- Within a bay
- Between bays

Bibliography

Jamil Abdo, Tariq Masood "New Technique to Develop Intelligent, Highly Reliable & Efficient Unmanned Power Generation Modular" published in 22nd International Conference on Renewable Energies and Power Quality (ICREPQ'24) Bilbao (Spain), 26–28 June 2024. https://www.icrepq.com/icrepq24/392-24-abdo.pdf.

P Chandhra Sekhar; RA Deshpande; V Sankar "Evaluation and Improvement of Reliability Indices of Electrical Power Distribution System" published in National Power Systems Conference (NPSC) in 2016.

Jing Xiao; Ning Wu; Shuai Han; Xiaorui Wu; Wenlan Gong "Electric Vehicle Support Distribution Grid Emergency Power Protection Strategy After a Major Power Outage" published in 3rd International Conference on Energy Engineering and Power Systems (EEPS) in 2023.

Ritu Malhotra "Management of Risk and Economics of LV Power System Protection for Industrial Facilities" published in IEEE Transactions on Industry Applications in 2016.

Montri Wiboonrat "Zone Protection Approach of Data Center System Reliability and Power Quality" published in International Conference on Advanced Robotics and Intelligent Systems (ARIS) in 2017.

Shijie Cui; Peng Zeng; Ende Hu "Research on Distributed Intelligent Protection Strategy of Distribution Network Based on Reinforcement Learning" published in IEEE 2nd International Conference on Information Technology, Big Data and Artificial Intelligence (ICIBA) in 2021.

Jibran Ali; Federico Silvestro; Mahmood Jamil "Virtual Power Plant for Improving Power System Protection Issues: Solution to the Problem of Power System

Reliability under Distributed Energy Resources" published in Australasian Universities Power Engineering Conference (AUPEC) in 2018.

Hadi Zayandehroodi; Azah Mohamed; Hussain Shareef; Marjan Mohammadjafari; Masoud Farhoodnea "A Novel Protection Coordination Strategy Using Back Tracking Algorithm for Distribution Systems with High Penetration of DG" published in IEEE International Power Engineering and Optimization Conference Melaka, Malaysia in 2012.

Seongil Kim; Byungchul Sung; Jeong-Hyeok Heo; Duck-Su Lee "Protection against Transient Overvoltages Caused by Intermittent Earth Faults in Unearthed Marine Power Distribution Network" published in IEEE International Conference on Power Systems Technology (POWERCON) in 2022.

JA Sa'ed; S Favuzza; MG Ippolito; F Massaro "Verifying the Effect of Distributed Generators on Voltage Profile, Power Losses and Protection System in Radial Distribution Networks" published in 4th International Conference on Power Engineering, Energy and Electrical Drives in 2013.

Kun Liu; Xin Cui; Bin Wang; Xinzhou Dong "Study on High Impedance Grounding Fault Protection for Personal Safety Guard in Distribution Grid" published in IEEE 8th International Conference on Advanced Power System Automation and Protection (APAP) in 2019.

Qiushi Li; Xianghua Tang; Xiongjie Shi "Demonstration and Application of AC/DC Hybrid Power Supply System in Building" published in 2nd IEEE Conference on Energy Internet and Energy System Integration (EI2) in 2018.

Luyao Pei; Cheng Xu; Jinsong Zhang; Bo Wang; Yining Ma "Artificial Intelligence Algorithm in Intelligent Power Supply and Distribution Network Design System" published in IEEE 3rd International Conference on Electronic Technology, Communication and Information (ICETCI) in 2023.

Jamil Abdo Tariq Masood "Operational Reliability – Production Best Practice Pragmatic Case Studies & Improvement" published in Proceedings of the 9th North American Conference on Industrial Engineering and Operations Management Washington D.C., USA, 4–6 June 2024.

Tariq Masood, Jamil Abdo "Maintenance Strategy Production Best Practices – Pragmatic Case Studies and Operations Improvement" published in Proceedings of the 9th North American Conference on Industrial Engineering and Operations Management Washington D.C., USA, 4–6 June 2024.

Jamil Abdo, Tariq Masood "Techno Economics Analysis to Reshape Global Clean and Green Energy" published in 22nd International Conference on Renewable Energies and Power Quality (ICREPQ'24) Bilbao (Spain), 26–28 June 2024. https://www.icrepq.com/icrepq24/415-24-abdo.pdf.

Nasser Alemadi, Tariq Masood, R Aggarwal, Suhail A Qureshi, D Kothari "Novel Selection Approach of STATCOM Device Tailored and Optimized Engineering Configuration to Implement in Between Qatar and Bahrain" published in

International Journal of Automation and Power Engineering 2014. 10.14355/ijape.2014.0302.01.

Tariq Masood, Jamil Abdo "New Technique to Enhance Hybrid Renewable Energy Resources' Operational Flexibility and Reliability" published in 22nd International Conference on Renewable Energies and Power Quality (ICREPQ'24) Bilbao (Spain), 26–28 June 2024. https://www.icrepq.com/icrepq24/396-24-masood.pdf.

Tariq Masood, Jamil Abdo "Review: Uninterrupted Liquid Hydrogen (E-Methanol) Production Framework to Deliver Clean & Green Energy" published in 22nd International Conference on Renewable Energies and Power Quality (ICREPQ'24) Bilbao (Spain), 26–28 June 2024. https://www.icrepq.com/icrepq24/414-24-masood.pdf.

Tariq Masood, Abdel Aty Edris, Suhail Aftab Qureshi, Muhammad Tajammal, Murtaza Hashmi, DP Kothari, Samer Kareem "SSSC Tailored to Optimization Performance in Between United Arab Emirates and Oman on the GCC Power Grid" published in the 13th IET International Conference on AC and DC Power Transmission Conference on 14–16 February 2017. http://ieeexplore.ieee.org/document/7934934.

Tariq Masood, Abdel Aty Edris, Suhail Aftab Qureshi, Muhammad Tajammal, Murtaza Hashmi, DP Kothari, Samer Kareem "Smart Grid Operations and Control Challenge by Implementing STATCOM Tailored to Optimization Performance at the GCC Power Grid" published in the 13th IET International Conference on AC and DC Power Transmission Conference on 14–16 February 2017. http://ieeexplore.ieee.org/document/7934988.

Tariq Masood, Muhammad Tajammal, Samer Kareem, Suhail Aftab Qureshi, Abdul Jabbar Khan, DP Kothari, Nair Jameel "Unified Power Flow Controller Modeling and Analysis Technique on the GCC Power Grid Operations and Control" published in 2018 IEEE International Energy Conference, Limassol Cyprus 7–9 June 2018. https://ieeexplore.ieee.org/document/8398834.

Tariq Masood, Muhammad Tajammal, Samer Kareem, Suhail Aftab Qureshi, Murtaza Hashmi, Nasir Jameel "Unified Power Flow Controller (UPFC) Modeling and Analysis Technique by using Matlab and EMTP" published in International Conference on Renewable Energies and Power Quality (ICREQP'18) Salamanca (Spain) 21–23 March 2018. http://www.icrepq.com/icrepq18/250-18-masood.pdf.

Tariq Masood, Muhammad Tajammal, Samer Kareem, Suhail Aftab Qureshi, Murtaza Hashmi, Naser Jameel "STATCOM Controller Performance Enhancement and Optimization Technique" published in International Conference on Renewable Energies and Power Quality (ICREQP'18) Salamanca (Spain) 21–23 March 2018. http://www.icrepq.com/icrepq18/249-18-masood.pdf.

Tariq Masood, Muhammad Tajammal, Samer Kareem, Suhail Aftab Qureshi, Murtaza Hashmi, DP Kothari "Smart Grid Operational Functions and Control Challenges by Implementing SSSC Tailored to Optimize Performance in Between

Qatar and KSA on the GCC Electrical-Power Grid" published in the ASTESJ Journal Advances in Science, Technology and Engineering Systems Journal Vol. 2, No. 1, 2017. https://astesj.com/special-issue-4.

Tariq Masood, Raj K Aggarwal, Suhail Aftab Qureshi, RAJ Khan "STATCOM Model Against SVC Control Model Performance Analyses Technique by MATLAB" published in International Conference on Renewable Energies and Power Quality (ICREPQ'10) 2010. https://icrepq.com/icrepq'10/655-Masood.pdf.

Tariq Masood, N Al-Emadi, SA Qureshi, YY Almulla, AJ Khan, SK Shah "FACTS Control Devices (Statcom, SSSC and UPFC) Re-Configuration Techniques By PSIM/MATLAB" published in 2007 International Conference on Electrical Engineering 2007. http://ieeexplore.ieee.org/abstract/document/4287295.

Tariq Masood, Raj K Aggarwal, SA Qureshi, RAJ Khan "STATCOM and SVC Control Operations and Optimization During Network Fault Conditions" published in 2010 IEEE International Symposium on Industrial Electronics 2010. https://ieeexplore.ieee.org/document/5636868.

Tariq Masood, RK Aggarwal, SA Qureshi, RAJ Khan "Statcom Control Reconfiguration Technique for Steady State and Dynamic Performance Optimization During Network Fault Conditions" published in International Conference on Renewable Energies and Power Quality (ICREPQ'10) Granada (Spain) 2010. https://www.icrepq.com/icrepq'10/656-Masood.pdf.

Tariq Masood, Samer Karim Shah, Muhammad Tajammal "Unified Power Flow Controller Modeling and Analysis Technique on the GCC Power Grid" published in 2018 IEEE International Energy Conference (ENERGYCON) 2018. https://ieeexplore.ieee.org/abstract/document/8398834.

Tariq Masood, Suhail A Qureshi, RK Aggarwal, DP Kothari "Multiple FACTS Devices Integration Through Distributed Control System and Its Application on the GCC Interconnection" published in 2013 7th IEEE GCC Conference and Exhibition (GCC) 2013. https://ieeexplore.ieee.org/abstract/document/6705750.

12

Smart Grid Operation & Control

12.1 Objectives

- What is a smart grid and its features?
- What are the distribution automation field devices?
- How to detect and locate a fault in a distribution system?
- What are the important volt-var devices, and the areas affected by these devices?
- What is meant by distributed automation?
- What are the key functions of the SCADA system?

12.2 Introduction

Smart grid is not just smart metering, but technologies to modernize the power distribution system, which includes wires, substations, transformers, switches, and much more. This approach aims to improve the power distribution systems' reliability, quality, and efficiency on the power network and for energy users.

Bi-directional electric grids and communication networks improve reliability, security, and efficiency of the electric system for small to large-scale generation, transmission, distribution, storage, and consumption. It includes software and hardware applications for dynamic, integrated, and interoperable optimization of electric system operations, maintenance, and planning; distributed energy resources interconnection and integration; feedback and controls at the *consumer level.*

IEEE Std 2030 establishes the Smart Grid Interoperability Reference Model (SGIRM) and provides a knowledge base addressing terminology, characteristics, functional performance, and evaluation criteria, and the application of engineering principles for smart grid interoperability of the electric power system with end-use

Protection of Electrical Power Distribution Systems: Smart grid, Microgrid, AI, and Cyber Security, First Edition. Tariq Masood, Jamil Abdo, and Atif Iqbal.

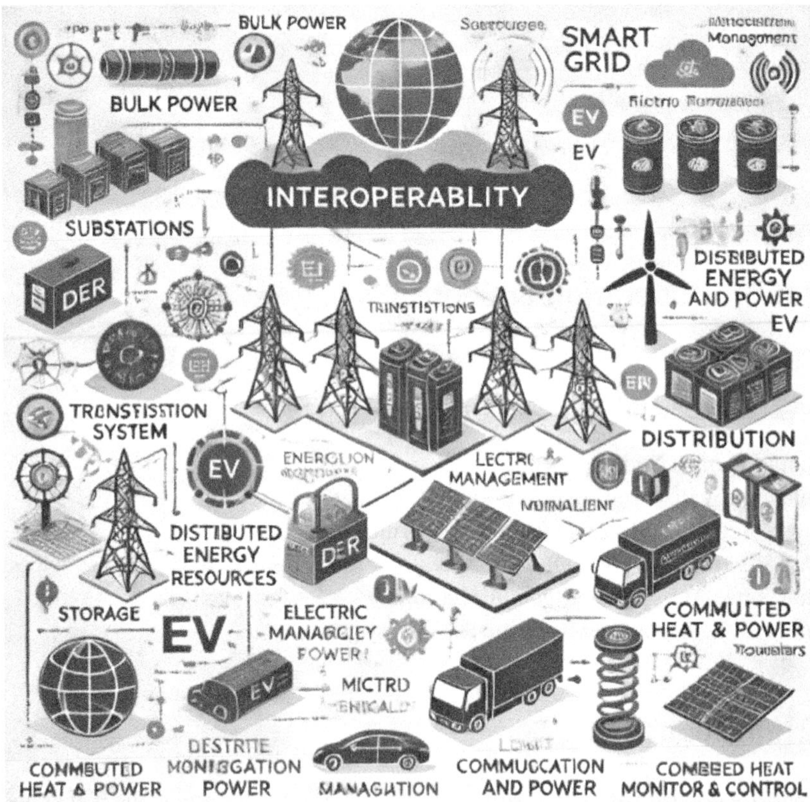

Figure 12.1 Smart grid concept.

applications and loads. The IEEE 2030 SGIRM defines three integrated architectural perspectives: power systems, communications technology, and information technology, as shown in Figure 12.1.

12.2.1 Distributed Intelligence

This is achieved by advanced sensors with processing and communications capabilities built into every element of the grid – switches, transformers, substations, distribution lines, etc.

12.2.2 Broadband Communication

This is the key infrastructure enabling comprehensive system-wide monitoring and coordination to enable applications such as distribution automation, demand response, outage management, and power quality monitoring. It needs low latency, high bandwidth, and QoS prioritization, which require a broadband network.

12.3 Distribution System

Power distribution, a critical component of the smart grid, is undergoing a significant transformation. The terminal units, integral to this transformation, play a crucial role in the smart grid. They are not just revolutionizing the system but also serving as the vital communication bridge between control centers and power grids, thereby amplifying the impact of these technological advancements. The distribution side is experiencing a wave of innovation, with the emergence of cutting-edge terminal units that signify a remarkable leap forward in power distribution, as shown in Figure 12.2.

Innovative terminal units have various functions: status monitoring, fault detection, location, restoration, network reconfiguration, information transmission, and switch control. Network reconfiguration, for example, is a key function that allows the system to adapt to changes in demand or supply, thereby maintaining grid stability and ensuring a reliable power supply.

Basic Principle: CTs are based on the principle of current induction, while PTs are based on voltage induction. Application areas: CTs are primarily used

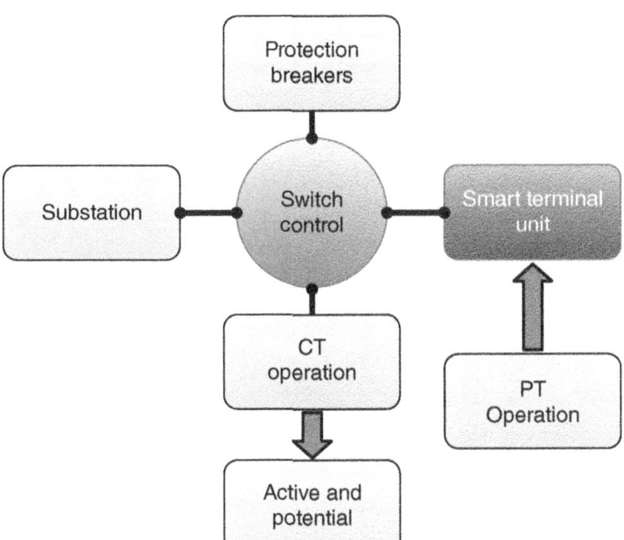

Figure 12.2 Smart terminal unit.

for current measurement and protection, while PTs are mainly used for voltage measurement and protection. Output type: CTs output current, whereas PTs output voltage.

12.4 The Smart Grid Ultimate Goal

A smart grid is an automated approach to a widely distributed energy delivery network characterized by a bidirectional flow of electricity and information. It can monitor and respond to changes in everything from power plants to customer preferences to individual appliances. **Changing** the entire energy conversion chain into a **smart infrastructure as shown in Figure 12.3.**

The smart infrastructure includes the following aspects of control and operations: first, the decentralized energy management system; second, communication solutions; and third, intelligent substation automation. Fourth, there is condition monitoring and asset management; fifth, there is the power transmission system. Sixth, the distribution automation systems; seventh, the smart metering; and last, the building automation system. These include different segments as shown in Figure 12.4.

The control strategies in smart grids are critical to ensuring the electricity network's stability, reliability, and efficiency. As smart grids integrate more renewable energy sources, distributed generation, and advanced technologies,

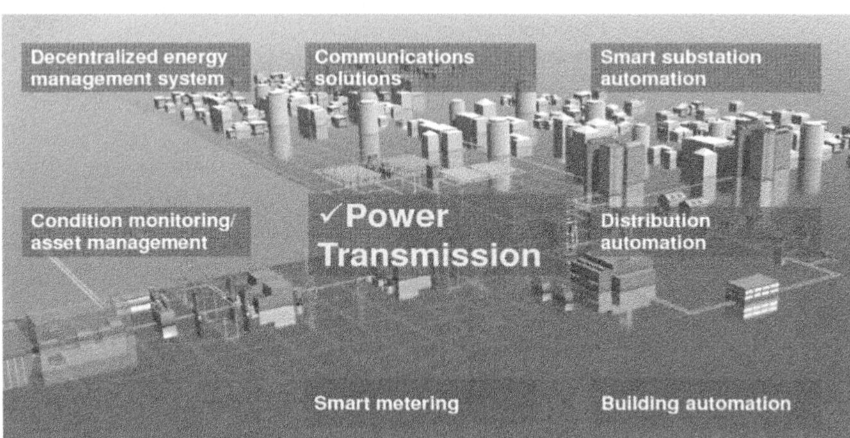

Figure 12.3 Smart infrastructure.

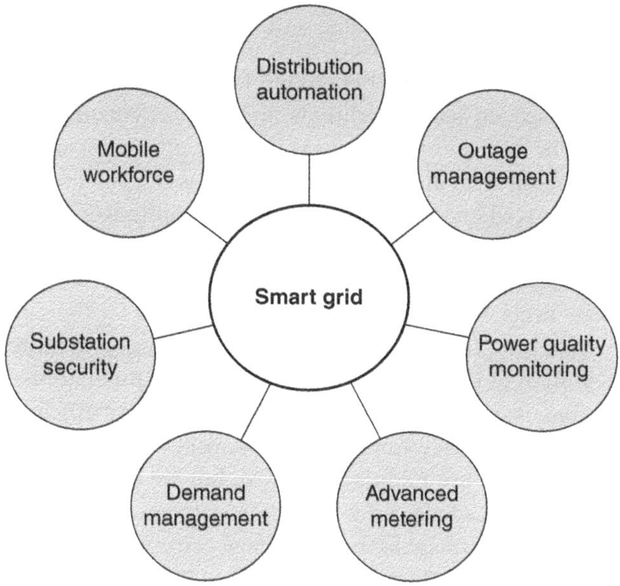

Figure 12.4 Smart grid components.

such as AI and machine learning, sophisticated control methods become necessary to manage their complex dynamics, reassuring us about the future of energy management.

12.5 Smart Grid Decentralized Versus Centralized Control

Centralized Control:

Traditionally, power systems have relied on centralized control, where a single control center manages the entire grid. This approach provides a comprehensive view of the grid's status, enabling optimal decision-making and coordination. Centralized control systems collect data from various parts of the grid, process it, and then send control signals back to different grid components. This model works well for large, interconnected grids but may face scalability and real-time response challenges, especially as the grid becomes more distributed. Challenges include the need for a robust communication infrastructure and potential vulnerabilities to cyberattacks.

Decentralized Control:
With the rise of distributed energy resources (DERs) and microgrids, decentralized control is gaining importance. In this model, local controllers make decisions independently based on local conditions and goals. Decentralized control not only reduces the communication overhead and allows for faster response times when managing local disturbances but also has the potential to significantly improve system resilience, giving us reason to be optimistic about the future of energy management.

Voltage and Frequency Control:
First, maintaining voltage levels within acceptable limits is crucial for the safe and efficient operation of the grid. In smart grids, voltage control becomes more complex due to the integration of DERs, which can cause voltage fluctuations. This task, while challenging, is of utmost importance for the safe and efficient operation of the grid. Techniques such as on-load tap changers (OLTC), capacitor banks, and voltage regulators are traditionally used, but smart grids also leverage real-time data and advanced algorithms to adjust voltage dynamically. Reactive power management is a crucial aspect of voltage control, with DERs and inverter-based resources playing a significant role.

Second, the frequency control ensures that the grid operates at a stable frequency, typically 50 or 60 Hz, depending on the region. Frequency deviations can occur due to imbalances between generation and load. Primary, secondary, and tertiary frequency controls are implemented in layers, with primary control providing immediate response through automatic generation control (AGC), secondary control adjusting the setpoints, and tertiary control handling economic dispatch and reserves. Smart grids enhance frequency control through demand response, where loads are adjusted to help balance frequency, and by integrating fast-responding energy storage systems.

Exercise 12.1 Describe the centralized and decentralized control.

12.6 Distribution Automation (DA)

Electrical power distribution systems automation, specifically distribution automation (DA), is a technology-driven solution that enhances essential distribution functions' speed, cost, and accuracy. A key component of DA is the use of digital sensors and switches, which, when combined with advanced control and communication technologies, automate feeder switching, voltage and equipment health monitoring, and outage, voltage, and reactive power management. This results in reliable and cost-effective services for customers, providing a sense of being informed about the technology.

Figure 12.6 indicates radial low-voltage network operations; Figure 12.7 demonstrates the network's voltage measurement; Figure 12.8 also indicates the current measurement; Figure 12.9 shows the feeder branch load measurement to maintain seamless operations.

Devices in the distribution automation field are listed below:

1) Remote fault indicators
2) Smart relays
3) Automated feeder switches and reclosers
4) Automated capacitors
5) Automated voltage regulators and load tap changers
6) Automated feeder monitors
7) Transformer monitors

To implement DA, advanced field devices are typically equipped with radio, wireless, or cellular communication to transmit data to collection points. These collection points then use backhaul communications networks to relay the data back to utility control centers. At the control center, the data is typically integrated into the supervisory control and data acquisition (SCADA) system, distribution management system (DMS), and outage management system (OMS) for processing, analysis, and action, either automatically or by operators.

12.6.1 Communication Networks

Communication networks are the backbone that allows utilities to connect devices to SCADA, DMS, and other information and control systems. This enables our skilled grid operators to confidently manage power flow and reliability issues. There are a variety of wired and wireless technologies available, and the choice of communication network is based on *factors such as bandwidth, latency, cost, cybersecurity, reliability, coverage, and backup power needs.*

12.6.2 Distribution Automation (DA) – FDIR

12.6.2.1 Fault Detection, Isolation, and Restoration (FDIR)

Fault detection recognizes that a fault occurred and that the grid is not operating as intended. Fault isolation identifies the fault, for example, what kind of fault has occurred, and localizes it, for example, finding where it happened. Fault diagnosis is the combined effort of both fault detection and isolation. But fault restoration, the process of reconfiguring the grid that is set in action after fault diagnosis, truly brings our system back to life, restoring it to its normal operating conditions. Figure 12.5 indicates the operation of FDIR in electrical power distribution systems.

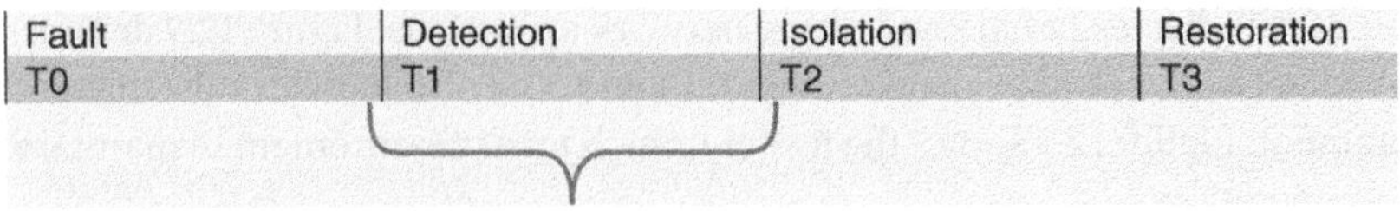

Figure 12.5 FDIR timeline schematic.

12.6.2.2 Radial Low-Voltage Networks

Low-voltage networks, the standard design for MV to LV distribution networks, are often used in sparsely populated areas. They are known for their simple structure, low initial cost, and reliability. As shown in Figure 12.6, a radial network has a single power source for a group of customers. It also indicates that a power failure, short-circuit, or downed power line will interrupt power to the entire line. However, if one power source fails, switches can be thrown to feed power to customers from another source. The cost-effectiveness of radial networks, both in terms of initial construction and ongoing reliability, makes them an attractive option for most customers.

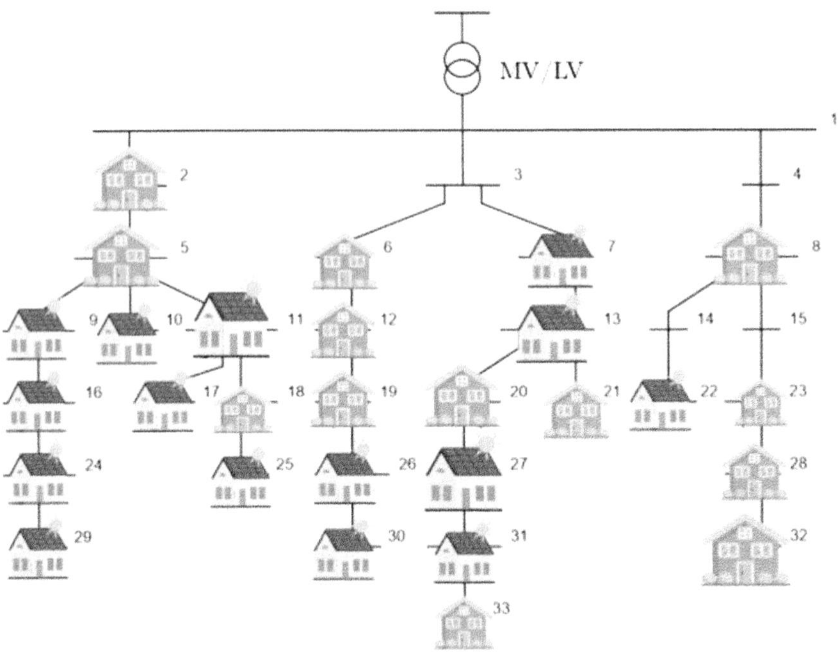

Figure 12.6 Radial low-voltage network.

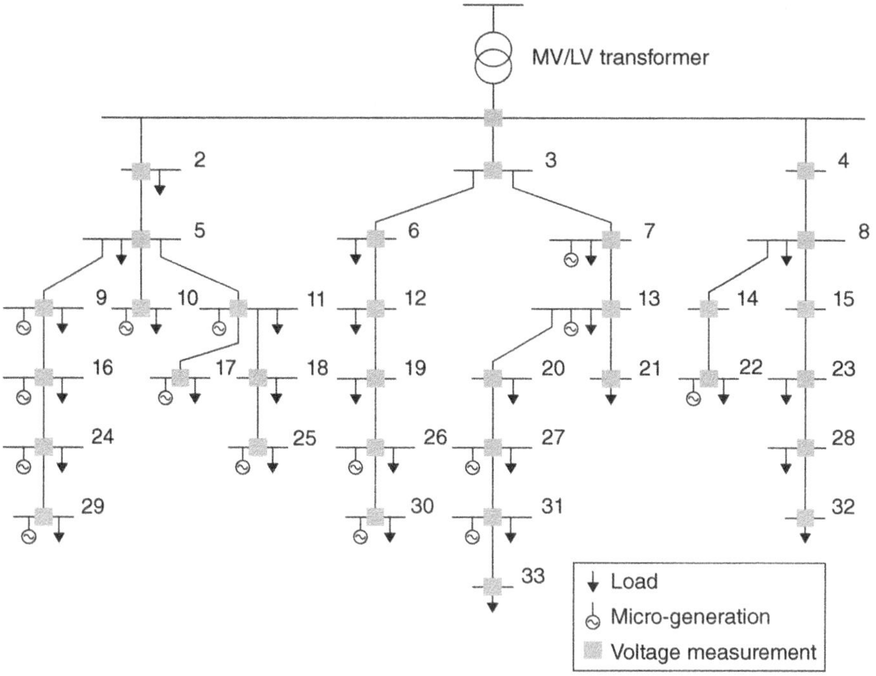

Figure 12.7 Voltage measurement schematic.

Exercise 12.2 Describe the voltage measurement schematically.

12.6.2.3 Voltage Measurement Schematic

Before measuring voltage, it is crucial to understand the voltage level and signal source. This knowledge will guide you in choosing the most suitable method for measuring voltage, be it ground-referenced or differential, depending on the type of signal. Figure 12.7 illustrates this concept. When determining the voltage's size, the process is made more convenient with digital testers. Start with the highest range and let the tester automatically select the range for you, making the process more efficient. It is common to get a negative voltage reading if you have accidentally reversed the probes. Do not worry, simply switching the probe placement can help you determine which point is ground and which carries a charge.

An ampere (AM-pir), or amp, is the international unit used for measuring current. It expresses the quantity of electrons (sometimes called "electrical charge") flowing past a point in a circuit over a given time (Figures 12.8 and 12.9).

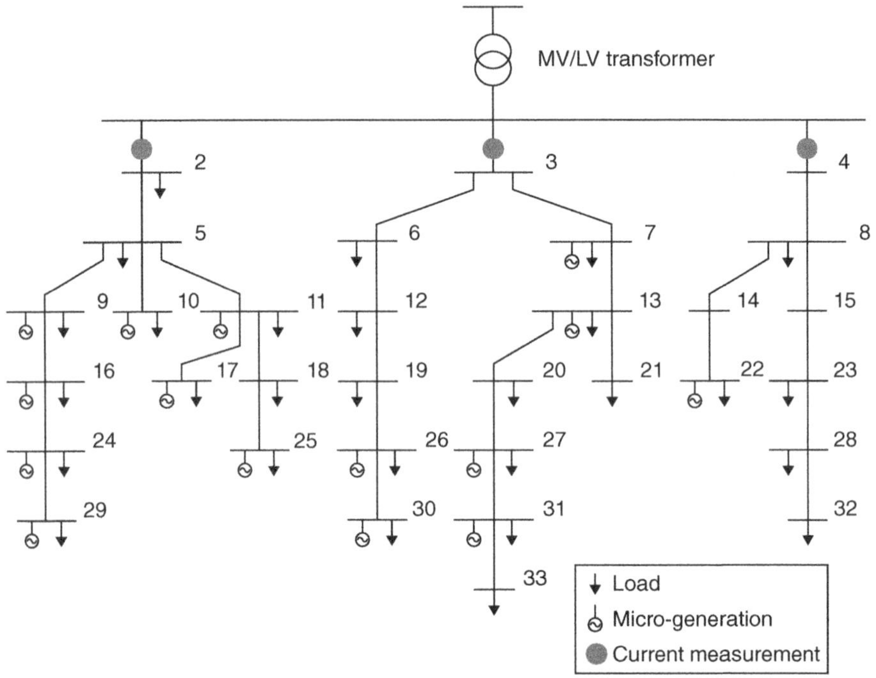

Figure 12.8 Current measurement schematic.

12.6.3 Fault Detection, Isolation, and Restoration (FDIR) and FLISR[1]

Figure 12.10 indicates the sequence of the fault detection.

Step 1: Fault detection: A fault occurrence is detected by a sudden increase in current at the beginning of the feeder.

Step 2: Identifying the faulty Feeder: You can easily identify the feeder under fault by comparing the currents at the beginning of each feeder. The feeder that shows an increase in the current in any phase is the one under fault, as shown in Figure 12.6. This process is straightforward and well within your capabilities, making you feel competent in your role.

Step 3: Faulty branch identification/Faulty sector localization: Identifying which branch is under fault inside a feeder is done using a simple criterion: the branch with the highest voltage drop is considered to be under fault. However, as fault resistance, which is the resistance introduced by the fault, increases to higher values, the voltage drop becomes comparable with a drop that could be part of normal grid operation, as shown in Figure 12.7. According to the

1 Fault location, isolation, and service restoration.

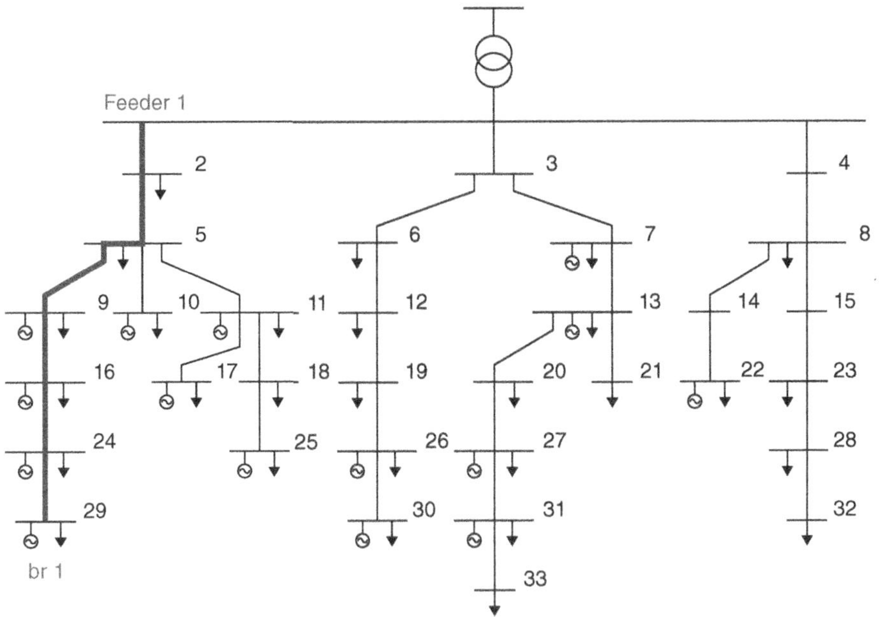

Figure 12.9 Feeder branch load measurement schematic.

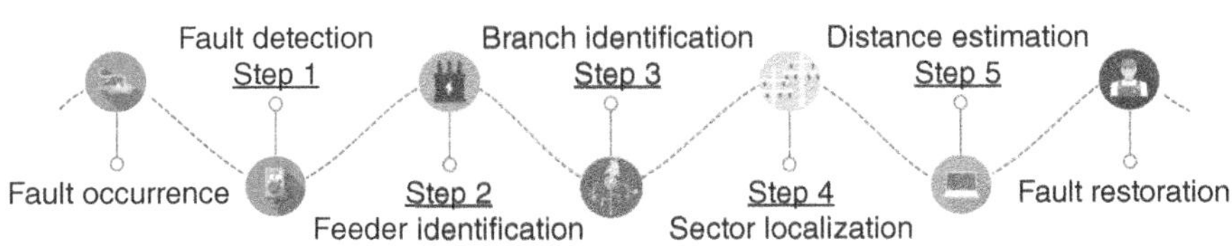

Figure 12.10 Fault detection timeline schematic.

EN50160 standard in LV grids, only a voltage drop higher than 10% of the one under normal operating conditions, such as when the feeder is at full capacity and all components are functioning optimally, is considered an anomaly (Figures 12.11 and 12.12).

12.7 Volt-Var Optimization (VVO)

VVO uses real-time information and online system modeling to provide optimized and coordinated control for unbalanced distribution networks with discrete controls. VVO uses load characteristics to achieve various objectives. Conservation voltage reduction (CVR) mode: Energy efficiency at system level. Peak shaving

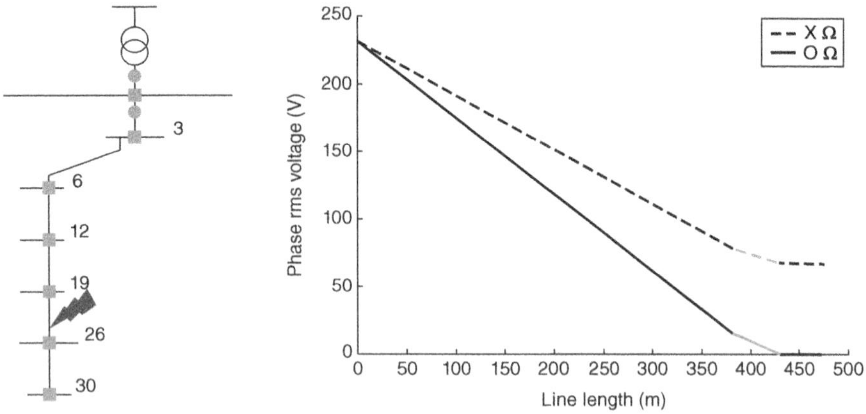

Figure 12.11 Fault detection and isolation/faulty sector localization.

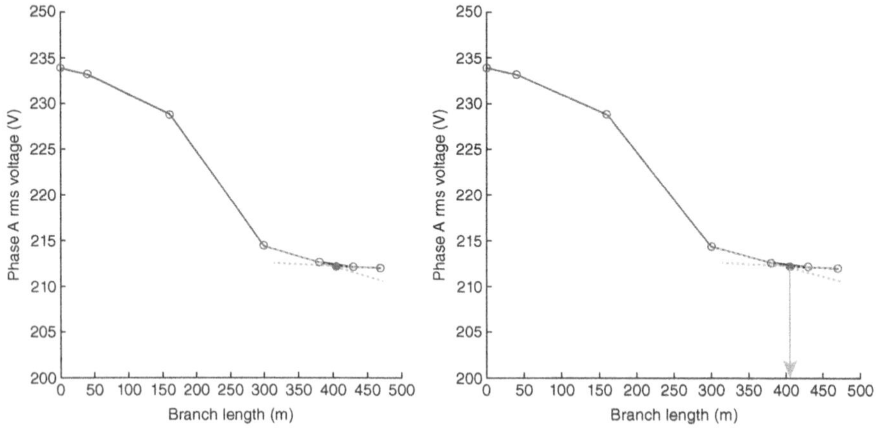

Figure 12.12 Fault distance estimation.

mode: Reduction of system capacity. Less capacity needs means less power generation needs. Power quality correction: Correct system voltage in real-time. Table 12.1 indicates the voltage control and equipment locations (Figure 12.13).

12.8 SCADA System

SCADA systems are used to control one or more remote stations from a master control station. They include digital control equipment, sensing equipment, and telemetry equipment, as well as two-way communications to and from the master

Table 12.1 Voltage control equipment and locations.

Equipment	Function	Area impacted
Load tap changer	Raise or lower voltage	Entire feeders
Voltage regulator	Raise or lower voltage	Downstream of connection point
Capacitor banks	Compensate reactive load Support voltage	Entire feeder with greatest effect closer to the load

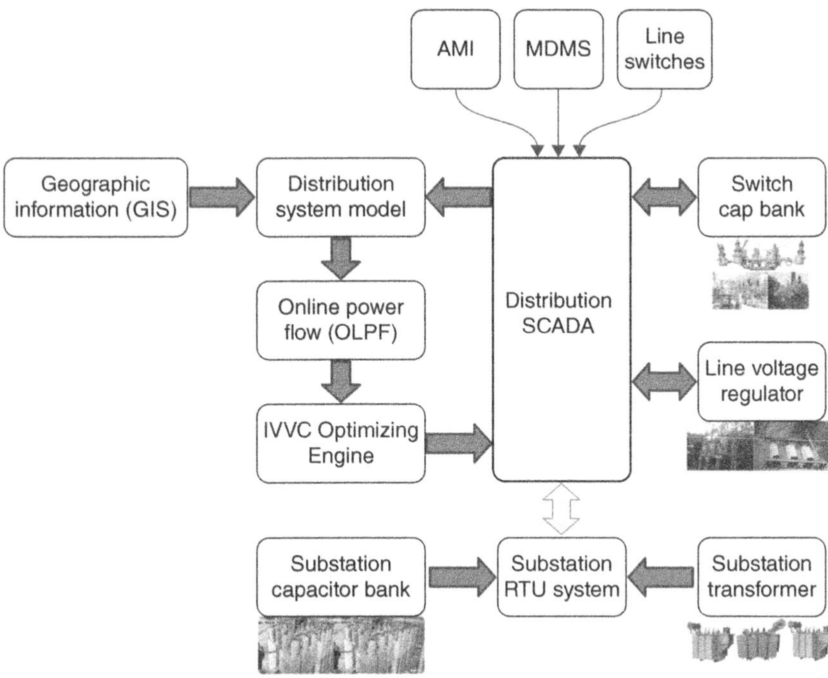

Figure 12.13 Centralized volt-var optimization configuration.

and remotely controlled stations. The sensing and telemetry equipment includes the sensors, DAC and ADC, actuators, and relays used at the remote stations to sense operation and alarm conditions and to activate equipment such as circuit breakers remotely.

The communication equipment includes modems for transmitting the digital data and the communication link (radio, phone link, microwave link, or power line communication).

12.8.1 SCADA System Functions

SCADA can perform the following functions:

1) Automatic bus sectionalizing
2) Automatic reclosing after a fault
3) Synchronism check
4) Protection of equipment in a substation
5) Fault reporting
6) Transformer load balancing
7) Voltage and reactive power control
8) Equipment condition monitoring
9) Data acquisition for sensors
10) Status monitoring data logging

12.8.2 SCADA System Outage Monitoring

Many utilities use advanced metering infrastructure called smart meters to implement outage monitoring. Utilities set up automated processes for pinging meters over large areas when large outages are reported. The affected feeders and neighborhoods are located, and service is restored. Table 12.2 indicates the DA's important terms.

Table 12.2 Distribution automation (DA) – important terms.

DA application	Specific smart grid functions
Reliability and outage management	Remote fault location and diagnostics Automated feeder switching Outage status monitoring and notification Optimized restoration dispatch
Voltage and reactive power management	Integrated voltage and volt-ampere reactive (VAR) controls (IVVC) Automated voltage regulation Conservation voltage reduction (CVR) Real-time load balancing Automated power factor corrections
Asset health management	Real-time or near real-time equipment health monitoring
Distributed energy resources (DER) integration	Integrated and automated DER dispatching and management Operation of customer-sited thermal energy storage systems

Exercise 12.3 Fault detection, isolation, and restoration (FDIR) or fault location, isolation, and service restoration (FLISR).

12.9 Case Studies

12.9.1 Northern Virginia Electric Cooperative (NOVEC)

Electric cooperative	Virginia	143,196 Customers
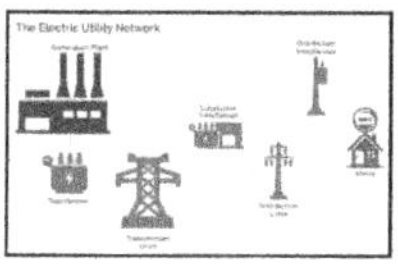		

Distribution circuit impacted: 105 (of 235) Distribution substation impacted 37(of 53)

DA communication network → IP-based communication links

Total cost of DA implementation under SGIG $10,000,000	Distribution automation devices deployed					
	Automated feeder switches	Yes	14	Remote fault indicators	No	
	Automated capacitors	Yes	164	Transformer monitors	Yes	56
	Automated regulators	Yes	340	Smart relays	Yes	25
	Feeder monitors	No		Fault limiters	No	
	Automated reclosers	Yes	117	Smart reclosers	Yes	19
	Substation battery bank monitors	Yes	33			

12.9.2 Electric Power Board of Chattanooga

Municipal/public utility	Tennessee, Georgia	172,079 Customers
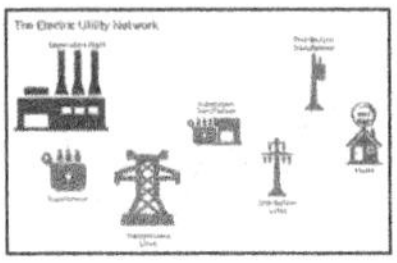		

Distribution circuit impacted: 232 (of 370)

DA Communication Network → Fiber optical network

Total cost of DA implementation under SGIG $49,878,568	Distribution Automation Devices Deployed				
	Automated feeder switches	Yes	1294	Remote fault indicators	No
	Automated capacitors	No		Transformer monitors	No
	Automated regulators	No		Smart relays	No
	Feeder monitors	No			

12.9.3 Duke Energy

Investor-owned utility	Ohio, Indiana, Kentucky North Carolina, South Carolina	4,514,000 Customers
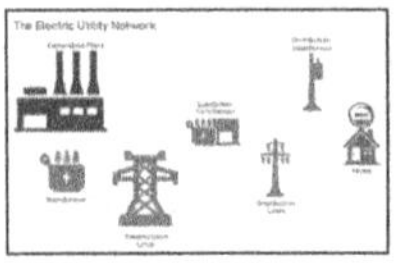		

Distribution circuit impacted: 105 (of 235) Distribution substation impacted 37(of 53)

DA communication network → cellular DA and SCADA network

Total cost of DA implementation under SGIG $189,471,768	Distribution automation devices deployed					
	Automated feeder switches	Yes	914	Remote fault indicators	Yes	4755
	Automated capacitors	Yes	2098	Transformer monitors	No	
	Automated regulators	Yes	914	Smart relays	Yes	4755
	Feeder monitors	Yes	83			

12.9.4 Consolidated Edison (Con Edison)

Investor-owned utility	New York, New Jersey	3,578,188 Customers
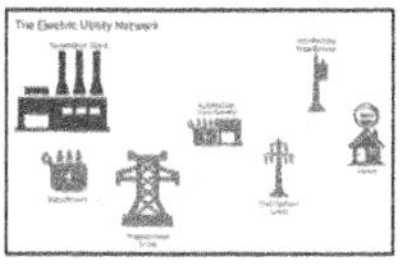		

Distribution circuit impacted: 840 (of 2297)

DA Communication Network → Master radio sites to upgrade SCADA wireless

Total cost of DA implementation under SGIG	Distribution Automation Devices Deployed					
$272,341,798	Automated feeder switches	Yes	797	Remote fault indicators	Yes	1851
	Automated capacitors	Yes	449	Transformer monitors	Yes	17,401
	Automated regulators	Yes	111	Smart relays	Yes	205
	Feeder monitors	Yes	617	Reclosers controls	Yes	307
	Remote battery monitors	Yes	17			

12.9.5 Centerpoint Energy

Investor-owned utility	Houston, TX	2,320,156 Customers
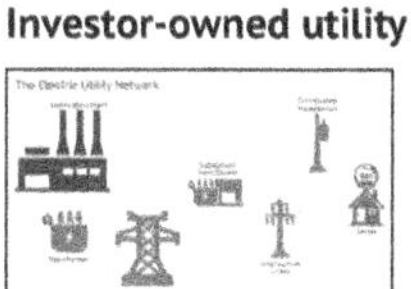		

Distribution circuit impacted: 188 (of 1516) Distribution substation impacted 31 of 240

DA Communication Network → RF, WiMAX, and cellular technologies

Total cost of DA implementation under SGIG	Distribution Automation Devices Deployed					
$120,604,288	Automated feeder switches	Yes	567	Remote fault indicators	No	
	Automated capacitors	No		Transformer monitors	No	
	Automated regulators	No		Smart relays	Yes	171
	Feeder monitors	No				

12.9.6 PPL Electric Utilities Corporation

Investor-owned utility	Pennsylvania	1,396,751 Customers
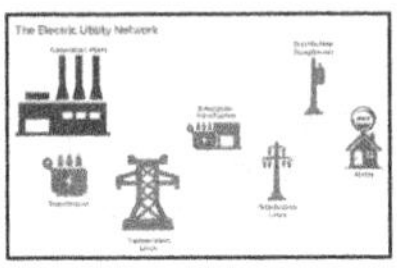		

Distribution circuit impacted: 50 (of 1153) Distribution substation impacted 10 of 376

DA Communication Network → WiMAX, cellular, and fiber optic cable

Total cost of DA implementation under SGIG $38,108,290

Distribution Automation Devices Deployed					
Automated feeder switches	Yes	214	Remote fault indicators	No	
Automated capacitors	Yes	195	Transformer monitors	No	
Automated regulators	No		Smart relays	No	
Feeder monitors	No		Automated reclosers	Yes	77

12.9.7 PEPCO - DC

Investor-owned utility	Washington, DC	249,059 Customers
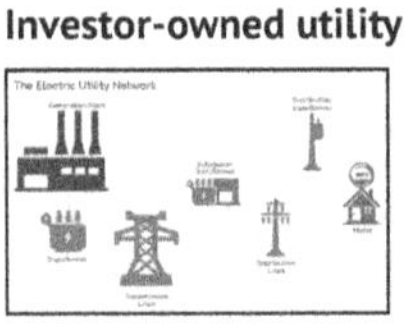	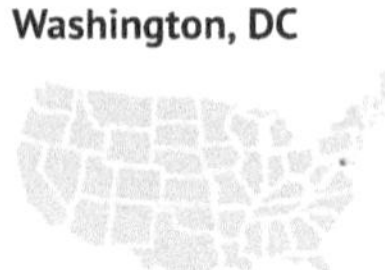	

Distribution circuit impacted: 19 (of 779)

DA Communication Network → Wireless mesh

Total cost of DA implementation under SGIG $8,308,800

Distribution Automation Devices Deployed					
Automated feeder switches	Yes	42	Remote fault indicators	No	
Automated capacitors	No		Transformer monitors	Yes	41
Automated regulators	No		Smart relays	Yes	306
Feeder monitors	No		Substation DRTUs	Yes	6
Transformers health sensors	Yes	14	Automated circuit Recloser switches	Yes	64

12.9.8 Southern Company

Investor-owned utility	Alabama, Florida, Georgia, Mississippi	4,395,000 Customers
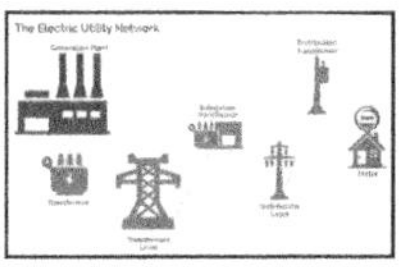		

Distribution circuit impacted 2081 of 4706 Distribution substation impacted 359 of 3325

DA Communication Network → Radio communication via SCADA platform

Total cost of DA implementation under SGIG $207,274,827	Distribution Automation Devices Deployed					
	Automated feeder switches	Yes	2193	Remote fault indicators	Yes	263
	Automated capacitors	Yes	1869	Transformer monitors	No	
	Automated regulators	Yes	3339	Smart relays	Yes	848
				Capacitors monitors	Yes	7876

12.9.9 Wisconsin Power and Light Company (WPL)

Investor-owned utility	Wisconsin	453,573 Customers
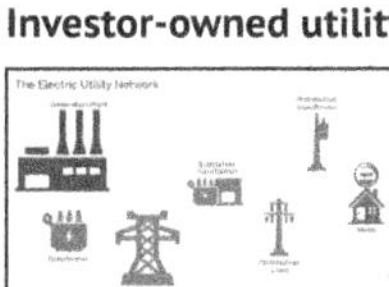		

Distribution circuit impacted: 298 (of 906)

DA Communication Network → Capacitor control radios, take out radio, router, hopper radios

Total cost of DA implementation under SGIG $6,306,738	Distribution Automation Devices Deployed					
	Automated feeder switches	No		Remote fault indicators	No	
	Automated capacitors	Yes	576	Transformer monitors	No	
	Automated regulators	No		Smart relays	No	
	Feeder monitors	No				

12.9.10 Central Lincoln People's Utility District

Municipal public utility	Oregon	38,465 Customers
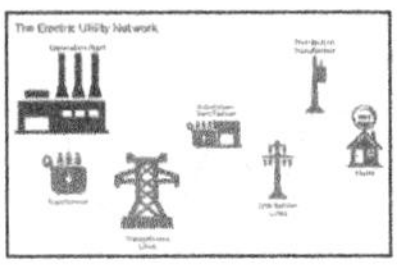		

DA Communication Network → Fiber optic cable and high-speed wireless

Total cost of DA implementation under SGIG
$2,561,406

Distribution Automation Devices Deployed

Automated feeder switches	Yes	17	Remote fault indicators	No
Automated capacitors	No		Transformer monitors	No
Automated regulators	Yes	2	Smart relays	No
Feeder monitors	Yes	14		

12.9.11 Florida Power & LIG, HT Company (FPL)

Investor-owned utility	Florida	4,524,000 Customers
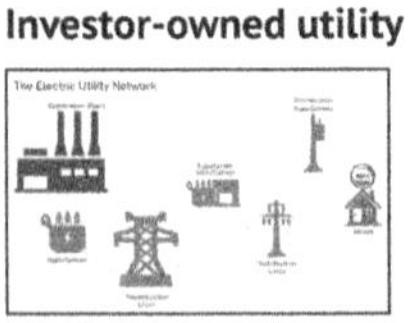		

Distribution circuit impacted: 476 (of 3124)

DA Communication Network IP-Based Communication Links

Total cost of DA implementation under SGIG
$84,667,288

Distribution Automation Devices Deployed

Automated feeder switches	Yes	285	Remote fault indicators	Yes	3879
Automated capacitors	Yes	1403	Transformer monitors	Yes	2716
Automated regulators	Yes	1806	Smart relays	Yes	1084
Feeder monitors	Yes	1014	Throw over sensors	Yes	745

Exercise 12.4 Describe the essential components of case studies as mentioned above.

12.10 Smart Grid System for PV and Wind

Transmission system: It incorporates power electronics for HVDC[2] and FACTS[3] applications. The integration of advanced monitoring and control technologies, such as phasor measurements, into SCADA[4] and EMS[5] is a crucial aspect of the system's operations. As a protection engineer, your role in this integration is significant and ensures the system's smooth functioning.

Distribution system: Implementing information technology and advanced metering infrastructure in the distribution system is critical to ensuring reliability. This, coupled with the implementation of FDIR[6] using DMS, distribution SCADA, communications to substations and feeders, and substation and distribution automation (SA and DA) with controllable devices, provides a robust framework for the system's operations.

12.11 R&D Challenges

- Cybersecurity and interoperability
- Modeling, visualization, and data analytics
- Advanced grid devices and power electronics
- Advanced demand-side solutions

Exercise 12.5 Describe a comparative analysis of transmission and distribution systems.

12.12 Standardization Efforts in Smart Grid Technologies

Importance of Standards:

Standards are critical for ensuring that different smart grid components can communicate and work together effectively. They help prevent the "vendor lock-in"

2 High-voltage direct current.
3 Flexible AC transmission systems.
4 Supervisory control and data acquisition.
5 Energy management system.
6 Fault detection, isolation, and restoration.

problem, where utilities might be forced to rely on a single vendor for all their equipment and systems. Standards also ensure that innovative grid technologies are secure, reliable, and interoperable, essential for overall performance and safety.

12.12.1 Key Organizations Involved in Standardization

Several international organizations are involved in developing standards for smart grids, underscoring **the importance of this work. The IEC**[7] has developed a comprehensive set of standards for smart grids, covering everything from communication protocols to cybersecurity. Essential standards include IEC 61850 for communication networks and systems in substations and IEC 62351 for power system security. **IEEE**[8] has published numerous standards relevant to smart grids, including IEEE 1547 for interconnecting distributed resources with the grid and IEEE 2030 for smart grid interoperability, all of which are crucial for **the advancement of smart grids. NIST**[9] plays a central role in coordinating the development of innovative grid standards, with its Smart Grid Interoperability Framework outlining the critical standards needed to ensure the smart grid's interoperability. Lastly, the **ISO**[10] works closely with other organizations to develop standards that address various aspects of smart grids, including energy management (ISO 50001) and environmental management (ISO 14001), highlighting the significance of these standards.

Interoperability and Communication Protocols: Interoperability is a cornerstone of smart grid operation, ensuring that different devices, systems, and networks can work together seamlessly. This is particularly important as the grid becomes more decentralized, with increasing numbers of DERs[11], microgrids, and IoT devices. Common communication protocols include the first protocol DNP3 (Distributed Network Protocol), widely used for communication between control centers and substations. The second protocol, Modbus, is used for communication between electronic devices, such as sensors and controllers, playing a crucial role in the grid's operation.

OpenADR (Open Automated Demand Response) is a standard for automated demand response, allowing utilities to communicate with consumers to adjust their energy usage during peak demand periods. Lastly, IEC 61850 is a standard for the design of electrical substation automation, allowing devices to communicate effectively.

---

7 The International Electrotechnical Commission.
8 The Institute of Electrical and Electronics Engineers.
9 The National Institute of Standards and Technology.
10 International Organization for Standardization.
11 Distributed Energy Resources.

Exercise 12.6 Describe energy policy on smart and microgrid operations.

12.12.2 Impact of Policies on Smart and Microgrid Operations

12.12.2.1 Renewable Energy Policies

Government policies play a pivotal role in promoting renewable energy sources like solar and wind, thereby driving the adoption of smart grid technologies. These policies, which often include subsidies, tax incentives, and RPS[12] that require utilities to source a certain percentage of their energy from renewable sources are instrumental in this transition. Smart grids, in turn, are indispensable for integrating renewable energy into the grid, providing the flexibility and control needed to manage the variable output of these sources. The regulatory support for smart grids is therefore critical for achieving renewable energy targets.

12.12.2.2 Energy Efficiency and Demand Response

Energy efficiency policies and smart grid initiatives are not just parallel tracks, but interconnected pathways toward a sustainable energy future. Regulators often set energy efficiency targets for utilities, incentivizing them to implement smart grid technologies that help consumers reduce their energy usage. However, it is the demand response programs, which rely on smart grid infrastructure, that truly shine in managing peak demand. These programs allow utilities to incentivize consumers to reduce their electricity usage during high-demand periods, thereby ensuring grid reliability. Regulatory support for demand response is crucial for its success, as it provides the framework for compensating participants and ensuring grid reliability.

12.12.2.3 Cybersecurity and Privacy Regulations

Cybersecurity and privacy have become significant concerns as smart grids rely heavily on digital communication and data exchange. Regulatory bodies are increasingly mandating robust cybersecurity measures to protect the grid from cyber threats and ensure consumer data privacy. These measures include, but are not limited to, regular security audits, encryption of sensitive data, and the implementation of multifactor authentication. Standards like IEC 62351 and the NIST[13] Cybersecurity Framework provide guidelines for securing smart grid infrastructure. Regulations may also require utilities to report cybersecurity incidents and implement measures to protect sensitive data.

---

12 Renewable portfolio standards.

13 National Institute of Standards & Technology.

12.13 Challenges in Regulatory and Standardization Implementation

12.13.1 Harmonization Across Regions

One of the challenges in standardization is ensuring harmonization across different regions and countries. While international standards provide a common framework, local regulations and market conditions can lead to variations in how these standards are implemented. This lack of uniformity can create barriers for technology developers and vendors, who must adapt their products to meet regional requirements. The urgency and necessity of harmonization efforts are clear, as they are needed to ensure smart grid technologies can be deployed globally without significant modifications.

12.13.2 Regulatory Uncertainty

The rapid pace of technological innovation in smart grids can outpace the development of regulations, leading to uncertainty for utilities and technology providers. This uncertainty can hinder investment in smart grid projects and delay the adoption of new technologies. Regulators face the challenge of developing flexible policies to accommodate new technologies while providing clear guidance for market participants. The importance of adaptive regulatory frameworks that can evolve with technological advancements cannot be overstated.

Exercise 12.7 Describe balancing innovation and regulations.

12.13.3 Balancing Innovation and Regulation

Regulators must balance promoting innovation and ensuring grid reliability and security. Overly stringent regulations can stifle innovation, while too much flexibility can lead to security vulnerabilities or reliability issues. Collaborative efforts between regulators, industry stakeholders, and technology developers are crucial in this regard. Industry stakeholders, including utilities and technology providers, play a significant role in shaping the regulatory environment, fostering innovation while maintaining high grid operation and security standards.

Exercise 12.8 Describe emerging technologies and standards.

12.14 Future Directions in Regulatory and Standardization Efforts

Emerging Technologies and Standards:
As the smart grid industry rapidly evolves with the introduction of new technologies like blockchain, artificial intelligence, and advanced energy storage systems, the need for new standards becomes increasingly urgent. These technologies present unique challenges and opportunities that must be addressed through the establishment of standards. For instance, blockchain technology for peer-to-peer energy trading necessitates transaction verification and data security standards. Similarly, AI-driven grid management systems require standards to ensure reliability and transparency.

International Collaboration:
International collaboration is not just important but also essential for the development and updating of standards that mirror the latest advancements in smart grid technology. Organizations like IEC, IEEE, and ISO, in close partnership with regional regulatory bodies, are at the forefront of this global effort. By participating in international forums and conferences, you become part of a global smart grid community, sharing best practices, discussing emerging challenges, and aligning regulatory approaches across different regions.

Regulatory Sandboxes:
Regulatory sandboxes are gaining recognition for their role in testing new technologies and business models in a controlled environment. These sandboxes provide a reassuring platform for regulators to observe how innovations perform in the real world and gather data to inform future regulations. They are particularly useful for testing smart grid technologies that involve significant uncertainty or potential risks, such as integrating large-scale DERs or deploying advanced cybersecurity measures, ensuring the safety and reliability of smart grid innovations.

12.15 Self-Assessment Questions

Self-Assessment Question 12.1

A) Describe the **Smart grid system for PV and Wind.**
B) Describe the Fault Detection, Isolation & Restoration (FDIR) and FLISR.[14]

Self-Assessment Question 12.2

Describe the following.

- SCADA system and its operational functions also discuss one or two case studies related to power distribution system
- Challenges in Regulatory and Standardization Implementation

Bibliography

Jamil Abdo, Tariq Masood "New Technique to Develop Intelligent, Highly Reliable & Efficient Unmanned Power Generation Modular" published in 22nd International Conference on Renewable Energies and Power Quality (ICREPQ'24) Bilbao (Spain), 26–28 June 2024. https://www.icrepq.com/icrepq24/392-24-abdo.pdf.

Jamil Abdo, Tariq Masood "Operational Reliability – Production Best Practice Pragmatic Case Studies & Improvement" published in Proceedings of the 9th North American Conference on Industrial Engineering and Operations Management Washington D.C., USA, 4–6 June 2024.

Tariq Masood, Jamil Abdo "Maintenance Strategy Production Best Practices – Pragmatic Case Studies and Operations Improvement" published in Proceedings of the 9th North American Conference on Industrial Engineering and Operations Management Washington D.C., USA, 4–6 June 2024.

Jamil Abdo, Tariq Masood "Techno Economics Analysis to Reshape Global Clean and Green Energy" published in 22nd International Conference on Renewable Energies and Power Quality (ICREPQ'24) Bilbao (Spain), 26–28 June 2024. https://www.icrepq.com/icrepq24/415-24-abdo.pdf.

M. Saeed Ullah, Tariq Masood, Umair Zahid, H. Abuzar Ahsan "Electrochemical Oxidation of Dissolved organic Dye in Aqueous Solution" published in IMASCON-2019.

Nasser Alemadi, Tariq Masood, R Aggarwal, Suhail A Qureshi, D Kothari "Novel Selection Approach of STATCOM Device Tailored and Optimized Engineering

14 Fault location, isolation, and service restoration.

Configuration to Implement in Between Qatar and Bahrain" published in International Journal of Automation and Power Engineering 2014. 10.14355/ijape.2014.0302.01.

Tariq Masood, Jamil Abdo "New Technique to Enhance Hybrid Renewable Energy Resources' Operational Flexibility and Reliability" published in 22nd International Conference on Renewable Energies and Power Quality (ICREPQ'24) Bilbao (Spain), 26–28 June 2024. https://www.icrepq.com/icrepq24/396-24-masood.pdf.

Tariq Masood and Jamil Abdo "Review: Uninterrupted Liquid Hydrogen (E-Methanol) Production Framework to Deliver Clean & Green Energy" published in 22nd International Conference on Renewable Energies and Power Quality (ICREPQ'24) Bilbao (Spain), 26–28 June 2024. https://www.icrepq.com/icrepq24/414-24-masood.pdf.

Tariq Masood, Abdel Aty Edris, Suhail Aftab Qureshi, Muhammad Tajammal, Murtaza Hashmi, DP Kothari, Samer Kareem "SSSC Tailored to Optimization Performance in Between United Arab Emirates and Oman on the GCC Power Grid" published in the 13th IET International Conference on AC and DC Power Transmission Conference on 14–16 February 2017. http://ieeexplore.ieee.org/document/7934934.

Tariq Masood, Abdel Aty Edris, Suhail Aftab Qureshi, Muhammad Tajammal, Murtaza Hashmi, DP Kothari, Samer Kareem "Smart Grid Operations and Control Challenge by Implementing STATCOM Tailored to Optimization Performance at the GCC Power Grid" published in the 13th IET International Conference on AC and DC Power Transmission Conference on 14–16 February 2017. http://ieeexplore.ieee.org/document/7934988.

Tariq Masood, Muhammad Tajammal, Samer Kareem, Suhail Aftab Qureshi, Abdul Jabbar Khan, DP Kothari, Nair Jameel "Unified Power Flow Controller Modeling and Analysis Technique on the GCC Power Grid Operations and Control" published in 2018 IEEE International Energy Conference, Limassol Cyprus, 7–9 June 2018. https://ieeexplore.ieee.org/document/8398834.

Tariq Masood, Muhammad Tajammal, Samer Kareem, Suhail Aftab Qureshi, Murtaza Hashmi, Nasir Jameel "Unified Power Flow Controller (UPFC) Modeling and Analysis Technique by using Matlab and EMTP" published in International Conference on Renewable Energies and Power Quality (ICREQP'18) Salamanca (Spain) 21–23 March 2018. http://www.icrepq.com/icrepq18/250-18-masood.pdf.

Tariq Masood, Muhammad Tajammal, Samer Kareem, Suhail Aftab Qureshi, Murtaza Hashmi, Naser Jameel "STATCOM Controller Performance Enhancement and Optimization Technique" published in International Conference on Renewable Energies and Power Quality (ICREQP'18) Salamanca (Spain) 21–23 March 2018. http://www.icrepq.com/icrepq18/249-18-masood.pdf.

Tariq Masood, Muhammad Tajammal, Samer Kareem, Suhail Aftab Qureshi, Murtaza Hashmi, DP Kothari "Smart Grid Operational Functions and Control Challenges by Implementing SSSC Tailored to Optimize Performance in Between Qatar and KSA on the GCC Electrical-Power Grid" published in the ASTESJ Journal Advances in Science, Technology and Engineering Systems Journal Vol. 2, No. 1, 2017. https://astesj.com/special-issue-4.

Tariq Masood, Raj K Aggarwal, Suhail Aftab Qureshi, RAJ Khan "STATCOM Model Against SVC Control Model Performance Analyses Technique by MATLAB" published in International Conference on Renewable Energies and Power Quality (ICREPQ'10) 2010. https://icrepq.com/icrepq'10/655-Masood.pdf.

Tariq Masood, N Al-Emadi, SA Qureshi, YY Almulla, AJ Khan, SK Shah "FACTS Control Devices (Statcom, SSSC and UPFC) Re-Configuration Techniques By PSIM/MATLAB" published in 2007 International Conference on Electrical Engineering 2007. http://ieeexplore.ieee.org/abstract/document/4287295.

Tariq Masood, Raj K Aggarwal, SA Qureshi, RAJ Khan "STATCOM and SVC Control Operations and Optimization During Network Fault Conditions" published in 2010 IEEE International Symposium on Industrial Electronics 2010. https://ieeexplore.ieee.org/document/5636868.

Tariq Masood, RK Aggarwal, SA Qureshi, RAJ Khan "Statcom Control Reconfiguration Technique for Steady State and Dynamic Performance Optimization During Network Fault Conditions" published in International Conference on Renewable Energies and Power Quality (ICREPQ'10) Granada (Spain) 2010. https://www.icrepq.com/icrepq'10/656-Masood.pdf.

Tariq Masood, Samer Karim Shah, Muhammad Tajammal, "Unified Power Flow Controller Modeling and Analysis Technique on the GCC Power Grid" published in 2018 IEEE International Energy Conference (ENERGYCON) 2018. https://ieeexplore.ieee.org/abstract/document/8398834.

Tariq Masood, Suhail A Qureshi, RK Aggarwal, DP Kothari "Multiple FACTS Devices Integration Through Distributed Control System and Its Application on the GCC Interconnection" published in 2013 7th IEEE GCC Conference and Exhibition (GCC) 2013. https://ieeexplore.ieee.org/abstract/document/6705750.

13

Microgrid Operation and Control

13.1 Objectives

- Microgrid operations and classification of microgrids
- AC and DC microgrid operations
- Hybrid microgrid
- Off-grid and different types of island operations
- Microgrid forwarding and forming inverters operations
- Advantages of the short-circuit study ratio and traditional short-circuit study methods
- Strong and weak microgrid
- Time-based current curve
- TCC operations in a microgrid

13.2 Introduction

The US Department of Energy (DOE) defines a microgrid as: *A group of interconnected loads and distributed energy resources (DERs) within clearly defined electrical boundaries that acts as a single controllable entity with respect to the grid. A microgrid can connect and disconnect from the grid to enable it to operate in both grid-connected and island mode, as shown in* Figure 13.1.

13.3 Classification of Microgrids

In the first layer, we delve into microgrid operation, providing a comprehensive understanding of the microgrid size: small <10 kW, medium <1 MW, and large microgrid >1 MW power operations. The discussion on microgrid application is

Protection of Electrical Power Distribution Systems: Smart grid, Microgrid, AI, and Cyber Security,
First Edition. Tariq Masood, Jamil Abdo, and Atif Iqbal.

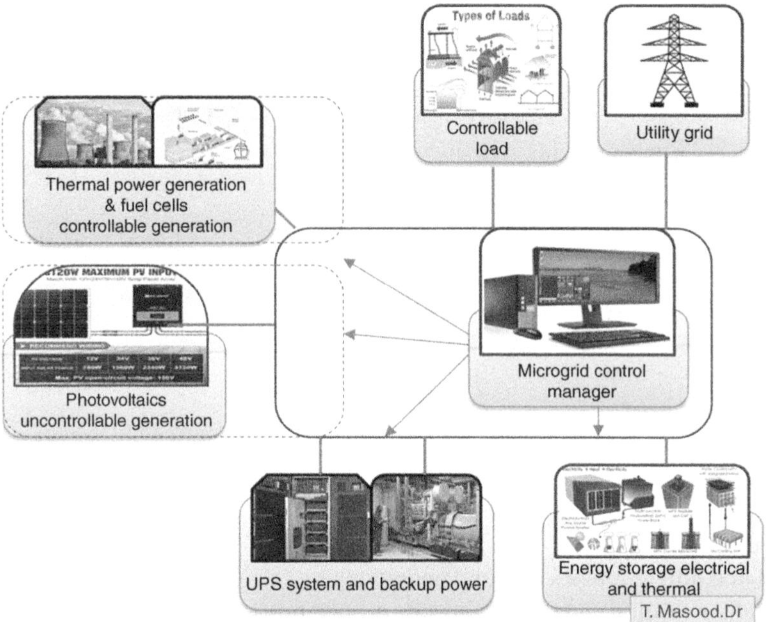

Figure 13.1 Microgrid schematic. Source: Landscape Photography of Factory / Pixabay / CC0, xfilephotos / Shutterstock, Christian Delbert / Adobe Stock Photos.

thorough covering three main components: premium, loss reduction, and resilience in power delivery. We ensure a comprehensive understanding of these concepts. The section also explores grid connection, handling the transient and standalone operations scenarios, leaving no aspect of microgrids unexplored, as shown in Figure 13.2.

In the second layer, the distributed system, as discussed in the second instance of microgrid operations, is a comprehensive solution that covers DC microgrid, AC microgrid, and hybrid microgrid system operations. Each of these microgrid architectures is designed with specific principles and operational advantages. The microgrid architecture, with its radial grid connection (providing a direct path for power flow), ring primary grid connection (allowing for power to be rerouted in case of a fault), and mesh-type hybrid connection (providing redundancy and reliability), is designed to deliver and maintain seamless power operations, ensuring a reliable power supply even in unforeseen situations, as shown in Figure 13.3. In the third layer, we discuss the distributed configuration, which covers single-phase, three-phase, and three-phase and neutral scenarios. The microgrid architecture, with its three main components, is designed to deliver and maintain seamless power operations. It provides comprehensive coverage for

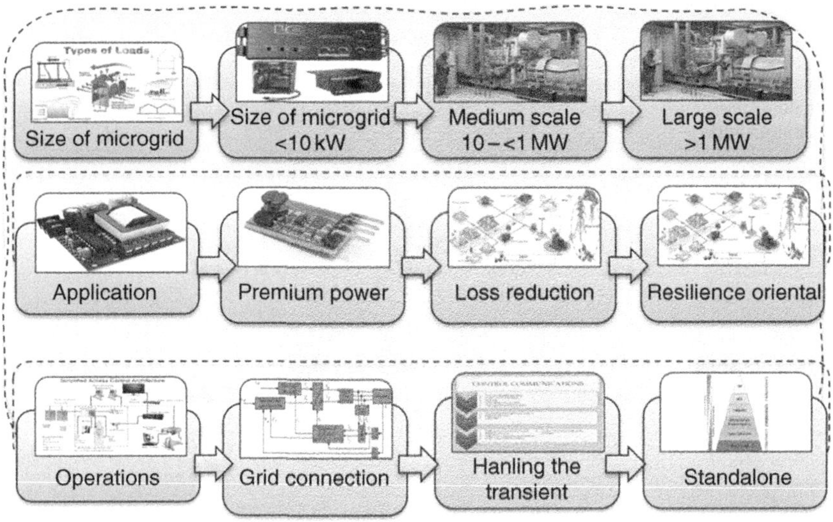

Figure 13.2 Microgrid size, application, and operations. Source: Christian Delbert / Adobe Stock Photos.

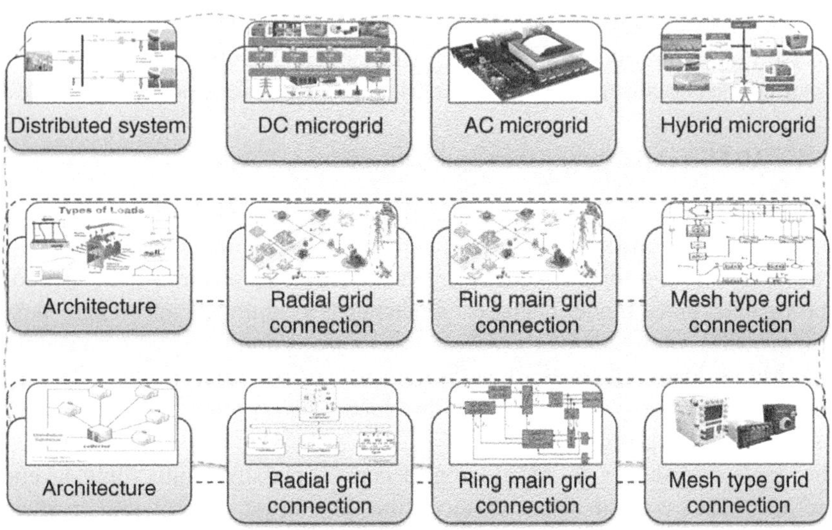

Figure 13.3 Microgrid distributed system, architecture and resources.

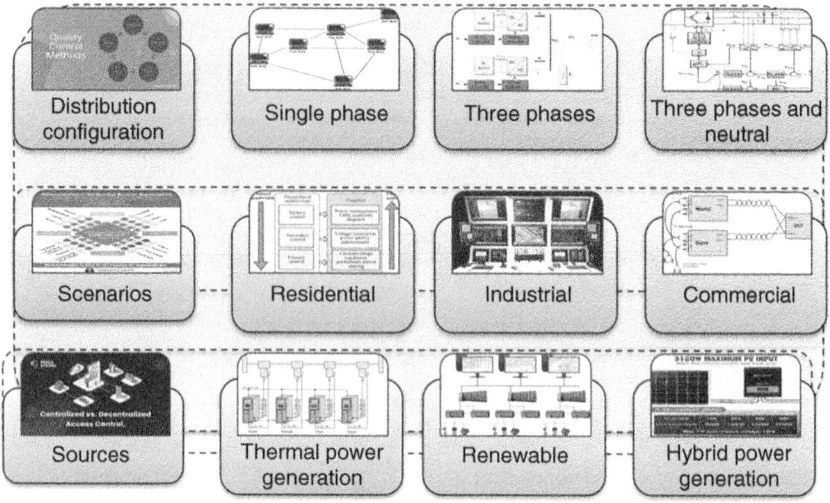

Figure 13.4 Microgrid distributed configuration, scenarios, and resources.

any type of operational contingency, ensuring the system's preparedness even in unforeseen situations, as shown in Figure 13.4.

Exercise 13.1 Describe AC microgrid operational and control challenges.
Describe the classification of the microgrid and its main components.

13.3.1 AC Microgrid Operations and Control

Figure 13.5 indicates an AC microgrid connecting various generation sources and loads using an AC bus system, which provides the following benefits. First, AC microgrids offer a promising solution by integrating with a conventional grid or operating in islanded mode. Their cost efficiency is a significant advantage, making them a vital component of the power system protection system. They can also handle higher loads. However, they do have some drawbacks, such as the cost of converters and the complexity of control methods. They may also have lower transmission efficiency.

The control strategies in AC microgrid can be classified into three layers: Inner and outer control layers that control the output current and manage the output active and reactive power of renewable energy sources (RESs), and the power-sharing control layer that adjusts the amplitude and frequency of the network.

Exercise 13.2 Describe DC microgrid operational and control challenges.

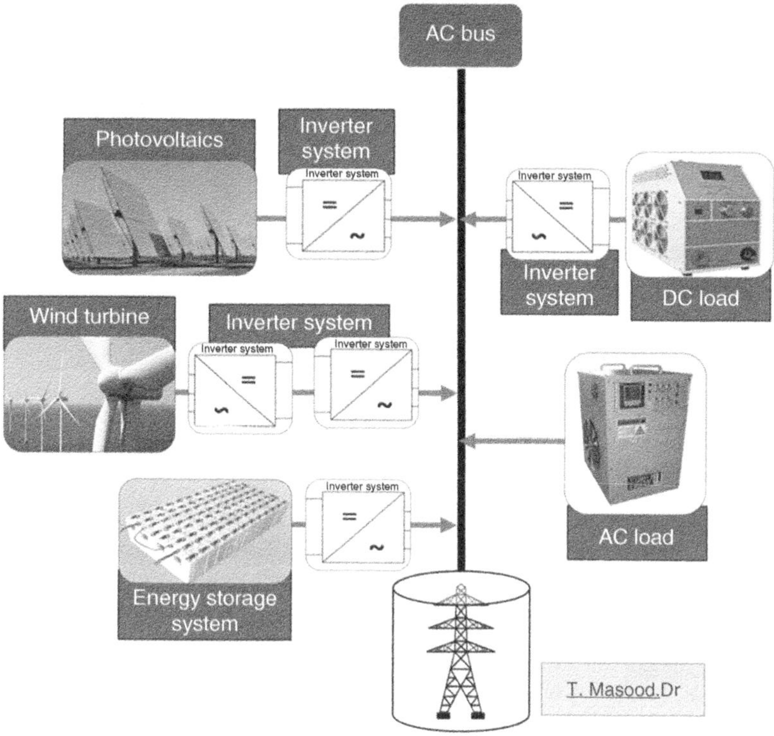

Figure 13.5 AC microgrid operations and control. Source: Tom Washington / Getty Images.

13.3.2 DC Microgrid Operations and Control

The distinguishing feature of the DC microgrid is its use of a DC bus network for interconnection, a departure from the AC bus used in traditional microgrids. This DC bus network serves as the interconnection point for the distributed generators and loads in the network.

Significant Advantages of the DC Microgrid: It notably enhances transmission efficiency and boasts a simple control system, ensuring seamless operation and high-power quality delivery. Despite its advantages, the DC microgrid **does** have some drawbacks. The higher initial investment required may be a deterrent for some. Additionally, the absence of reactive power sources in the DC power network poses a challenge in controlling voltage variation, and the power protection system can be complex as shown in Figure 13.6. Table 13.1 shows DC and AC microgrid's comparative performance analysis.

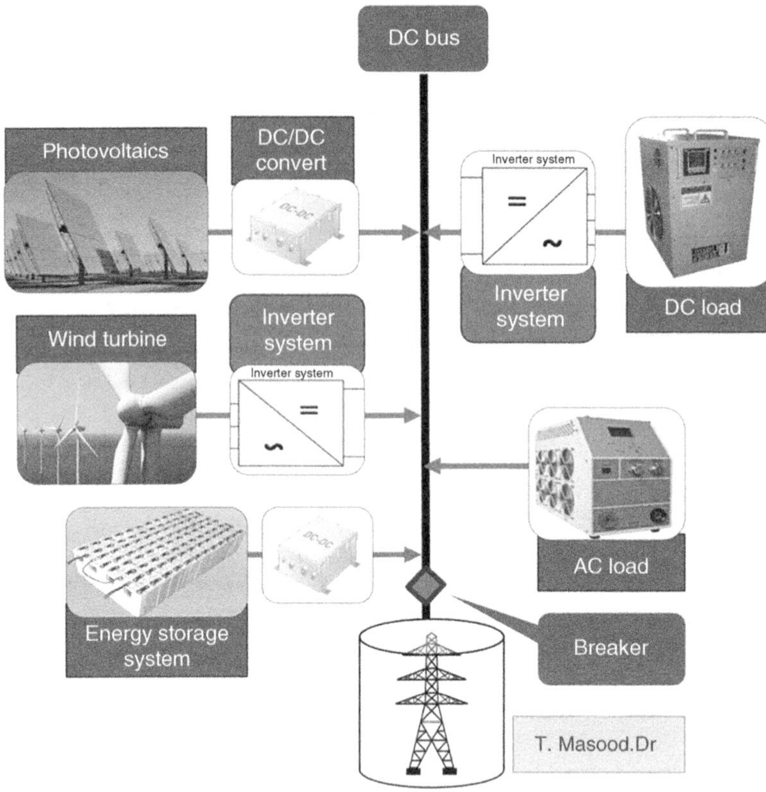

Figure 13.6 DC microgrid operations and control schematic. Source: Tom Washington / Getty Images.

Exercise 13.3 Describe the DC/AC system operational flexibility to allow for reconfiguration of existing facilities. Which DC and AC systems are vulnerable or complex to interface power electronics devices.

Exercise 13.4 Describe a static transfer switch (STS) that is connected to the primary grid. What is main purpose of the switch.

13.3.3 Hybrid Microgrid Operations and Control

A static transfer switch (STS) connects it to the primary grid. The power electronic converter interface controls the power flow between the networks and the utility grid. The power direction is subject to the balance between load and generation. The interface can minimize conversion stages, increase reliability, and reduce interfacing devices. Most importantly, it significantly reduces energy costs, making it a financially prudent choice for power system management, as shown in Figure 13.7.

Table 13.1 Performance comparison of DC and AC microgrid.

Factor affecting	DC microgrid	AC microgrid
Series-connected power converters	Yes	No
Possibility of reconfiguration of existing facilities	No	Yes
Stability	Free from external effects	Affected by external disturbances
Microgrid control	Simple control approach	Control process is complex due to frequency
Grid-compliant devices	No	Yes
Required number of power electronics interfaces	Medium	High
Complexity of the required power electronics interfaces	Simple	Complex
Quality of the energy in the microgrid	High	Low
Easy management of the energy storage	Yes	No
Protection system	Complex, costly and immature protection components	Simple, cheap, and mature protection schemes

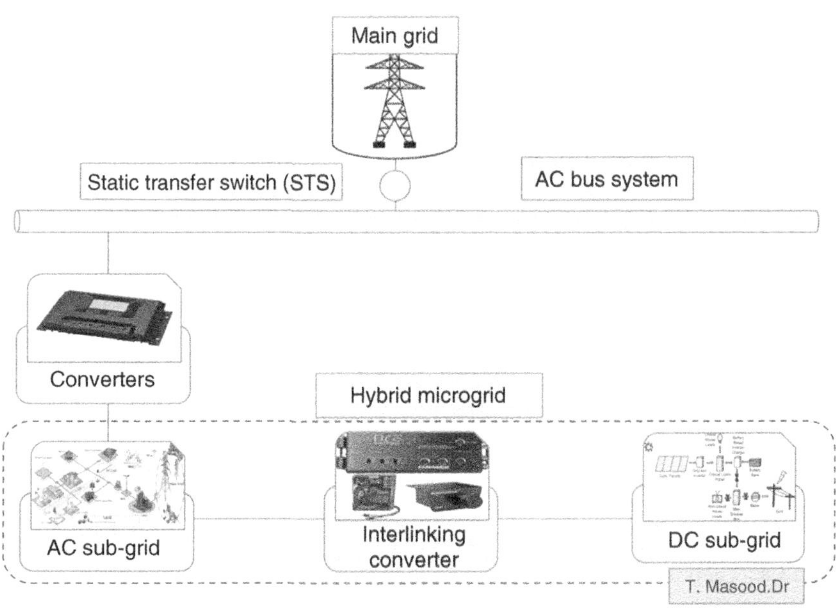

Figure 13.7 Hybrid microgrid operations and control schematic.

13.4 Microgrid Control

The microgrid control objectives, which are crucial but pivotal for the system's efficient functioning, include independent active and reactive power control, correction of voltage sag and system imbalances, and fulfillment of the grid's load dynamics requirements. The microgrid control flow is meticulously and systematically divided into four control boundaries: the control architecture, control method, control mode, and control level. This meticulous approach, further divided into three levels: primary, secondary, and tertiary, provides a clear and structured organization to the process that enhances efficiency, as shown in Figures 13.8 and 13.9 (centralized and decentralized models).

Tertiary control, a key player in the system, is used to control the operations optimally, significantly enhancing the overall efficiency. It works with global-level control to receive feedback from primary and secondary control and power flow control in grid-tie mode. Secondary control is designed to compensate for the frequency deviation caused by the primary control and local level control. Primary control is integrated to adjust the voltage stability provision, frequency stability preserving, power sharing, and protection devices, as shown in Figure 13.10.

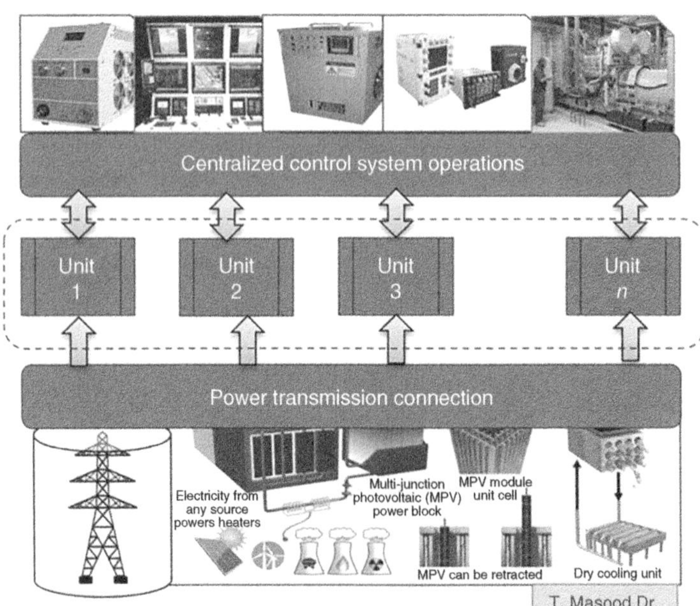

Figure 13.8 Centralized control. Source: Christian Delbert / Adobe Stock Photos.

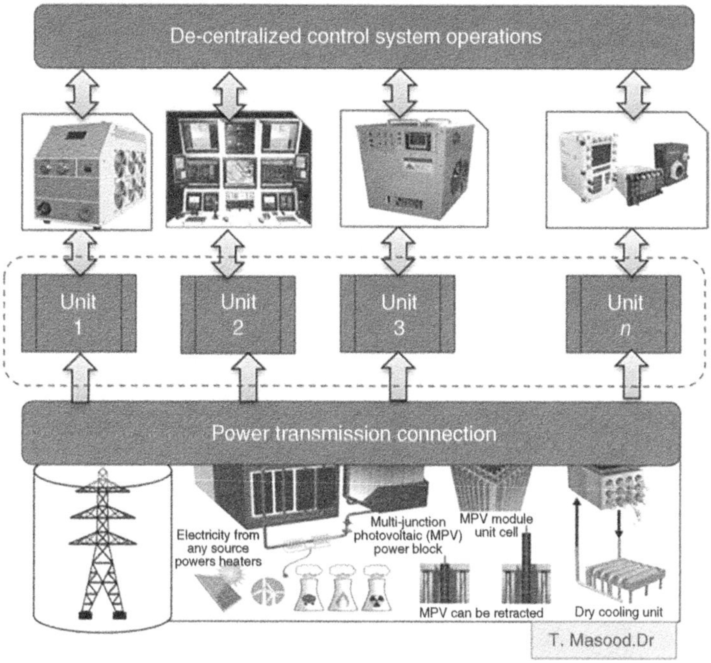

Figure 13.9 Decentralized control.

Exercise 13.5 Describe the centralized and decentralized control system operations.

Figure 13.8 indicates the Centralized Control: Traditionally, power systems have relied on centralized control, where a single control center manages the entire grid. This approach, with its comprehensive view of the grid's status, enables optimal decision-making and coordination. The centralized control systems collect data from various parts of the grid, process it, and then send control signals back to different grid components. This model, effective for large, interconnected grids, may face scalability and real-time response challenges, especially as the grid becomes more distributed. Challenges include the need for a robust communication infrastructure and potential vulnerabilities to cyberattacks.

Figure 13.9 indicates the Decentralized Control: With the rise of distributed energy resources (DERs) and microgrids, decentralized control is gaining importance. In this model, local controllers make decisions independently based on local conditions and goals, showcasing their adaptability. Decentralized control can improve system resilience, as local issues can be resolved without the need for

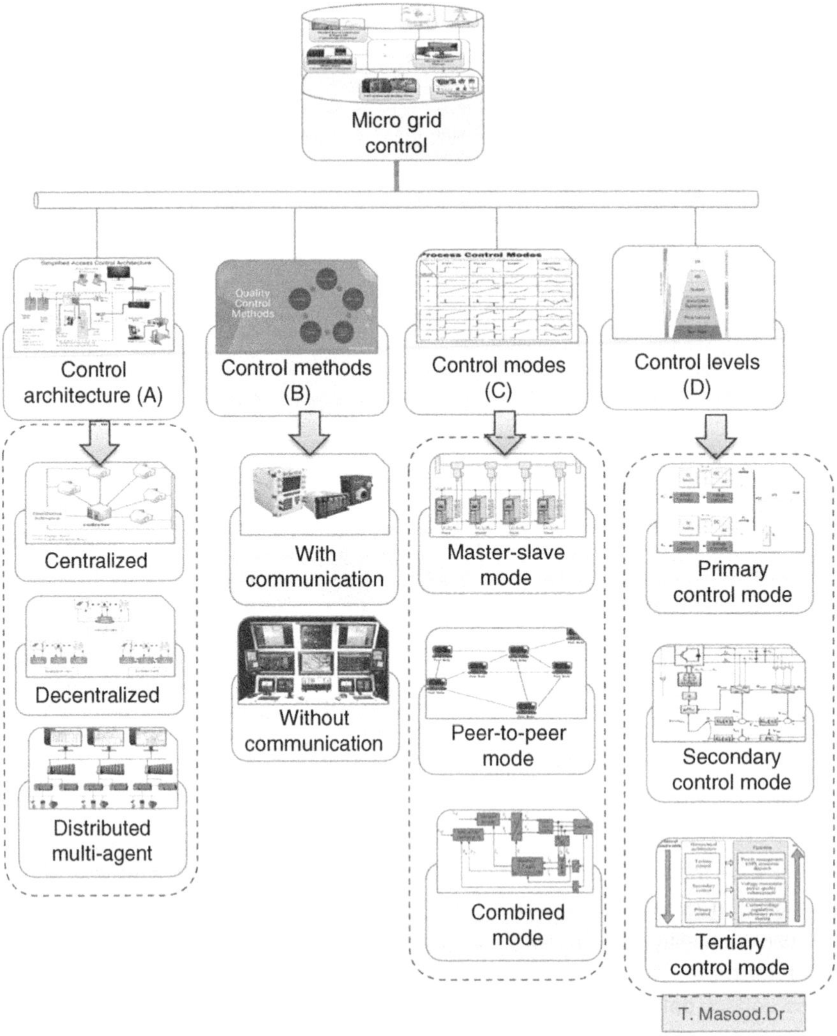

Figure 13.10 Microgrid operations and control schematic.

centralized intervention. It also reduces the communication overhead and allows for faster response times in managing local disturbances. However, it is important to be aware that achieving global optimization in a decentralized system can be challenging, as it requires coordination among multiple independent controllers and understanding the complexities involved.

13.5 Droop Control – Parallel Inverter in UPS System

Figure 13.11 illustrates the versatile operations of the droop control system, a technology that can be used to control AC electrical power generators integrated with both conventional and renewable power resources. The power output of a generator is reduced as the line frequency increases, showcasing the adaptability of this system. It is commonly used as the speed control mode of the governor of a prime mover driving a synchronous generator connected to an electrical grid.

Equally important is the parallel inverter in a UPS (uninterruptible power supply) system, a control technique that connects the outputs of two or more inverters to a standard AC busbar to supply a load, providing extra capacity or redundancy.

This section highlights the significant advantages of droop control, such as its ability to allow inverters to mimic the droop characteristics of synchronous generators and enable proper sharing of absolute power. However, it is important to note that the disadvantage is that it deteriorates reactive power sharing among distributed generators.

Exercise 13.6 Describe the concept of crucial coupling and IEEE standard for improvement.

The concept of standard coupling (PCC) is a crucial one in the power system. It refers to a common point or location where multiple customers and their equipment are connected to a utility power grid. This point, as defined by IEEE[1] is the

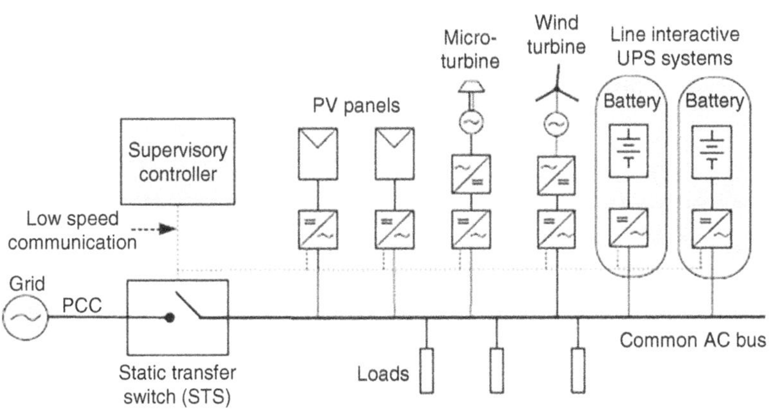

Figure 13.11 Droop control schematic.

1 IEEE Standard 519-2014 [5].

point on a public power supply system electrically nearest to a particular load at which other loads could be connected, ensuring efficient and reliable power distribution. The supervisory controller is a key component in the power system, continuously monitoring diagnostic parameters and defining the fault status. Its role is crucial in ensuring a high level of security and reliability in the power system, instilling confidence in its capabilities.

13.5.1 Bidirectional DC/DC Converters

Figure 13.12 indicates a bidirectional direct current (DC)/DC converter is not only vital, it is a game-changer in many new applications, particularly power distribution systems. However, it is essential to be aware of the limitations of the buck-boost bidirectional converter.

These limitations, such as hard switching, high duty-cycle operation, voltage-conversion ratio limitation, and high voltage stress on the switching device, underscore the urgent need for further research and development. This need is essential and urgent and cannot be overstated. The buck converter is a straightforward DC–DC converter that produces an output voltage less than its input. It is so named because the inductor always "bucks" or acts against the input voltage. Boost converters can increase the voltage above the input.

Exercise 13.7 Describe the role of buck-boost converters for bidirectional power flow in Distribution System.

Furthermore, the buck-boost converter, an efficient DC-to-DC converter, showcases a remarkable potential. It can produce an output voltage magnitude greater than or less than the input voltage magnitude, equivalent to a flyback converter using a single inductor instead of a transformer. This converter, with its two different topologies, can produce a range of output voltages that is significantly more significant (in absolute magnitude) than the input voltage, which is almost zero. This potential is a reason for optimism in the field of power distribution systems.

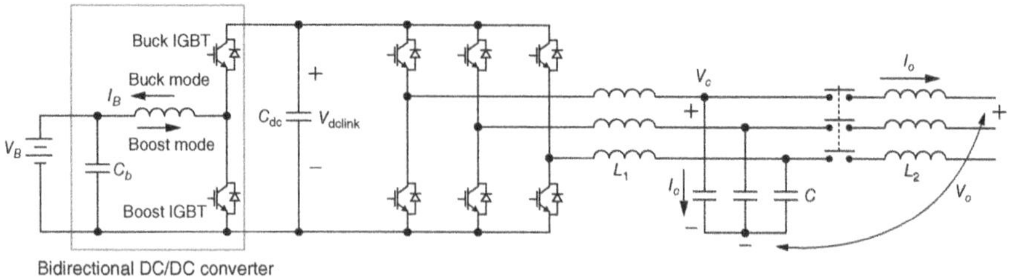

Figure 13.12 Bidirectional DC/DC control schematic.

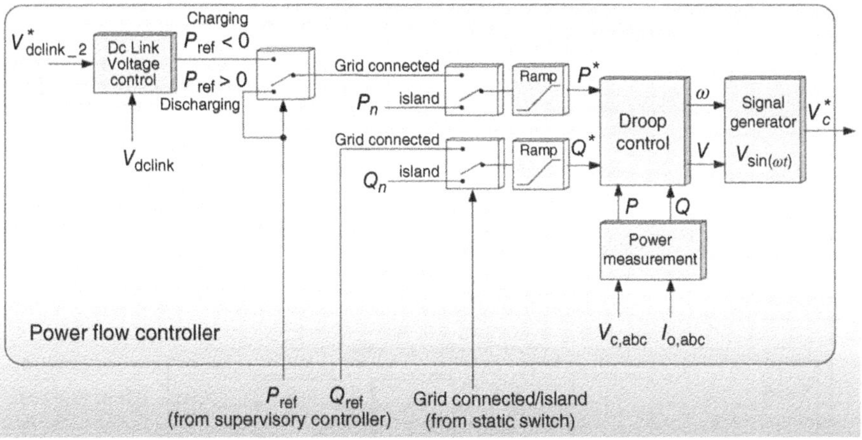

Figure 13.13 Power flow control by using the droop control technique.

13.5.2 Droop Control – Parallel Inverter in UPS System

Figure 13.13 underscores the controllers' significant role in designing SPV generation systems. The DC-link voltage control during low-voltage ride-through (LVRT) for SPV (Solar Photovoltaic system) generation systems is of utmost importance. This control is used for a grid-connected voltage source converter (VSC) of a microgrid, and is one of two controllers designed for this application, along with active/reactive power (PQ) control, as shown in the schematic. This also indicates the charging and discharging of the active power to control the reference value of the bus voltage and power flow in the network.

The static transfer switch (STS) is a critical component in uninterruptible power supply (UPS) systems. It provides a reliable backup source of power, ensuring that large loads and facilities are protected from short-term power disturbances. Its role in maintaining an uninterrupted power supply is paramount, instilling confidence in the system's reliability. Engineers and professionals in the field play a crucial role in the island mode of operations. Their expertise is vital in connecting active and reactive powers to adjust their requirements. Active and reactive power grids are connected to droop control, which is associated with a power measurement block showing the measured voltage and current delivered to the consumer.

13.5.3 Droop Control – Battery Charging and Discharging Systems

The top part of Figure 13.14a vividly demonstrates the pivotal role of the PI (proportional integral) control mechanism in managing the buck converter's

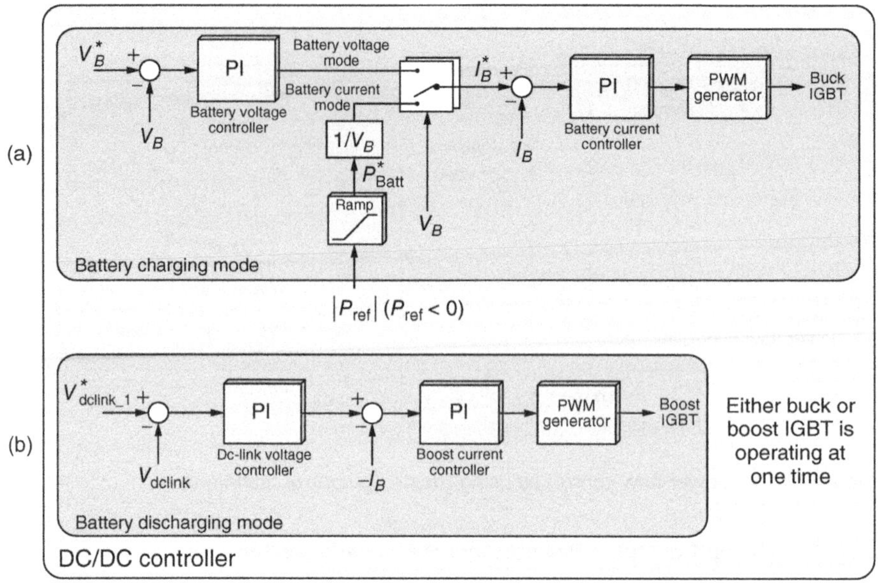

Figure 13.14 (a) and (b) Power flow control by using the droop control technique.

step-down output voltage. This mechanism, a vital feature of a buck converter, a type of DC-to-DC converter, excels at reducing voltage while amplifying current from its input (supply) to its output (load). Its sophisticated control, a testament to the technical advancement in power management systems, is instrumental in faster battery charging, a task even a linear regulator cannot accomplish. This means that with the PI control mechanism, batteries can be charged more quickly, reducing downtime and increasing the efficiency of the system. As DC-to-DC converters, buck converters outperform linear regulators in terms of power efficiency, a key advantage in power management systems. Unlike linear regulators, which dissipate power as heat without stepping up the output current, buck converters can achieve high efficiency, often over 90%. This superior performance, a testament to the advancements in power management, makes them ideal for converting a computer's primary supply voltage, usually 12 and 24 V, providing a significant advantage in power management.

In Figure 13.14a, the bottom part underscores the critical role of the PI control mechanism in managing the battery discharging system, a crucial aspect of power electronics. The PI[2] controls and adjusts the DC link voltage from the battery output to consume the power. The DC-link voltage control during low-voltage

2 Proportional and Integral control.

ride-through (LVRT) for SPV[3] generation systems is of utmost importance, highlighting the significant role of the PI control mechanism in our field of work.

Figure 13.14b also demonstrates the pivotal role of the PI control mechanism in managing the boost converter's step-up output voltage. This mechanism, a vital feature of a boost converter, a DC-to-DC converter, excels at increasing voltage while reducing current from its input (supply) to output (load). Its sophisticated control is instrumental in achieving faster battery discharging, a task that linear regulators cannot accomplish.

13.5.4 Off-Grid and Island Microgrid

Islanded microgrid solutions offer access to grid-quality power in remote locations, typically at a lower electricity cost and with a significantly lower environmental impact than conventional grids. This makes them a cost-effective and eco-friendly choice for remote power needs.

Figure 13.15 shows an off-grid power system, a symbol of independence in the energy world. It harnesses the power of alternative sources like solar, wind, or micro-hydro, converting it into usable electricity. This system stands independently, not relying on local utility, and is immune to blackouts. It is a powerful tool that allows you to break free from the central power system most people are tied to, giving you a sense of control and empowerment over your energy needs.

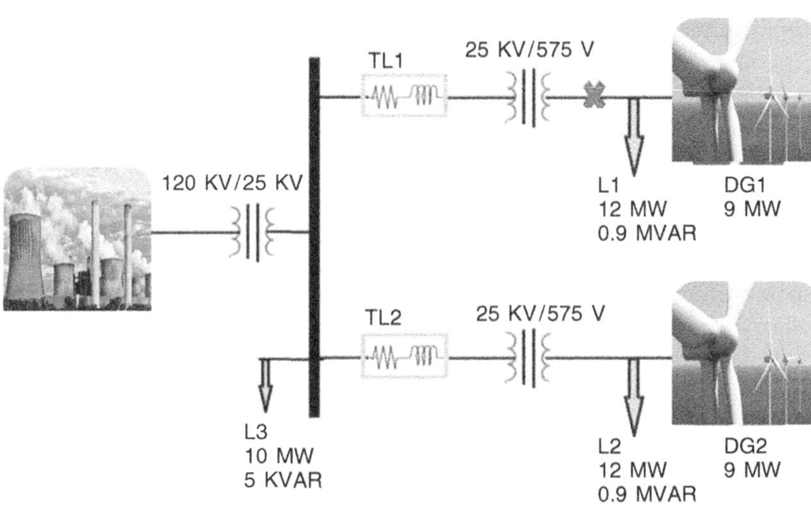

Figure 13.15 Off-grid and island microgrid. Source: Landscape Photography of Factory / Pexels / CC0.

3 (Solar Photovoltaic system).

The island power system is a reliable solution that ensures your home remains powered, even when the electrical utility is no longer supplying electricity. This system, powered by distributed generators like solar panels or wind turbines, operates independently of grid conditions, including during weather-related outages. It is a secure and dependable power source for your home, providing you with a sense of security and peace of mind.

13.5.5 Different Types of Islanding

A planned island is the ability to decide when to get disconnected from the grid; second, it is an unplanned or dynamic island. The microgrid decides the islanding operation automatically when it detects a problem with the grid. The dynamic island is the efficiency of the microgrid to adapt to unplanned island situations, such as the sudden loss of generation or a fault. Third, a seamless island is the microgrid's ability to form a planned or unplanned island without any interruption to loads, ensuring a smooth and uninterrupted transition, as shown in Figure 13.16.

13.5.6 Grid Following Inverters

Inverter-based technologies have traditionally been limited to a grid-following control strategy. However, the future holds promise with the emergence of alternative control strategies that could revolutionize grid integration. In a grid-following topology, inverters receive voltage and frequency measurements from the grid and active and reactive power setpoints from the power plant controller.

Grid-following inverters, by their nature, are limited in their ability to provide additional support to the grid strength. They act as a controllable current source synchronized with the grid, following any grid voltage and frequency changes. However, they are unable to enhance the grid strength, highlighting the need for alternative control strategies. The power plant controller in grid-following inverters plays a pivotal role in monitoring the grid conditions. It heavily relies on external setpoints from the supervisory system that monitor the power system network. This reliance underscores the current limitations and the need for improvement in control strategies, as shown in Figure 13.17. Reliance on a higher supervisory control system introduces communication delays; therefore, an immediate response that is inherently provided by synchronous generators cannot be mimicked by the grid-following inverter in terms of a fast-reacting generator, not to mention the lack of ability in voltage and frequency regulation in this control approach.

Figure 13.17 also shows that most inverter-based resources utilize a PLL (phase-lock loop) to synchronize with the grid and function as a constant current source. The figure illustrates a generic inverter schematic, highlighting the role of the PLL as a critical component in the overall control system that connects

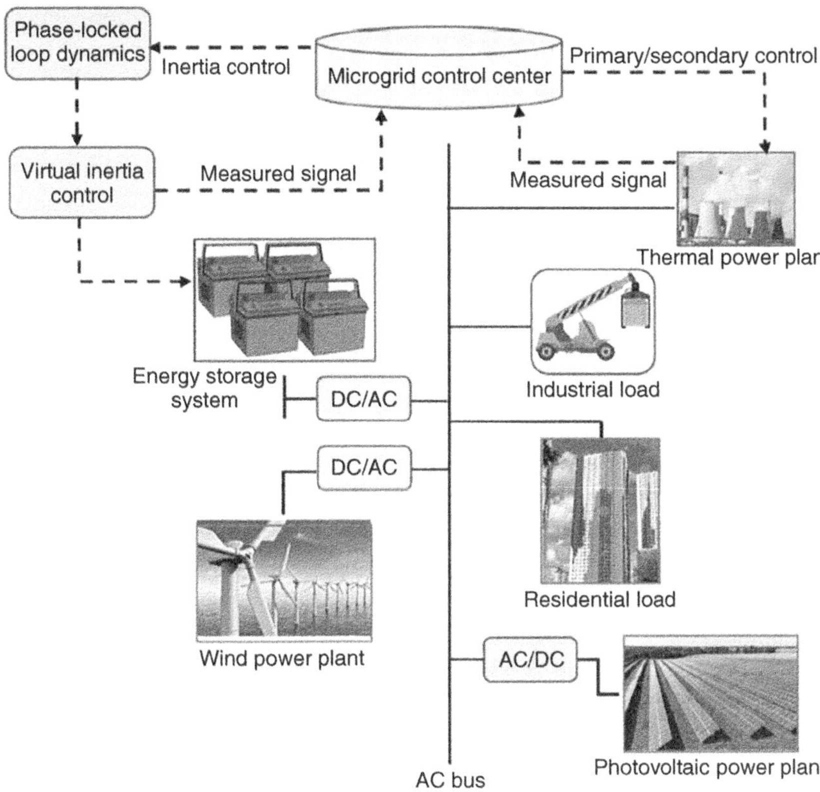

Figure 13.16 Different types of islands. Source: miklyxa / Adobe Stock Photos, Kingarion / Shutterstock, Piotr / Adobe Stock Photos.

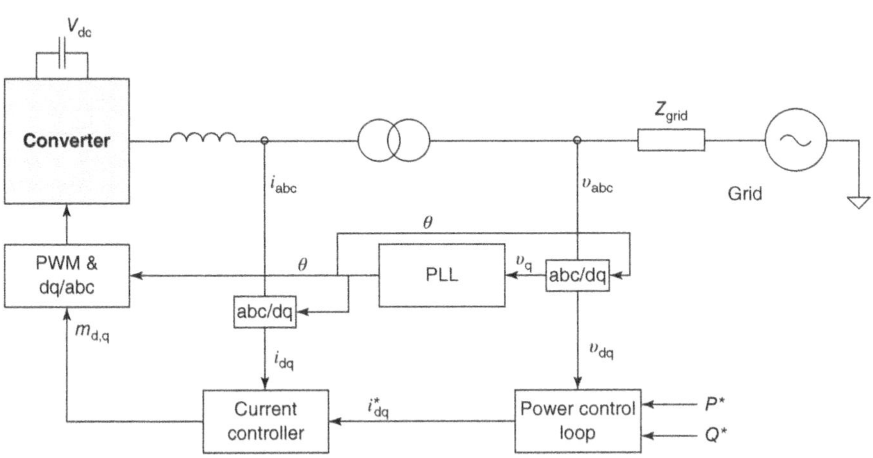

Figure 13.17 Grid following inverters.

the network and the inverter. The inverter's "grid-following" operation, which involves deriving the grid phase and frequency using a sophisticated closed-loop control system, underscores the complexity and sophistication of the system you are working with.

13.5.7 Grid-Forming Inverters

A grid-forming inverter receives voltage and frequency setpoints, which should be maintained in the grid. Simultaneously, its active and reactive powers injection or absorption are proportional to the grid voltage and frequency deviation from their nominal values. Therefore, a grid-following inverter operates as a controllable voltage source, as shown in Figure 13.18.

While grid-forming inverters are adept at forming and maintaining a grid voltage and frequency, they face a significant challenge in fully mimicking synchronous generator dynamic behavior when changes occur in the network. Sudden changes in the loading condition or the occurrence of contingency events can push the grid voltage and frequency to deviate from their nominal operating values, a struggle that these inverters must grapple with. The rapid deviation of grid voltage and frequency is a prevalent issue in weak networks. In these conditions, the lack of inertia means that a standard grid-forming inverter can only provide a partial increase in network robustness. This limitation underscores the urgent need for innovative solutions in the field of power systems and renewable energy.

Figure 13.18 shows that based on virtual synchronous generator (SG) concepts, SG behaviors in P and Q are emulated using a voltage source (amplitude and frequency) instead of a current source. Auxiliary energy storages are used to support the functions.

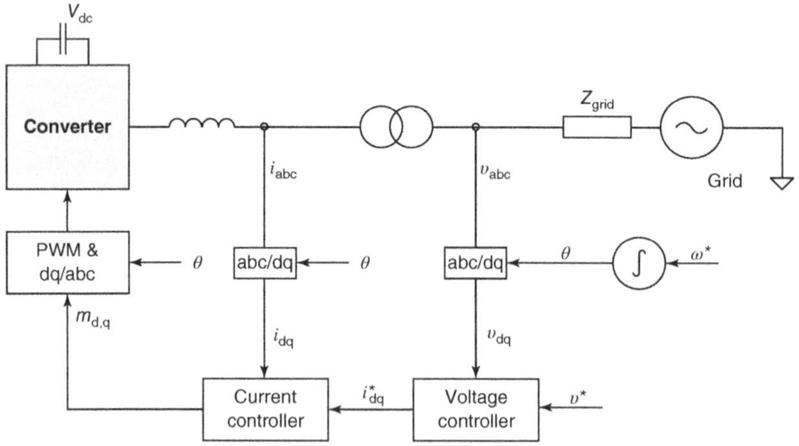

Figure 13.18 Grid-forming inverters.

13.6 Short-Circuit Study

Short-circuit studies identify worst-case fault currents under N-X conditions or other contingency scenarios from which an SCR value representing the strength of the POI is derived. The nominal PPC (power plant control) parameters, such as active and reactive power rise times, ramp rates, and step-change magnitudes, are not arbitrarily set. They are meticulously calculated according to the worst-case conditions, underscoring the gravity and importance of your work.

Adjusting power plant control parameters is not a one-time, static process. Instead, it is a dynamic and continuous regulation that takes into account SCR detection feedback from a power meter near the POI and bounded gain factor droop laws. This dynamic regulation is a key factor in ensuring system reliability. Regular and alarm-based monitoring and review of detected SCR trends are not just recommended but crucial. Your role in this process is instrumental in identifying the potential need for updated site studies, adjustments to nominal PPC parameters, and gain factors, thereby ensuring the system's reliability.

13.6.1 Advantages of Short-Circuit Ratio Study

- Short-circuit studies derive an SCR (short-circuit ratio) value representing "strength" of POI (point of interconnection).
- Power plant control parameters, active and reactive, are fixed according to the worst-case conditions.
- Power Plant control parameters remain unchanged.

13.6.2 Traditional Short-Circuit Ratio (SCR) Method

SCR is defined as the ratio between short-circuit apparent power (SCMVA) from a (line-line-line-ground) 3LG fault at the given location in the power system to the rating of the inverter-based resources at that location

13.7 Strong and Weak Microgrid

A strong grid, as indicated by a High SCR, is a robust system that can handle fluctuations in power injections or consumptions without significant changes in voltage magnitude and phase angle. This means that even with varying power demands, the grid can maintain a stable voltage, ensuring a reliable power supply.

A weak grid, as demonstrated by a low SCR, represents a system with a high sensitivity of voltage to changes in active and reactive power injection. This high sensitivity can have significant implications for power system stability and reliability.

Issues associated with a weak grid-fault ride-through are

1) HVRT – High voltage ride through
2) Unintentional islanding issues
3) Control instability

13.8 Types of Circuit Breakers in the AC and DC Microgrid

As defined in CSA C22.2 No. 5[4], a circuit breaker device is designed to open and close a circuit by nonautomatic means and to open the circuit automatically on a predetermined overcurrent without damage to itself when correctly applied within its rating. These versatile devices, which include molded-case circuit breakers, insulated-case circuit breakers, low-voltage power circuit breakers, medium-voltage circuit breakers, and high-voltage circuit breakers, offer a wide range of options to suit various electrical system requirements, as shown in Figure 13.19.

Figure 13.20 Cut-off Characteristics indicate the available peak short-circuit current 198 kA, available RMS short-circuit current 86 kA, Peak let-through current of the fuses 49 kA, and RMS let-Through current of fuse 21 kA. In addition, it also shows the fuse melt, arc, and clearing period. Figure 13.21 illustrates the ANSI relays, highlighting their critical role in power distribution system functionality. Meanwhile, Figure 13.22 provides a comprehensive overview of the operations and availability of both low-voltage and medium-voltage circuit breakers, underscoring their importance in ensuring reliable electrical performance.

Moulded case CB	Insulated case and low voltage CB	Medium voltage CB	High voltage CB

Figure 13.19 Different types of circuit breakers.

4 UL 489 – CSA C22. 2 No. 5 – ANCE NMX-J-266, Standard for Safety for Molded-Case Circuit Breakers, Molded-Case Switches and Circuit Breaker Enclosures. Testing.

Table 13.2 Circuit breakers rating.

Control parameters	Rating
Interrupting rating	25 kA, 31.5 kA, 63 kA
Rated interrupting time	None for MCCB or LVPCB
	(But typical values are 1.5 cycles for MCCB, 3 cycles for LVPCB). 5 cycles for MV CB.3 cycles optional for MV CB
Short-term rating	None for MCCB.
	65 kA for 30 cycles (LVPCB)
	40 kA for 2 s (MV CB)
Close and latch rating	None for MCCB or LVPCB
	104 kA for MV CB

Table 13.3 Fuses rating.

Control parameters	Rating
Voltage rating	600 V
Interrupting ratting	10 kA, 50 kA, 200 kA
Current limiting features	None for MCCB

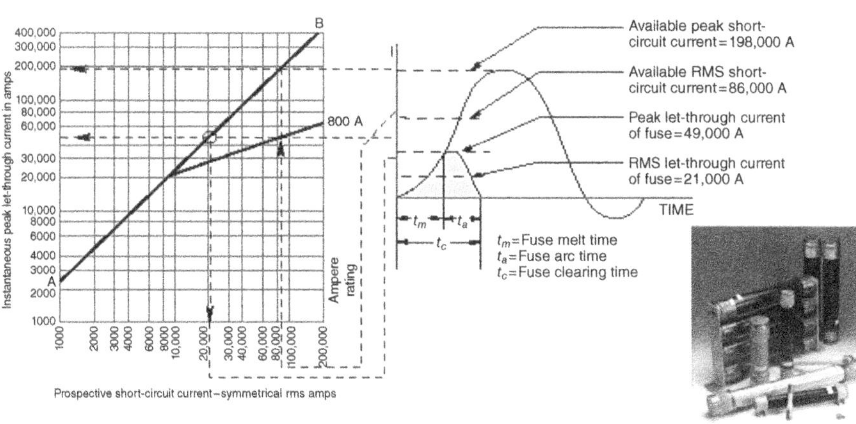

Figure 13.20 Cut-off characteristics.

Table 13.4 Switchgear, MCCs, and panelboards.

Control parameters	Rating
Rated voltage	600 V
Rated current	1200 A, 5000 A
Rated short-term withstand	65 kA, 250 MVA

Table 13.5 Current transformers.

Control parameters	Rating
Rated voltage	600 V
Current ratio	1200/5 A, 50/5 A
Continuous thermal	2 @ 30 °C
Current rating factor	1.5 @ 55 °C
ANSI metering class	0.3B1.8
	(0.3% accuracy with 1.8-ohm burden)
ANSI relay class	C200
	200 V on secondary at 20 × 5 A = 100 A
	(20 times secondary current)
	200 V/100 A = 2 ohm "C" means 10% accuracy with max 2-ohm burden

Figure 13.21 ANSI (American National Standards Institute) relays.

13.8.1 Time Current Curve

Time Current Curve – TCC: A time current curve (TCC) is a meticulously crafted graphical representation that plots the interrupting time of an overcurrent device based on a given current level. This curve, provided by the manufacturers

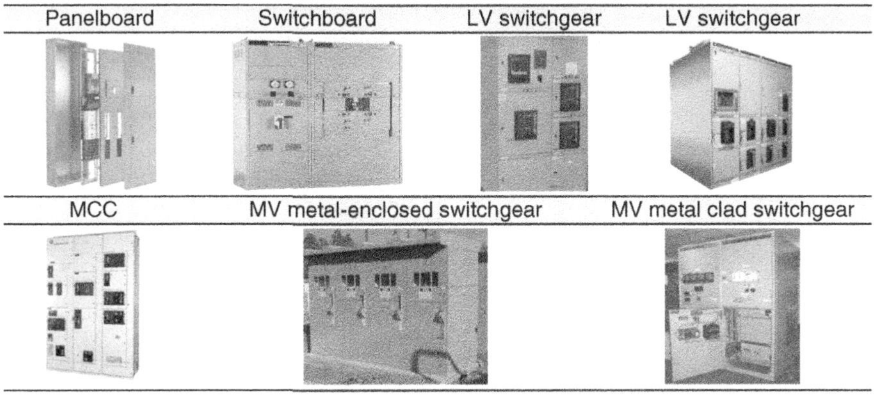

Figure 13.22 Grouping equipment.

of electrical overcurrent interrupting devices such as fuses and circuit breakers, shows the relationship between current (x-axis) and time (y-axis) on a log–log scale as shown in Figure 13.23. It is valid for a specific voltage and scale, ensuring its precision and reliability, as discussed in Section 3.4.

13.8.1.1 Typical Example of TCCs

Figure 13.24 indicates two sections of typical examples of TCCs. On the left-hand side, the CB with multiple functions and features is shown: LTPU, LTD, STPU, STD, and IPU will operate according to the CBs' defined settings. On the right-hand side, the Relay with 50/51 protection operations is shown. As professionals involved in power system protection and control, it is your responsibility to understand and implement these TCCs and ANSI[5] codes.

In protection, the ANSI code 50 refers to instantaneous operation, which means that the action should be carried out as fast as possible with no delay. This is particularly useful in situations where immediate action is required. Meanwhile, ANSI code 51 relates to delayed operation times, which is beneficial in scenarios where a delay is acceptable.

Exercise 13.8 Describe Terminology of LSI in details.

13.8.1.2 Terminology of LSI (Long Time, Short Time, and Instantaneous)

First, this is also known as the current rating long time; this setting has a practical range of 0.5–1000 seconds. This information equips you with the knowledge to provide a time-based delay before a circuit breaker trips for low-level current faults, giving you control over the system's operation. Second, this type of setting,

5 American National Standards Institute.

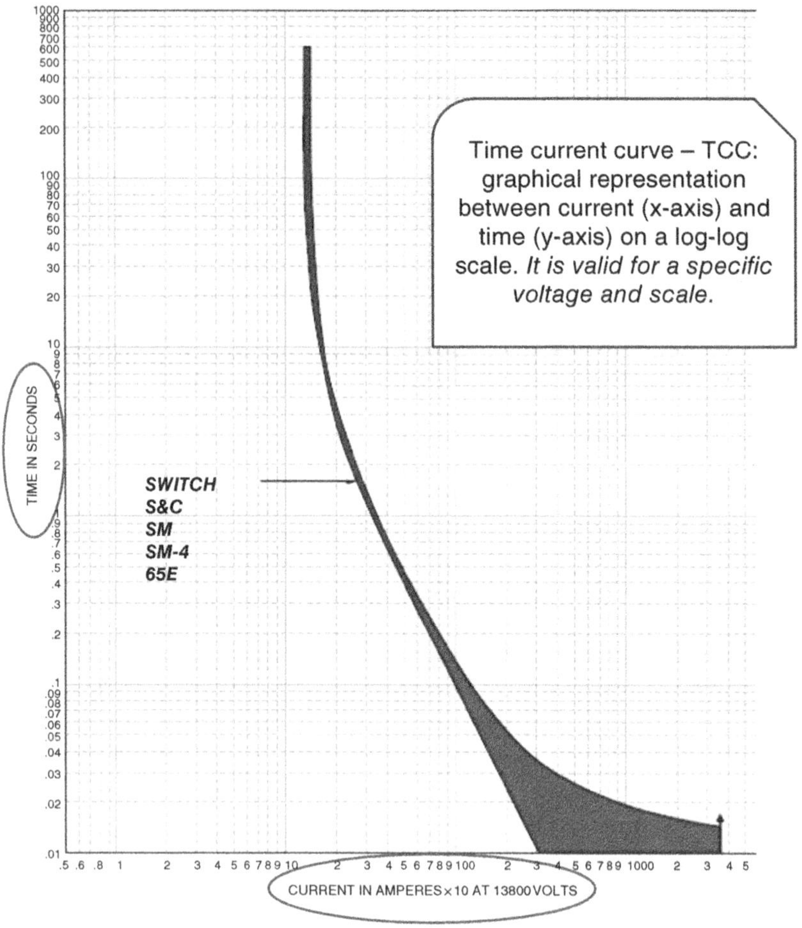

Figure 13.23 Time Current Curve - TCC.

calculated by multiplying the short-time pickup setting by the long-time pickup setting, plays a significant role in providing additional protection. It significantly enhances system safety by responding to short-duration overcurrent, such as a sudden surge in power demand or a short circuit, ensuring a secure electrical system. Third, this is also known as monostable relays; these relays operate without any intentional time delay. They change from a nonworking to a working position when the coil is energized and return to a nonworking state when the coil is no longer energized. This immediate response, with operating times usually less than three cycles, assures a quick and efficient operation, instilling confidence in the efficiency of the system. Examples of instantaneous relays include solenoid-type relays and balance beam relays as shown in Figure 13.25.

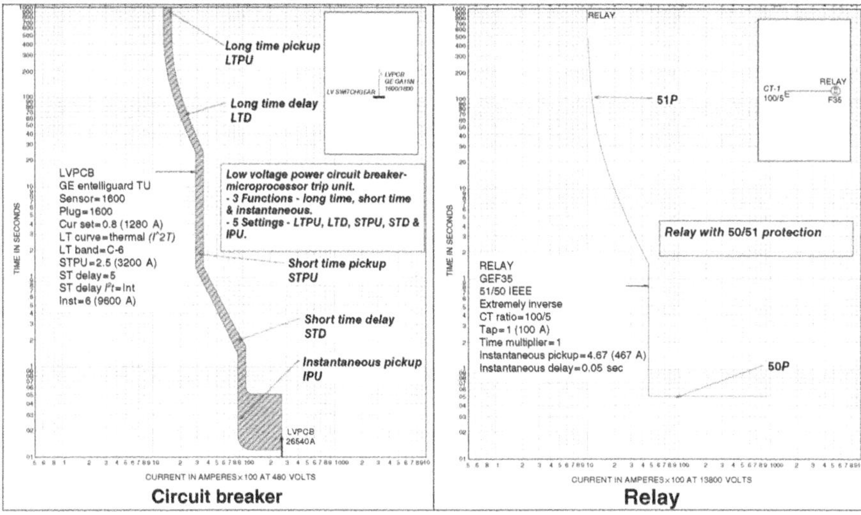

Figure 13.24 Typical example of TCC.

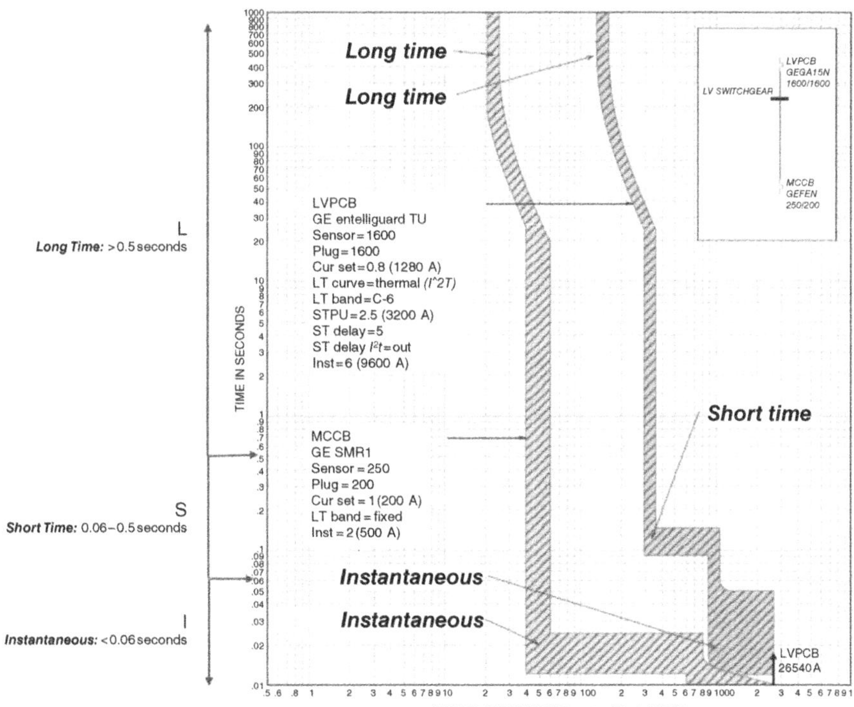

Figure 13.25 TCC – LSI (Long, short Instantaneous) Terminology.

Exercise 13.9 Describe the TCC time band features and functionalities.

13.8.1.3 Time Current Curve – TCC-Time Band

The LVPCB's long-time pickup value is equally significant. It denotes the current level at which the long-time delay function is activated, playing a crucial role in the circuit breaker's operation. The long-time delay feature, on the other hand, determines the duration the circuit breaker can withstand a sustained overcurrent (beyond the long-time pickup value) before triggering a trip signal.

The short-time delay (STD) is the second key feature of the CB. It is a circuit breaker setting that dictates how long a circuit breaker will carry a short circuit before initiating a trip signal. The appropriate STD setting, which can be 6, 12, 18, 24, or 30 cycles, depends on factors such as the type of equipment being protected and the expected short circuit current. LVPCBs function for LTD, STD, and ITD. The bottom of the band is pickup or commit time.

The Top of the band is precise time – when the current stops as shown in Figure 13.26.

Exercise 13.10 Describe Terminology of the MCCB and LVPCB.

13.8.1.4 Terminology of MCCB and LVPCB

MCCB (molded case circuit breaker): This is a key player in circuit control, housing two functions, long time and instantaneous. These functions, operating in

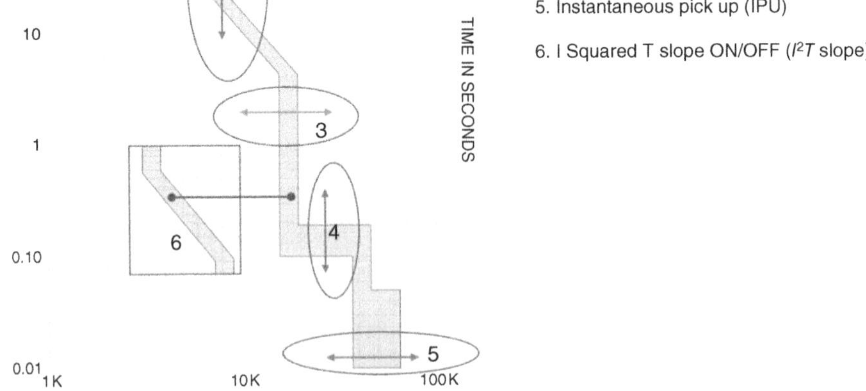

Figure 13.26 TCC – time band.

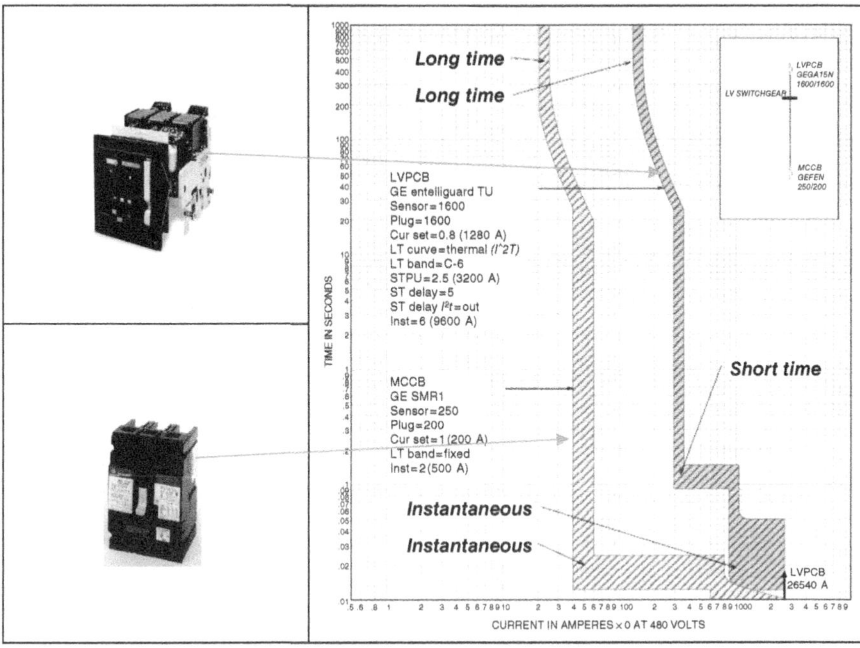

Figure 13.27 TCC – terminology MCCB and LVPBC.

the downstream of the circuit, are crucial in maintaining the circuit's integrity, underscoring its significance in circuit control. It is also known as the LI circuit breaker.

LVPCB (Low voltage power circuit breaker): This versatile equipment, known as low voltage power circuit breaker, contains four functions (3 + 1): time, which is the delay before the breaker trips, short term, which is a shorter delay, instantaneous time, which is the fastest response, and ground, also known as LSIG. Its adaptability allows it to be used upstream of the circuit, making it a valuable asset in circuit control, as shown in Figure 13.27.

13.8.1.5 TCC for MCCBs Normally 2 Settings

First, LTPU (long-time pickup), whereas essentially, the breaker's continuous ampere rating has been determined. Long-time protection is time-dependent when operating during contingency. The long-pickup (Ir) value (multiplied by the MCCB's ampere rating [In]) sets the maximum current level that the circuit breaker will carry continuously.

On a graph, instantaneous pickup is represented by a vertical line and a rectangular block on the right side at the bottom. Changing the setting moves the line and block along the x-axis, with higher currents moving it to the right and lower currents moving it to the left. This practical application of IPU is a circuit breaker

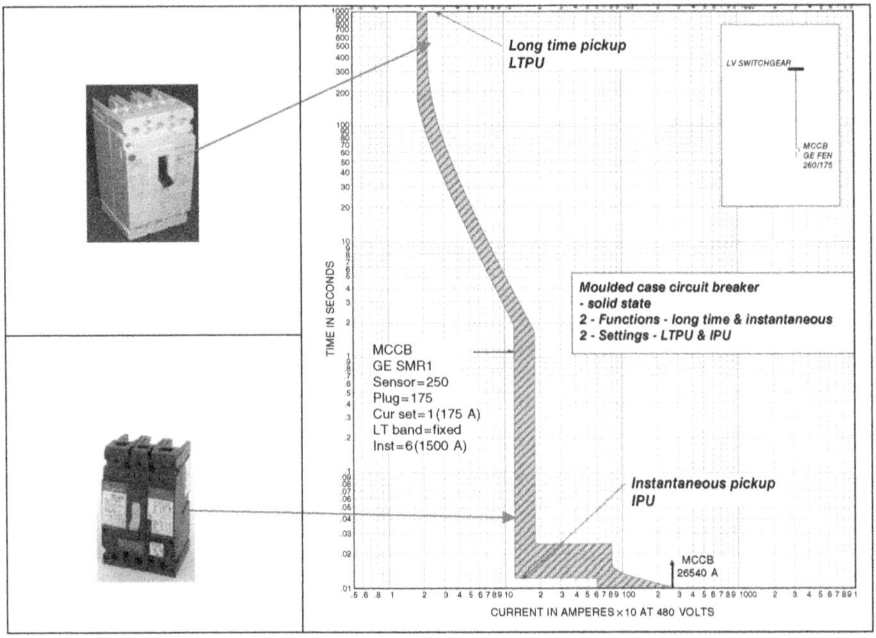

Figure 13.28 TCC – MCCB two settings.

setting that determines how strong a current surge must be to cause the breaker to trip immediately without any intentional time delay. It simulates the MCCBs as given below, as shown in Figure 13.28.

13.8.1.6 TCC for LVPCBs. Normally 5 Settings

Figure 13.29 indicates that the long-time pickup (LTPU) setting on a circuit breaker is of utmost importance as it determines the maximum current level the breaker can carry continuously. This setting also marks the initiation of the circuit breaker's long-time delay function.

Long-time delay (LTD) is a circuit breaker setting that allows a breaker to carry a sustained overcurrent before tripping. This setting enables weaker current surges, such as when a motor starts, to pass before the breaker trips. The longer the delay, the slower the breaker will trip, and the higher the current, the faster it will trip.

The short-time pickup (STPU) setting on a circuit breaker plays a crucial role in providing short-circuit protection. It determines the current level at which the breaker's short-time delay function begins to time, allowing the breaker to carry a certain amount of current for a short period of time before tripping. The STD is a circuit breaker setting that empowers the user with more control. It allows the breaker to carry a short-circuit current for an extended period before tripping. This setting gives the user more control over when a short-time pickup trip occurs

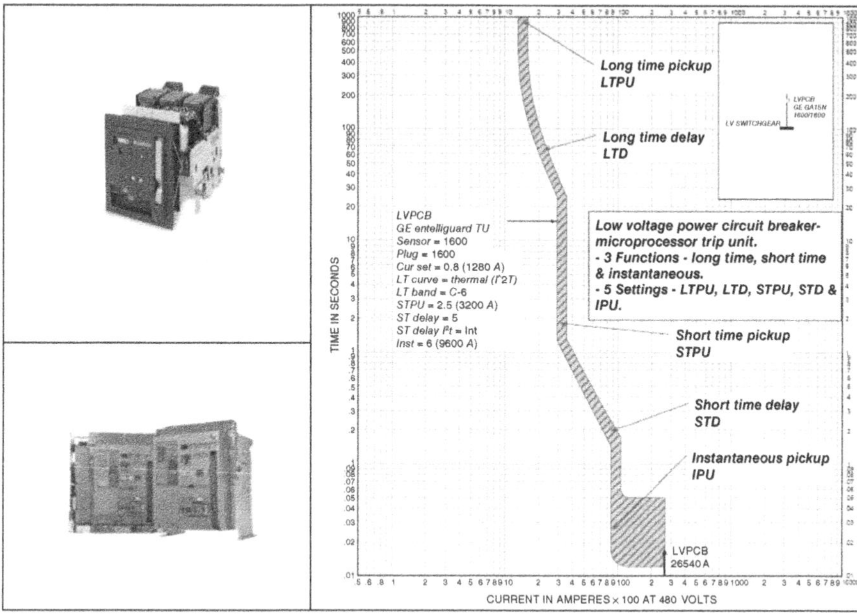

Figure 13.29 TCC- LVPCBs, normally 5 settings.

and can override other functions if needed, enhancing the user's ability to manage circuit breaker operations.

The instantaneous pickup (IPU) setting on a circuit breaker serves a vital function. It determines the strength of a current surge required to cause the breaker to trip immediately without any intentional time delay. This setting simulates the magnetic characteristics of a thermal-magnetic circuit breaker.

13.8.1.7 TCC for Fuse

Figure 13.30 indicates the fault current values; current limiting fuses are noncurrent limiting (left side figure). However, for high-fault current values,

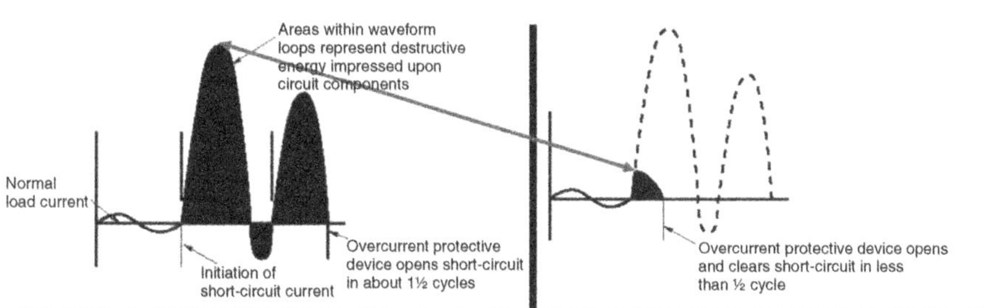

Figure 13.30 TCC – fuses operations.

current-limiting fuses are crucial in preventing system damage by isolating the faulted circuit before the fault current has reached its maximum value. This comprises current limitations, which include significantly limiting the fault current in the system and limiting the energy let-through downstream, underscoring the impact of your decisions on the system's integrity as discussed in Section 3.3.

Exercise 13.11 Describe fuses operational region and TCC.

Figure 13.31 indicates the fuses have two regions of operation, depending on the value of the current flowing through them, first the noncurrent limiting region,

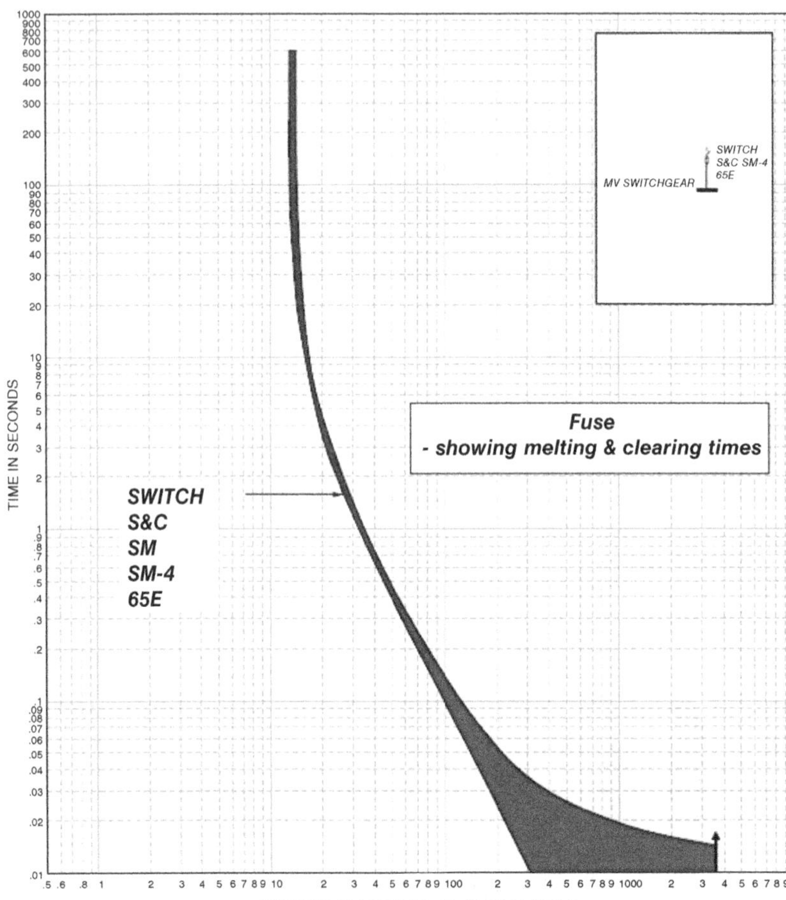

Figure 13.31 TCC – fuses operations.

and second current limiting region. When fuses operate in the noncurrent limiting region, their TCC looks like this.

Exercise 13.12 Describe the peak let-through of the fuses.

13.8.1.8 TCC – Peak Let-Through for Fuses

As electrical engineers, technicians, or professionals involved in circuit design and safety, your role in determining the peak let-through current is crucial. Start at the prospective short-circuit current, 86,000 A (point A), and proceed up to the 800-ampere fuse curve, as listed below, as shown in Figure 13.32.

1) Follow left to the instantaneous peak let-through current (point D).
2) Read the peak let-through current: 49,000 A.
3) If a fuse had not been used, the peak current would have been 198,000 A (point C).
4) The fuse has limited the peak current in the circuit from 198,000 A to 49,000 A.

Current limitation Curves-Bussmann® LOW-PEAK® time-delay Fuse KRP C800SP

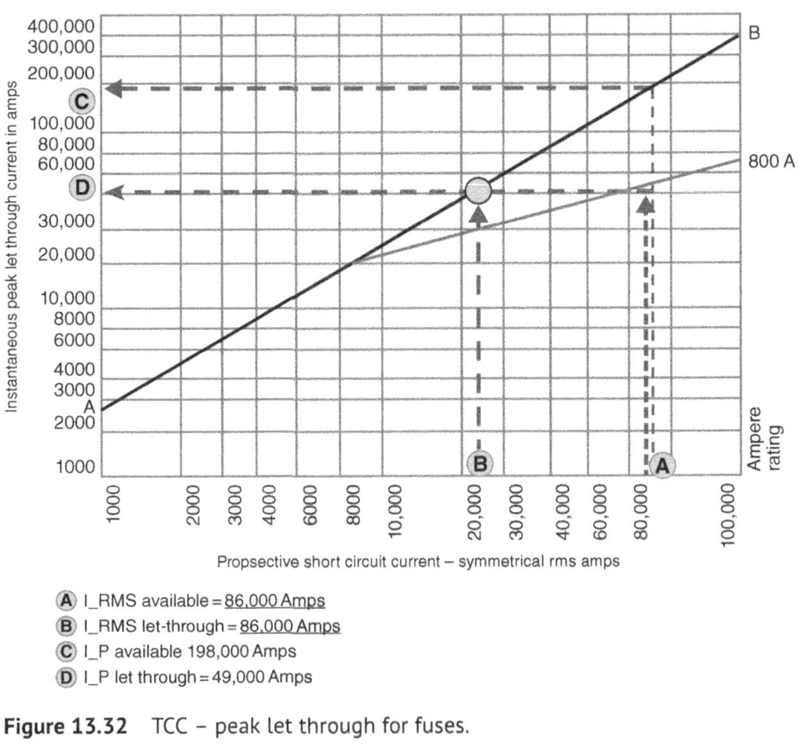

Figure 13.32 TCC – peak let through for fuses.

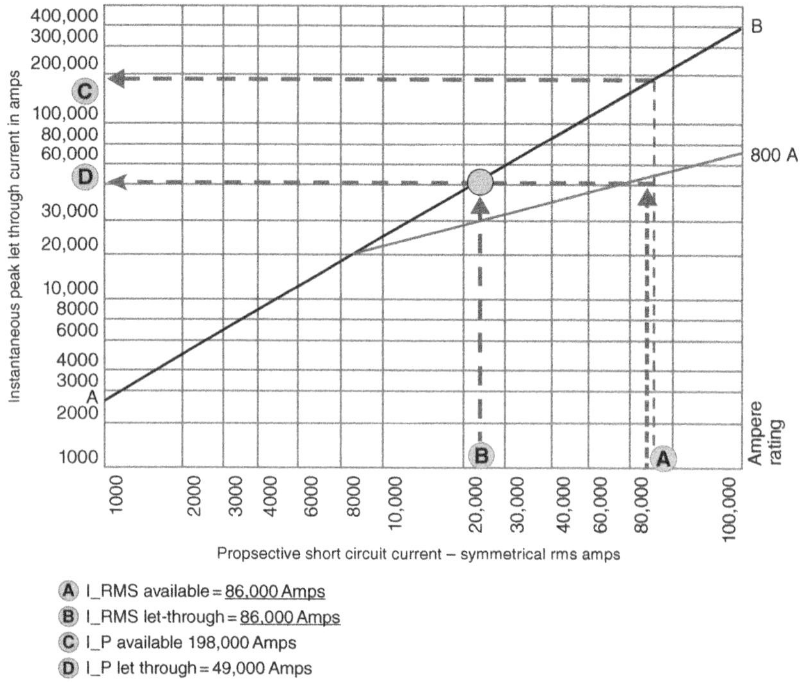

Figure 13.33 TCC – peak let through for fuses.

To determine the apparent RMS let-through current, start at the prospective short-circuit current, 86,000 A (point A), and proceed up to the crucial 800-ampere fuse curve. This step is of utmost importance as it guides us in understanding the let-through current, as shown in Figure 13.33.

- Follow left to the A–B line.
- Proceed down to the symmetrical RMS current scale (point B).
- Read the apparent RMS let-through current: 21,000 A.
- If a fuse had not been used, the RMS current would have been 86,000 A.

Current limitation Curves-Bussmann® LOW-PEAK® time-delay Fuse KRP C800SP

13.9 Low Voltage Trip Unit

As per CSA C22.2 No. 5[6], a trip unit is a user-replaceable part of a circuit breaker. The user, unless a rating plug is used, is responsible for activating the circuit

6 UL 489 – CSA C22. 2 No. 5 – ANCE NMX-J-266, Standard for Safety for Molded-Case Circuit Breakers, Molded-Case Switches and Circuit Breaker Enclosures. Testing.

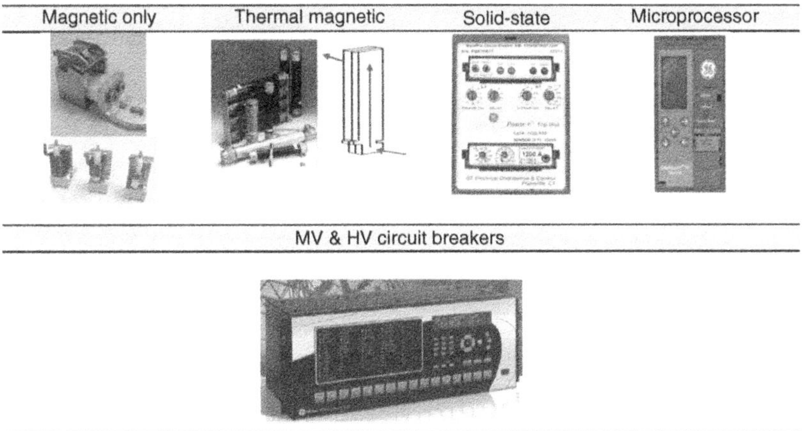

Figure 13.34 Different types LV (low voltage) trip units.

breaker release mechanism and, importantly, setting the rated current (I_n) of the circuit breaker. This active role of the user is a crucial part of the trip unit's functionality. See also Rating Plug. These trip units, including magnetic only, thermal-magnetic, solid-state, and microprocessor (digital), are versatile in their ability to trip low-voltage units, each showcasing unique functionalities and applications. They are also used in relays on MV and HV circuit breakers, as shown in Figure 13.34.

13.10 Symmetrical and Asymmetrical

Fault current is a combination of an alternating symmetrical current and a direct-current component. The sum of these two elements yields the total asymmetrical current. The alternating symmetrical current component is determined by system impedance, while the direct-current component, a significant determinant, is influenced by X/R or power factor at fault time. The asymmetry factor is determined by X/R & C. The formula for Total asymmetrical current is symmetrical current × asymmetry factor × 1.414. Figure 13.35 also indicates the asymmetrical current, generated voltage, and short-circuit current.

$$\text{Asymmetry factor} = \sqrt{1 + 2\epsilon^{\frac{-4\pi C}{X/R}}}$$

C = time in cycles and $\varepsilon = 2.71828$

When short-circuit studies are done, you can get 3 ratings: first, the momentary or ½ cycle, second, the interrupting or 5 cycle, and third, withstand or 30 cycle.

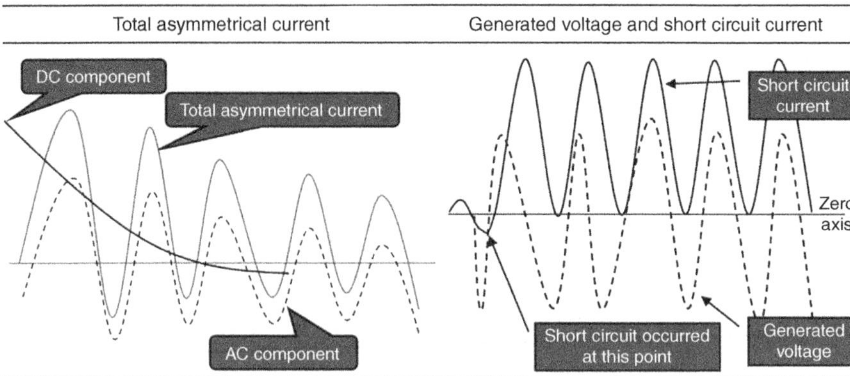

Figure 13.35 Total asymmetrical current.

For Example: Symmetrical short-circuit current = 23.55 kA and X/R = 26.523

- Momentary (½ cycle) = 1.606 × 23.55 × 1.414 = 53.48 kA.
- Interrupting (5 cycle) = 1.218 × 23.55 × 1.414 = 50.56 kA.
- Withstand (30 cycle) = 1.046 × 23.55 × 1.414 = 34.83 kA.

Based on these calculations, the following operations can be observed: first, the momentary (½ cycle) = close and latch; second, the interrupting (5 cycles) = interrupting; and third, the withstand (30 cycles) = short-time withstand.

$$\text{Asymmetry factor} = \sqrt{1 + 2\epsilon^{\frac{-4\pi C}{X/R}}} \tag{13.1}$$

C = time in cycles.
$\varepsilon = 2.71828$
$X/R = 26.523$

13.11 CB Selectivity or Coordination for Grid Operations

Figure 13.36 indicates the total asymmetrical schematic. Isolating a faulted circuit from the remainder of the electrical system is a key strategy to eliminate unnecessary power outages. This is achieved by the precise operation of the overcurrent protective device closest to the fault. Figure 13.36 presents the process flow schematics, which feature upstream and downstream circuit breakers on both sides of the schematic. It also demonstrates the direction of current flow from CB-M to CB-1 side, from upstream to downstream.

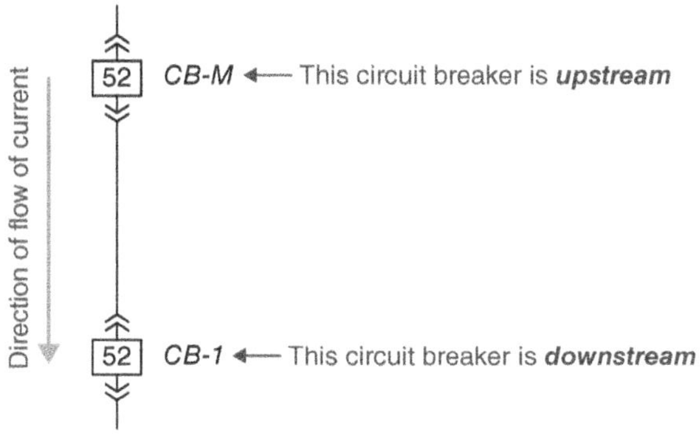

Figure 13.36 Total asymmetrical current.

The system on the right is a ***nonselective*** operation. If a fault has occurred, as shown in the figure, to overcome the situation, all cascaded circuit breakers open, causing a complete loss of power to the whole system. The system on the left is ***selective*** in its operation. This means that only the circuit breaker closest to the fault opens and clears the fault, while all upstream circuit breakers remain closed. It is important to note that the system is ***selective*** as shown in Figure 13.37.

Exercise 13.13 Describe selectivity or coordination of the CBs.

Figure 13.38 *indicates the operations coordinates the thermal and electronic CB. In the second scenario, the CB did not coordinate with fuse. To coordinate or be selective, you must see "daylight" between the curves.*

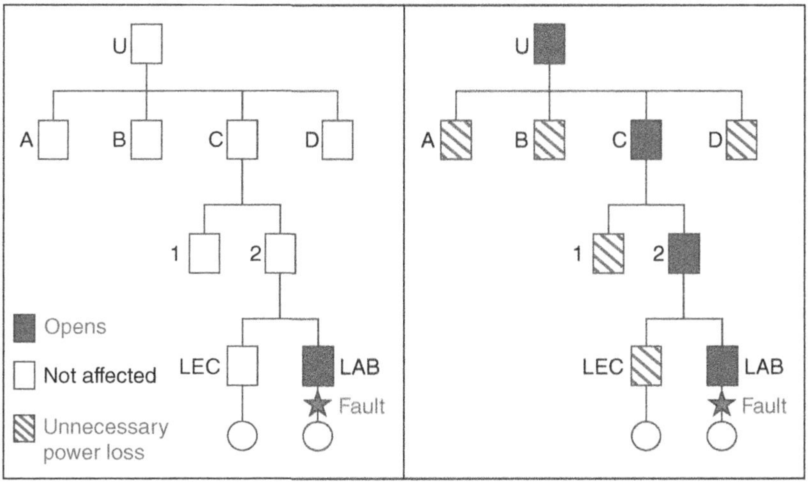

Figure 13.37 Selective and nonselective power control schematics.

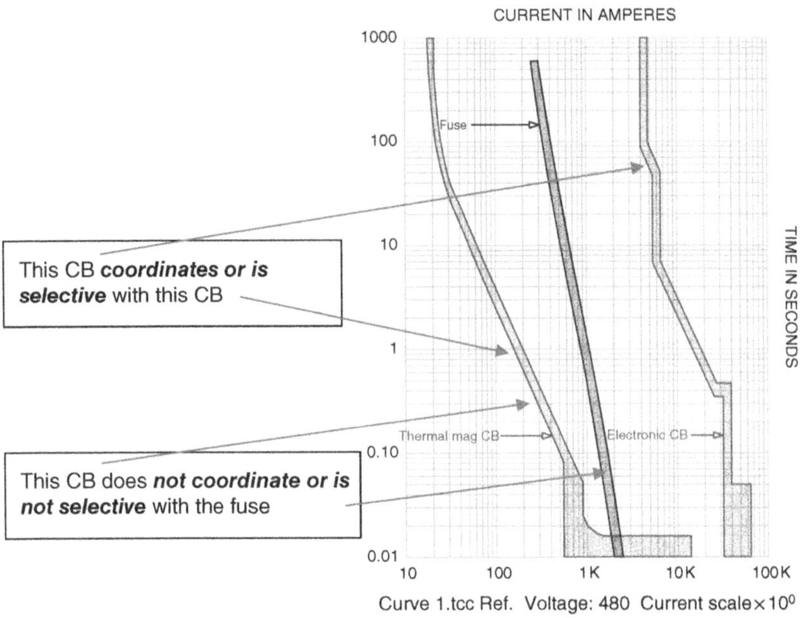

Figure 13.38 Selective or coordinate CB and fuses.

Exercise 13.14 Describe concept of upstream and downstream CBs to understand their operations and selectivity.

Figure 13.39 indicates the CBs. it is important to grasp the concept of upstream and downstream CBs to understand their operations and selectivity. When the power flow is from CB-1 to CB-2, CB-1 is **upstream**, and CB-2 is **downstream**. This arrangement does not imply selectivity or coordination.

These CBs are selective only in the large time region, and their selectivity can be enhanced by increasing the short-time pickup **(STPU)**, increasing the short-time delay **(STD)**, or enabling $\boldsymbol{I^2t}$ **IN.**

Figure 13.39 indicates the CBs. It is important to grasp the concept of upstream and downstream CBs to understand their operations and selectivity. When the power flow is from CB-1 to CB-2, CB-1 is **upstream**, and CB-2 is **downstream**. This arrangement does not imply selectivity or coordination. These CBs are selective only in the large time region, and their selectivity can be enhanced by increasing the short-time Pickup **(STPU)**, increasing the short-time delay **(STD)**, or enabling $\boldsymbol{I^2t}$ **IN.**

Q: *What was done to make them selective?*

A: ***I^2t IN was enabled*** *for CB-1*

Q: What can be done to make them selective in the instantaneous (I) region?

A: Disable Instantaneous (I) on CB-1 or use Zone Selective Interlock (ZSI)

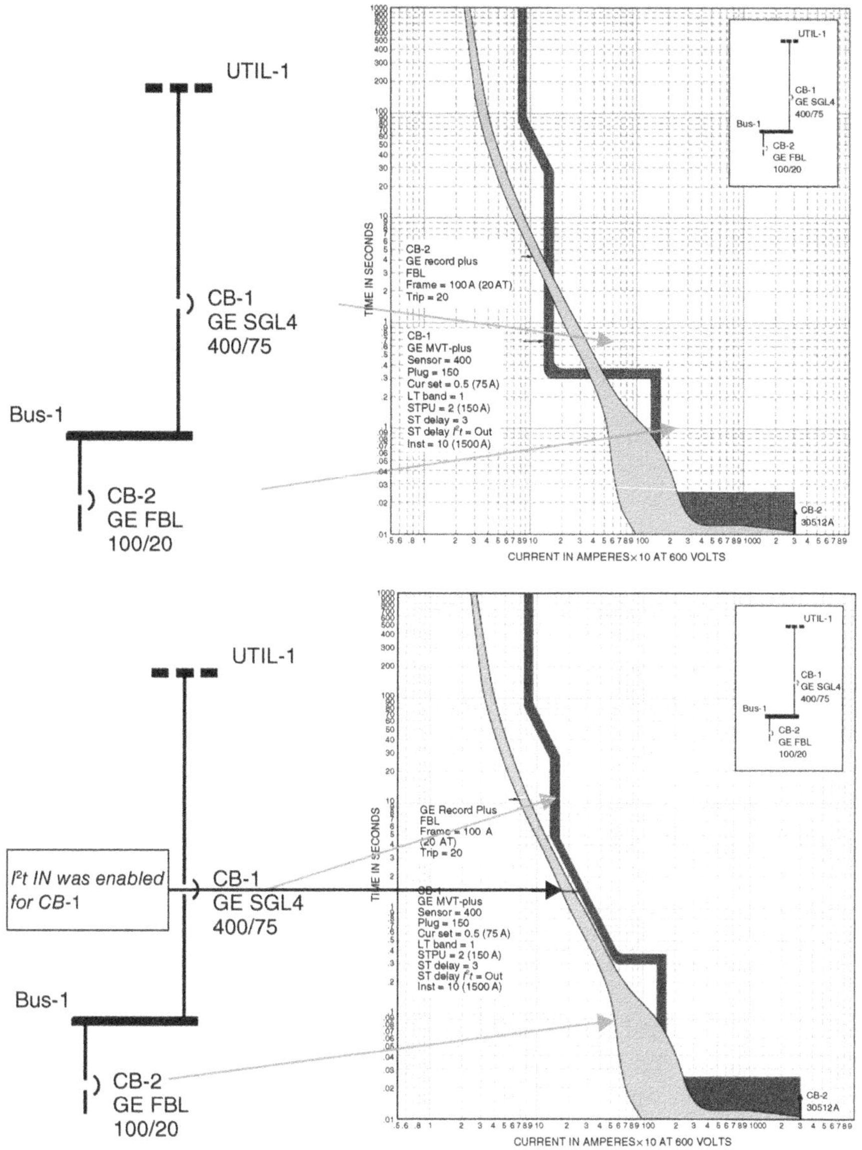

Figure 13.39 Selective or coordinate CBs.

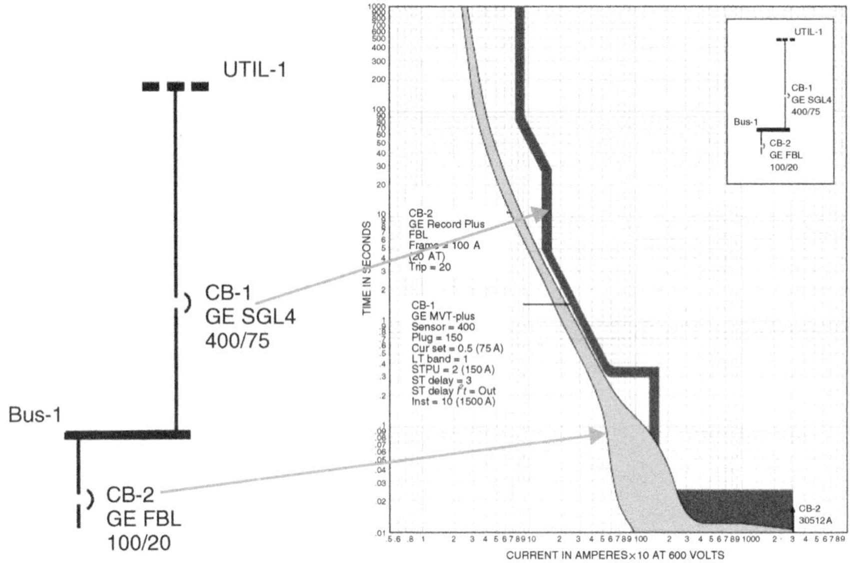

13.12 Microgrid Adaptive and Predictive Control Techniques

Adaptive Microgrid Control: Adaptive control systems can adjust their parameters in real-time based on changing grid conditions. This is particularly useful in smart grids, where the variability of renewable energy sources like wind and solar can cause rapid changes in generation. These systems use real-time data and machine learning algorithms to continuously optimize control actions, ensuring stability and efficiency despite the dynamic nature of the grid.

Predictive Microgrid Control: Predictive control involves forecasting future grid conditions and taking proactive control actions to prevent issues before they occur. This approach is highly reliant on accurate forecasting models for load, generation, and grid conditions. Model predictive control (MPC) is a popular method, where a model of the grid is used to predict future states, and control actions are optimized over a prediction horizon. MPC can be used for various applications, including voltage control, frequency regulation, and congestion management.

13.12.1 Control of Distributed Energy Resources (DERs)

13.12.1.1 Integration of DERs

As DERs like rooftop solar panels, wind turbines, and battery storage become more prevalent, controlling these resources is essential for maintaining grid stability. DERs can be both a challenge and an asset, depending on how they are managed. Virtual power plants (VPPs) aggregate multiple DERs to act as a single entity, providing control over a diverse set of resources and enabling participation in energy markets.

13.12.1.2 Inverter-Based Control

Many DERs are connected to the grid through power inverters, which convert DC output from renewable sources into AC for grid compatibility. Advanced inverter control strategies, such as droop control, grid-forming, and grid-following inverters, are used to manage power quality and support grid stability. These inverters can provide services, such as voltage support, frequency regulation, and reactive power compensation, making them crucial components in smart grid control strategies.

13.13 Artificial Intelligence (AI) and Machine Learning (ML) in Grid Control

13.13.1 AI and ML

AI and ML techniques are increasingly being integrated into smart grid control systems to handle the complexity and variability of modern power grids. AI algorithms can optimize control strategies by learning from historical data, predicting future states, and making real-time adjustments. Applications include fault detection, predictive maintenance, load forecasting, and energy management. Reinforcement learning, a subset of AI, is particularly promising for real-time grid control, where the system learns optimal actions through continuous interaction with the environment.

13.13.2 Data Analytics

Advanced data analytics tools process vast amounts of data generated by smart meters, sensors, and other IoT devices. This data-driven approach allows for more accurate decision-making, predictive maintenance, and optimization of grid operations. Big data analytics help in understanding patterns, identifying potential issues before they escalate, and enhancing overall grid performance.

13.14 Coordination of Multiple Control Levels

13.14.1 Hierarchical Control Structures

Smart and microgrids often employ hierarchical control structures that coordinate actions across different levels, from individual DERs to the entire grid. This structure allows for efficient management of local and global objectives. Primary control deals with immediate responses to disturbances, secondary control handles coordinated adjustments, and tertiary control focuses on economic optimization and long-term planning.

13.14.2 Interoperability and Standardization

Effective control strategies require seamless communication and interoperability between various components and systems within the grid. Standardization of protocols and interfaces is crucial for achieving this. Interoperability also ensures that new technologies and resources can be integrated into the grid without causing disruptions or requiring significant changes to existing infrastructure.

13.15 Regulatory and Standardization Aspects

The successful implementation and operation of smart grids depend significantly on a well-defined regulatory framework and adherence to industry standards. These aspects are crucial for ensuring interoperability, security, reliability, and the smooth integration of emerging technologies within the power grid.

13.15.1 Overview of Regulatory Frameworks

13.15.1.1 National and Regional Regulations

Smart grid deployment is heavily influenced by national and regional regulations that dictate how electricity is generated, distributed, and consumed. These regulations are often shaped by government policies aimed at promoting renewable energy, enhancing grid reliability, and protecting consumer interests. For example, the European Union has established the Clean Energy for All Europeans package, which includes directives on renewable energy, energy efficiency, and electricity market design. Similarly, in the United States, the Federal Energy Regulatory

Commission (FERC) plays a pivotal role in regulating the interstate transmission of electricity and setting guidelines for smart grid implementation. Regulations may also mandate the adoption of specific technologies, such as smart meters or energy storage systems, and set targets for reducing greenhouse gas emissions and increasing the share of renewable energy.

13.15.1.2 Market Structures and Economic Regulation

The shift toward smart grids has led to the evolution of electricity markets, where market structures and economic regulation are crucial for ensuring fair competition and protecting consumers. In many regions, regulators oversee the operation of wholesale electricity markets, where electricity is bought and sold between generators and distributors. The introduction of smart grids allows for more dynamic pricing, where electricity prices can vary in real-time based on supply and demand conditions. Regulators also ensure that grid operators and utilities provide nondiscriminatory access to the grid for all market participants, including small-scale DERs. This is essential for fostering innovation and competition in the energy market.

13.16 Self-Assessment Questions

Self-Assessment Question 13.1

A) Describe the classification of microgrid, AC microgrid, and control operations.
B) Describe the grid following and forming inverters.

Self-Assessment Question 13.2

Describe the following:

- Advantages of short-circuit ratio study
- Strong and weak microgrid
- Traditional short-circuit ratio study

Self-Assessment Question 13.3

Describe the following:

- Types of circuit breakers in the AC and DC microgrid
- Discuss time-current curve
- Terminology of LSI (long time, short time, and instantaneous)

Bibliography

Jamil Abdo & Tariq Masood "New Technique to Develop Intelligent, Highly Reliable & Efficient Unmanned Power Generation Modular" published in 22nd International Conference on Renewable Energies and Power Quality (ICREPQ'24) Bilbao (Spain), 26–28 June 2024. https://www.icrepq.com/icrepq24/392-24-abdo.pdf.

Jamil Abdo Tariq Masood "Operational Reliability – Production Best Practice Pragmatic Case Studies & Improvement" published in Proceedings of the 9th North American Conference on Industrial Engineering and Operations Management Washington D.C., USA, 4–6 June 2024.

Tariq Masood Jamil Abdo "Maintenance Strategy Production Best Practices – Pragmatic Case Studies and Operations Improvement" published in Proceedings of the 9th North American Conference on Industrial Engineering and Operations Management Washington D.C., USA, 4–6 June 2024.

Jamil Abdo Tariq Masood "Techno Economics Analysis to Reshape Global Clean and Green Energy" published in 22nd International Conference on Renewable Energies and Power Quality (ICREPQ'24) Bilbao (Spain), 26–28 June 2024. https://www.icrepq.com/icrepq24/415-24-abdo.pdf.

M. Saeed Ullah, Tariq Masood, Umair Zahid, H Abuzar Ahsan "Electrochemical Oxidation of Dissolved Organic Dye in Aqueous Solution" published in IMASCON-2019.

Nasser Alemadi, Tariq Masood, R Aggarwal, Suhail A Qureshi, D Kothari "Novel Selection Approach of STATCOM Device Tailored and Optimized Engineering Configuration to Implement in Between Qatar and Bahrain" published in International Journal of Automation and Power Engineering 2014. 10.14355/ijape.2014.0302.01.

Tariq Masood & Jamil Abdo "New Technique to Enhance Hybrid Renewable Energy Resources' Operational Flexibility and Reliability" published in 22nd International Conference on Renewable Energies and Power Quality (ICREPQ'24) Bilbao (Spain), 26–28 June 2024. https://www.icrepq.com/icrepq24/396-24-masood.pdf.

Tariq Masood and Jamil Abdo "Review: Uninterrupted Liquid Hydrogen (E-Methanol) Production Framework to Deliver Clean & Green Energy" published in 22nd International Conference on Renewable Energies and Power Quality (ICREPQ'24) Bilbao (Spain), 26–28 June 2024. https://www.icrepq.com/icrepq24/414-24-masood.pdf.

Tariq Masood, Abdel Aty Edris, Suhail Aftab Qureshi, Muhammad Tajammal, Murtaza Hashmi, DP Kothari, Samer Kareem "SSSC Tailored to Optimization Performance in Between United Arab Emirates and Oman on the GCC Power Grid" published in the 13th IET International Conference on AC and DC Power Transmission Conference on 14–16 February 2017. http://ieeexplore.ieee.org/document/7934934.

Tariq Masood, Abdel Aty Edris, Suhail Aftab Qureshi, Muhammad Tajammal, Murtaza Hashmi, DP Kothari, Samer Kareem "Smart Grid Operations and Control Challenge by Implementing STATCOM Tailored to Optimization Performance at the GCC Power Grid" published in the 13th IET International Conference on AC and DC Power Transmission Conference on 14–16 February 2017. http://ieeexplore.ieee.org/document/7934988.

Tariq Masood, Muhammad Tajammal, Samer Kareem, Suhail Aftab Qureshi, Abdul Jabbar Khan, DP Kothari, Nair Jameel "Unified Power Flow Controller Modeling and Analysis Technique on the GCC Power Grid Operations and Control" published in 2018 IEEE International Energy Conference, Limassol Cyprus, 7–9 June 2018. https://ieeexplore.ieee.org/document/8398834.

Tariq Masood, Muhammad Tajammal, Samer Kareem, Suhail Aftab Qureshi, Murtaza Hashmi, Nasir Jameel "Unified Power Flow Controller (UPFC) Modeling and Analysis Technique by using Matlab and EMTP" published in International Conference on Renewable Energies and Power Quality (ICREQP'18) Salamanca (Spain) 21–23 March 2018. http://www.icrepq.com/icrepq18/250-18-masood.pdf.

Tariq Masood, Muhammad Tajammal, Samer Kareem, Suhail Aftab Qureshi, Murtaza Hashmi, Naser Jameel "STATCOM Controller Performance Enhancement and Optimization Technique" published in International Conference on Renewable Energies and Power Quality (ICREQP'18) Salamanca (Spain) 21–23 March 2018. http://www.icrepq.com/icrepq18/249-18-masood.pdf.

Tariq Masood, Muhammad Tajammal, Samer Kareem, Suhail Aftab Qureshi, Murtaza Hashmi, DP Kothari "Smart Grid Operational Functions and Control Challenges by Implementing SSSC Tailored to Optimize Performance in Between Qatar and KSA on the GCC Electrical-Power Grid" published in the ASTESJ Journal Advances in Science, Technology and Engineering Systems Journal Vol. 2, No. 1, 2017. https://astesj.com/special-issue-4.

Tariq Masood, Raj K Aggarwal, Suhail Aftab Qureshi, RAJ Khan "STATCOM Model Against SVC Control Model Performance Analyses Technique by MATLAB" published in International Conference on Renewable Energies and Power Quality (ICREPQ'10) 2010. https://icrepq.com/icrepq'10/655-Masood.pdf.

Tariq Masood, N Al-Emadi, SA Qureshi, YY Almulla, AJ Khan, SK Shah "FACTS Control Devices (Statcom, SSSC and UPFC) Re-Configuration Techniques By PSIM/MATLAB" published in 2007 International Conference on Electrical Engineering 2007. http://ieeexplore.ieee.org/abstract/document/4287295.

Tariq Masood, Raj K Aggarwal, SA Qureshi, RAJ Khan "STATCOM and SVC Control Operations and Optimization During Network Fault Conditions" published in 2010 IEEE International Symposium on Industrial Electronics 2010. https://ieeexplore.ieee.org/document/5636868.

Tariq Masood, RK Aggarwal, SA Qureshi, RAJ Khan "Statcom Control Reconfiguration Technique for Steady State and Dynamic Performance Optimization During Network Fault Conditions" published in International Conference on Renewable Energies and Power Quality (ICREPQ'10) Granada (Spain) 2010. https://www.icrepq.com/icrepq'10/656-Masood.pdf.

Tariq Masood, Samer Karim Shah, Muhammad Tajammal, "Unified Power Flow Controller Modeling and Analysis Technique on the GCC Power Grid" published in 2018 IEEE International Energy Conference (ENERGYCON) 2018. https://ieeexplore.ieee.org/abstract/document/8398834.

Tariq Masood, Suhail A Qureshi, RK Aggarwal, DP Kothari "Multiple FACTS Devices Integration Through Distributed Control System and Its Application on the GCC Interconnection" published in 2013 7th IEEE GCC Conference and Exhibition (GCC) 2013. https://ieeexplore.ieee.org/abstract/document/6705750

14

Artificial Intelligence and Machine Learning's Industrial Application

14.1 Objectives

At the end of this section, you will have had an introduction to the general principles of the artificial intelligence (AI) components used for industrial applications:

- Thinking Probabilistic
- AI Machine Power and Disruption
- Designing Reliable Systems
- AI system Reliable Technology
- AI Machine Prediction Rules
- Machine Learning in Microgrid
- Tutorial, How to Simulate AI Models: Probability
- Challenges and Future Direction

14.2 Thinking Probabilistic

AIs introduced probabilistic thinking into the system. When we investigate a car accident, industrial incident, or accident, as far as protection is concerned. We ask whether the Panelboard Operator/protection or control engineers saw the incident/accident or catastrophic failure. We expect a yes or no answer. We are less accustomed to dealing with "sort of" or "bit." Yet, that is precisely the response the AI provides. It saw something it thought was a human approaching to take action, say 0.01% likelihood. When introducing AI into a system, we often transform that system from deterministic to probabilistic. Sometimes, the existing system is well-designed to accommodate probabilistic input. Other times, this creates an opportunity for greatly enhanced productivity via system redesign.

Protection of Electrical Power Distribution Systems: Smart grid, Microgrid, AI, and Cyber Security,
First Edition. Tariq Masood, Jamil Abdo, and Atif Iqbal.

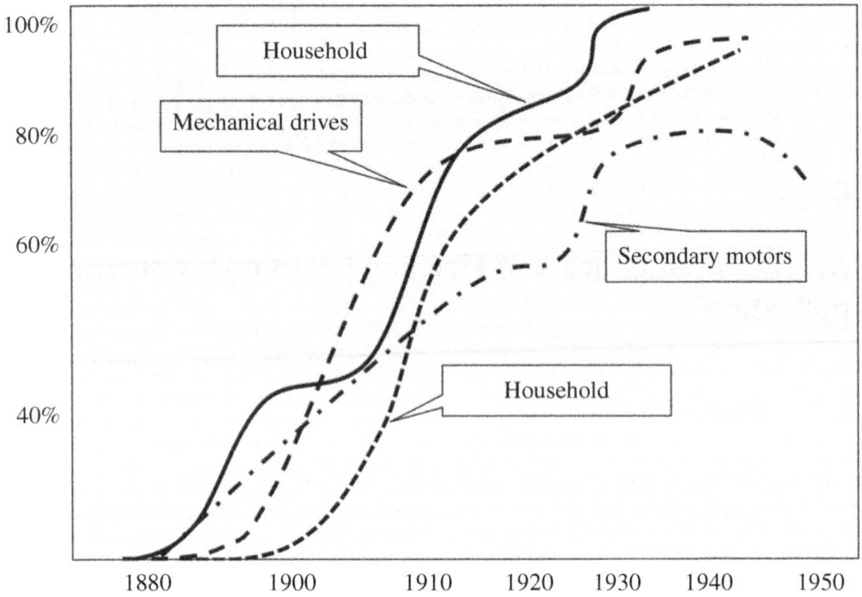

Figure 14.1 Modern productivity paradox.

To fully harness the power of AI, many industrial applications will require a shift to newly designed system-level solutions. These solutions, incorporating prediction, judgment, and a regulatory function, are not just beneficial but crucial for ensuring seamless operation during the transition from deterministic to probabilistic systems. They are a necessity in the age of AI integration. Prediction and judgment are the bedrock of decision-making. In a decision tree, prediction calculates the likelihood of each branch's occurrence, while judgment assesses the payoffs linked to the outcomes. These two elements, prediction and judgment, are unique inputs from the decision-maker's perspective as discussed in Figure 14.1.

Despite the challenges of doing system-level innovations with AI, there is a good reason to initiate sooner rather than later. The more it learns, the better it gets in terms of prediction and accuracy. The better it gets, the more influential the new system will be. Feedback control loops can have significant implications for system design. For an AI to learn, it must have access to the outcome data, which helps to design feedback as frequently as possible.

While machines cannot make decisions, AI can simulate decision-making. The AI generates a judgment to execute an action/decision, but it is important to remember that AI predictions are imperfect. To mitigate the risk of error, we rely on two lines of defense. First, we work through contingency before deploying AI, deciding what action the machine should take for each contingency. Second, after deployment, we rely on humans to step in when AI is unable to predict with high confidence or when the AI predicts a scenario for which we have not codified the judgment. This "human in the loop" approach underscores the crucial role of humans in AI decision-making.

Exercise 14.1 Describe the transformative power of AI in industrial applications and discover practical strategies for seamlessly transitioning to innovative system-level solutions.

14.3 AI Machine Power and Disruption

AI productions are imperfect; to mitigate the risk of being wrong, we embark on two lines of attack. First, before deploying AI, we work through contingencies and arrive at a conclusion as to what action the machine should take for each contingency. Second, after deploying the AI, we rely on humans to step in when the AI is unable to predict with high confidence or the AI predicts a scenario for which we have not codified the judgment Although machines do not have the power, they cannot create power through scale and relocate power by shifting whose judgment is used where and when for decision-making. Systems dedicated to AI can decouple the judgment from the decisions, such that it can be provided at a different time and place if judgment shifts from individually deployed by people for each decision to instate codify into systems, then this can lead to a number:

- Scaling and consequently a shift in power
- Changing who makes the decisions and consequently power in shift

Machines can decide when we can codify judgment. The AI generates predictions, and the machine draws upon human judgment to execute an action. Despite the challenges of doing system-level innovation with AI, there is a good reason to initiate it sooner rather than later. AI confers an advantage on first movers because AI learns quickly once it is deployed; the sooner it begins to learn and the more it learns, the better it gets, and a more effective system is obtained, as shown in Figures 14.2 and 14.3.

Exercise 14.2 Describe the systems dedicated to AI which can decouple the judgment from the decisions such that it can be provided at a different time and place if judgment shifts from individually deployed by people for each decision to instate codify into systems, then this can lead to number.

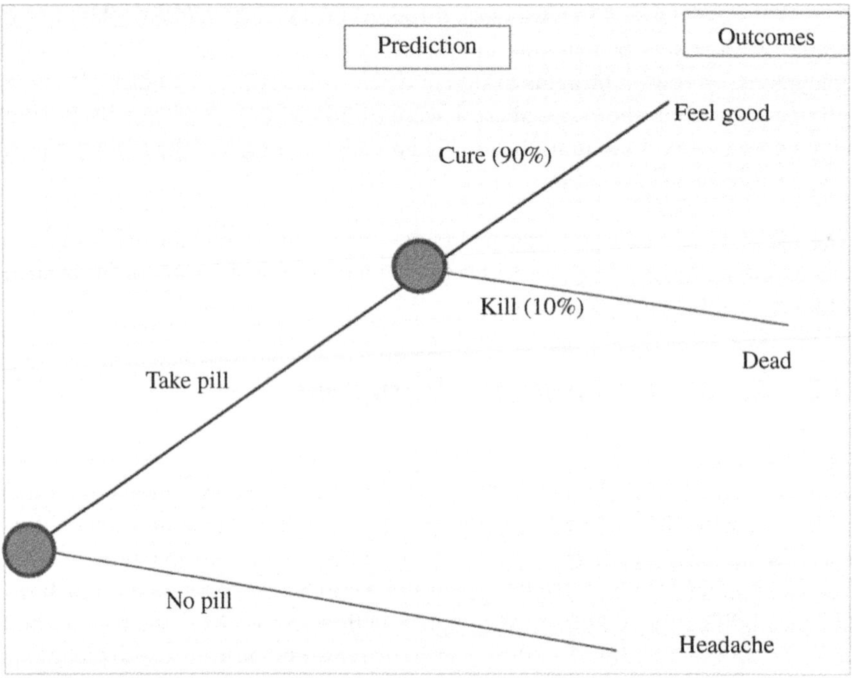

Figure 14.2 Pills decision tree.

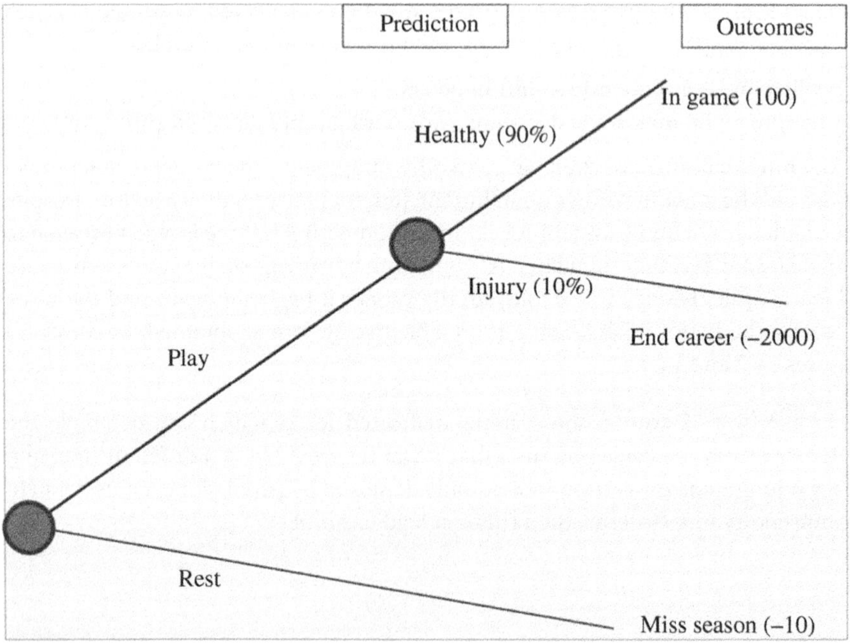

Figure 14.3 Play versus rest decision tree.

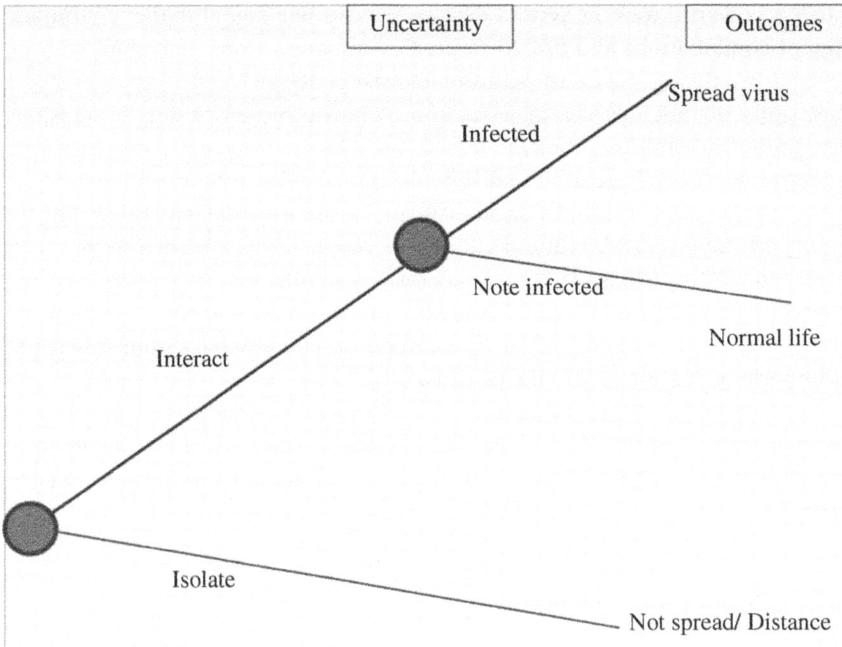

Figure 14.4 Isolate or interact decision tree.

14.4 Designing Reliable Systems

Understanding the two main system designs and approaches is not just important, it is crucial for addressing the reduced reliability that results from introducing AI-based decision-making. The weight of this responsibility should not be underestimated in two terms as listed below.

- Coordination
- Modularity

Modularity, however, involves building a wall around AI to enhance decision-making and avoid the costs associated with the lack of alignment between them and others within the organizations. It reduces coordination costs but comes at the expense of synergies. Systems are a combination of decisions that interact with each other. Consider a set of binary choices. Three decisions lead to eight different combinations. Ten interacting decisions generate 1024 combinations, and 20 interacting decisions generate 1,048,576 combinations. Systems with independent choices can get very complex quickly, underscoring the necessity of the tools and approaches we are discussing. That's why simulation is such a powerful tool for system design. We can use digital twins to simulate different combinations and use AI to predict outcomes from each combination, offering a promising and exciting future for system design.

Decisions do not operate in a vacuum. The outcomes of single decisions often influence multiple choices or actions. That is why we sometimes use predetermined rules instead of real-time decisions: rules enhance reliability. So, we accept the worst localized choices in exchange for excellent reliability for system performance. Reliability, as you can see, is the key feature of a system with interdependent decisions. This underscores the significance of your role in the system design. Your expertise and contributions are crucial to the success of the system, and we want to emphasize how important you are to us. Your role is not just important; it is integral to the process.

Exercise 14.3 Describe how to use digital twins to simulate different combinations and use AI to predict outcomes from each combination, offering a promising and exciting future for system design.

14.5 Embracing Uncertainty

Thinking in bets means embracing uncertainty and examining the likelihood of something happening. If it is sufficiently likely, we go left; if not, we go right. In other words, we express our decision rule contingent on the prediction as the cutoff point. This practical approach works well if forecasts are exact, highlighting the crucial role of accurate predictions in the decision-making process. The idea of basing decisions on thresholds that enumerate confidence is attractive. For instance, consider the process by which refugees have been assessed for entry into countries, a decision fraught with uncertainty. This practical approach can be seen in the Testimony of a refugee claimant as shown in Figure 14.4.

14.6 AI System Prediction Technology

One of the most crucial aspects of AI production is the distinction between correlations and causation. While correlations can be sufficient for certain applications, when it comes to AI informing causal relationships, we must rely on randomized experiments to gather the necessary operational data from power utility plants. These experiments are the gold standard in statistics, enabling us to accurately determine cause-and-effect relationships.

Imagine the potential of the power utility operation recommendation engine in the production machine. What if it could significantly enhance its accuracy?

Initially, it would excel at recommending various operational scenarios. But what if it reached a point where it could predict load with such precision that it could anticipate power utilities' needs for power quality and delivery? This could revolutionize the way we operate, shifting from a reactive to a proactive approach.

The AI that offers superior recommendations on the current platform is a game-changer. However, implementing this new model would necessitate a comprehensive redesign of our systems, particularly in how we manage the energy management system (EMS) supply and demand and its critical operating parameters. This redesign is not just a suggestion, but an urgent necessity to keep pace with the rapidly evolving landscape of power utility operations.

Exercise 14.4 Describe AI system and prediction technology.

14.7 AI Prediction Machine Rules

Turning rules into decisions to leverage prediction machines is a common necessity. However, the system must be capable of adapting to such changes. One rule is linked to another for the system to be reliable, so making decisions within that system may seem futile.

We present a COVID-19-related example where we developed a small oil system, initially involving 12 large companies. Here, the CEO played a crucial role in guiding their senior leadership on managing employee infectiousness from rapid antigen testing. This successful implementation allowed these 12 companies to continue their operations in an environment where the existing system would likely have led to a deadlock. The success of this demonstration subsequently inspired over 2000 more organizations to adopt this system and shift from rules to decisions, highlighting the importance of strong leadership in such transitions.

Exercise 14.5 Describe AI system prediction machine rules and prediction technology.

14.8 AI Model Architecture

Matrikon Process-NET gathers data from all production stations at Qatar Energy fields (ABB-DCS, Honeywell-DCS, SCADA, SOLAR, Bentley Nevada, and Allen Bradley), centralizes the data, and provides a means of presenting the information in numerous formats using Internet Explorer to manage and perform operations activities through AI models as shown in Figure 14.5.

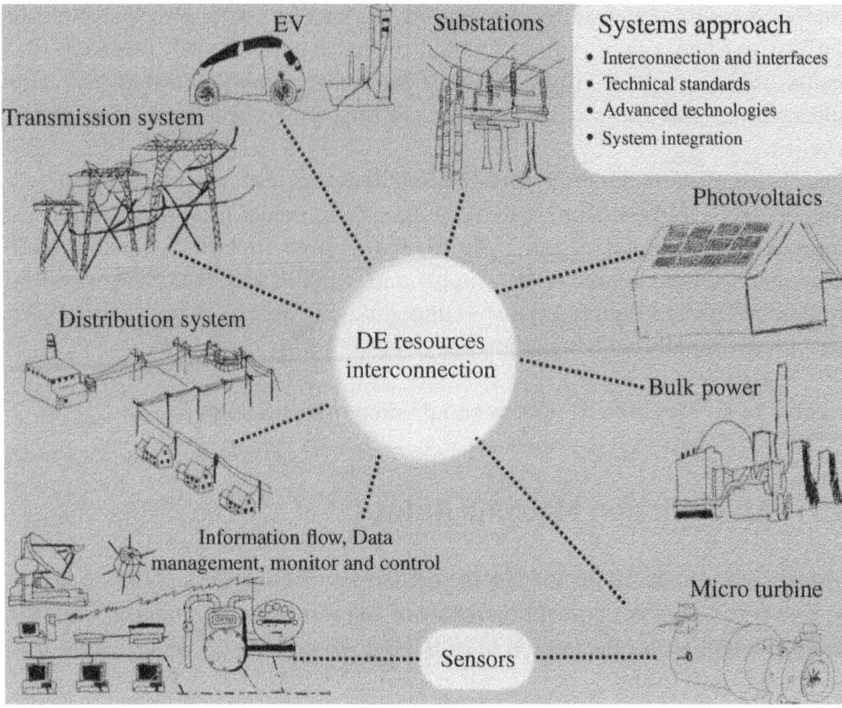

Figure 14.5 AI-virtual schematic to collect and compile daily operations data.

14.9 AI Models Tutorials

14.9.1 AI Industrial Assessment Models

Advanced Process AI – Machine/Deep Learning models are constituted by incorporating **(real-time and big historical data) with three layers of** supervised, unsupervised, and enforced forms of data. These layers represent different stages of data processing and model training. The supervised layer involves training the model on labeled data, the unsupervised layer deals with unstructured data to identify patterns, and the enforced layer ensures the model adheres to predefined rules and regulations. As shown in Figure 14.6. Machine/deep Learning is unconventional programming to develop different logic (through linear, nonlinear performance equations, and neural networks). To resolve complex engineering problems, equipment design issues, and operational bottlenecks, trends, and patterns to predict current and potential downsides and deliver as per business needs in compliance with health, safety, and environment with a degree of precision. They developed and customized real-time actionable production

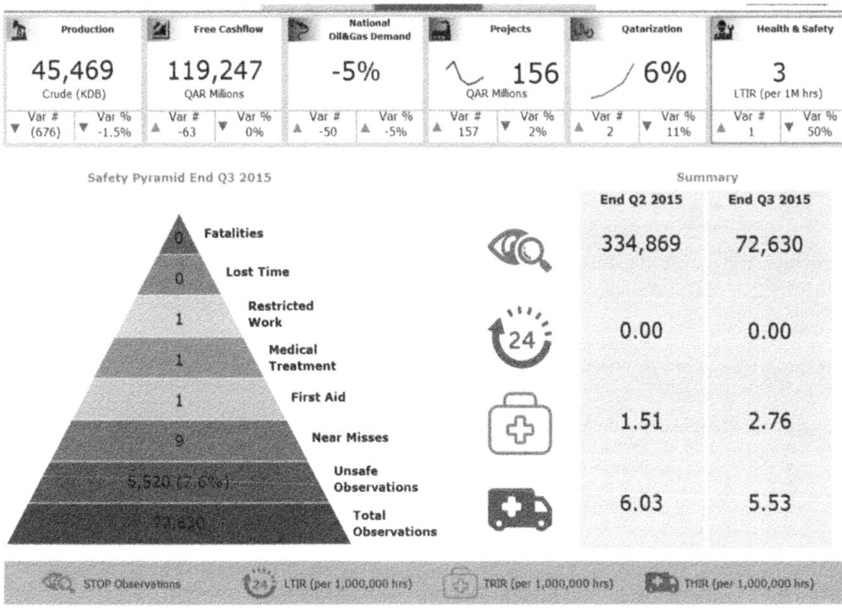

Figure 14.6 Actionable maintenance and operational reliability dashboard.

dashboards and equipment performance AI models to optimize, predict, and plan seamless operations as shown in Figure 14.7. To maximize production and uptime of the equipment, while ensuring product quality, health, safety, and environmental compliance, in line with consent-to-operate (CTO) government agency set legislation.

The following are the data collection points that can be considered as essential fuel for AI models.

1) Total tags: 50,000 (as per attached Big Data CSV (Comma Separated Values))
2) Plants: K.Main, North and South, F.Main, North, South, Jaleha Main, Diyab
3) 24 Injection manifolds
4) 600 Gas lift injection wells
5) 700 Production wells
6) 500 Power water injection wells
7) 150 Geothermal wells/renewable energy
   - Intelligent Actionable Production Dashboard (achieve zero trip incidents, to maximize uptime of the different Plant Operations)
   - Figure 14.8 shows the Intelligent Actionable Daily Production Reports to surveillance and monitor daily operational activities (achieve set production target, with minimum operator intervention)

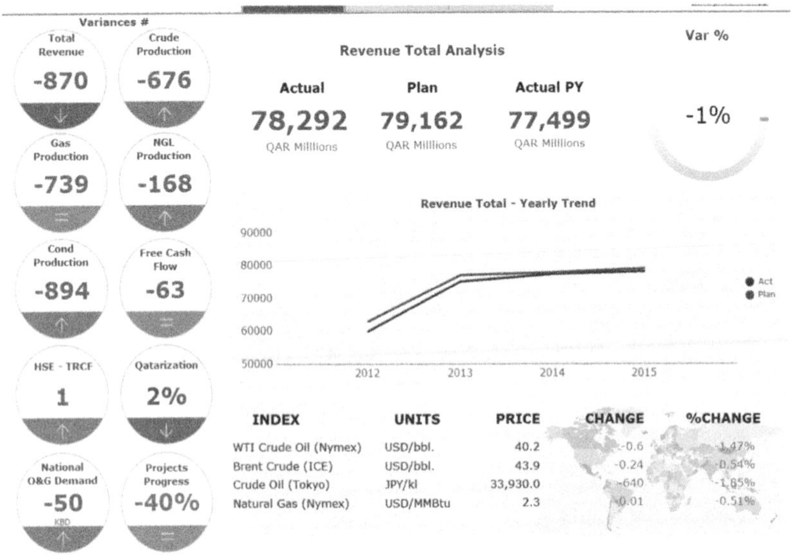

Figure 14.7 Actionable production dashboard – AI generated.

- Intelligent Flare Surveillance Models (to monitor and measure each molecule of hydrocarbon releases during normal or abnormal operations without any measuring device)
- Figure 14.9 indicates the Intelligent Actionable CPM (controller performance monitoring, enhances equipment performance to achieve zero trip incidents).

14.9.2 Introduction to Machine Learning in Microgrid Operations

Machine learning (ML) represents a significant advancement in the field of microgrid operations, offering a range of techniques that enhance decision-making, automation, and optimization within complex energy systems. ML algorithms enable microgrids to handle dynamic environments where energy production and consumption are fluctuating continuously.

- **Supervised Learning:** Supervised learning, which uses labeled datasets to train models, is pivotal in predicting energy consumption, identifying equipment failures, and optimizing grid operations. For instance, regression models can predict future energy demand based on historical consumption data and external factors like weather conditions. Classification algorithms, such as decision trees and support vector machines, can categorize different types of grid disturbances, enabling faster and more accurate responses.

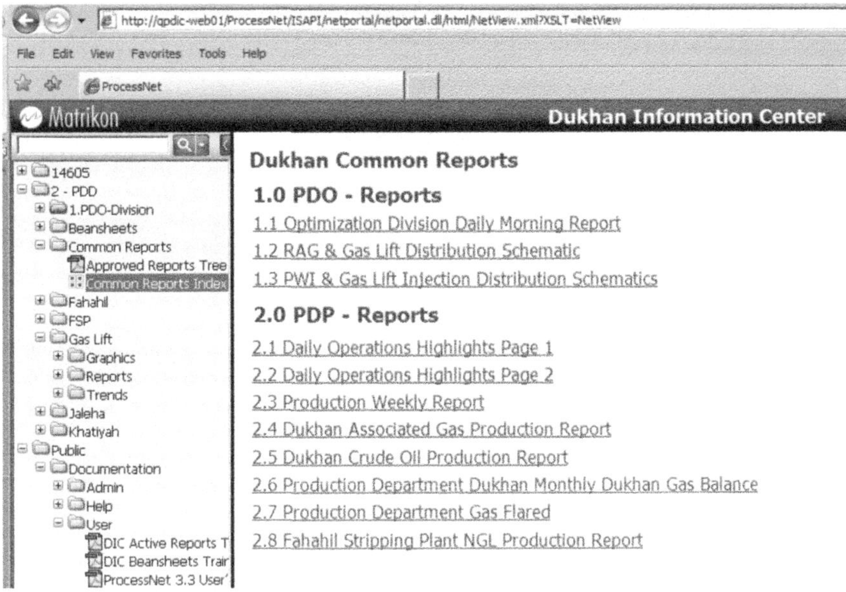

Figure 14.8 Actionable daily production reports senior management level – AI-based.

- **Unsupervised Learning:** Unsupervised learning does not rely on labeled data and is essential for uncovering hidden patterns in energy data. Techniques like clustering can group similar consumption patterns, helping to identify normal and abnormal behaviors in the grid. This is particularly useful for anomaly detection, where deviations from the norm can indicate potential issues such as equipment malfunctions or security breaches.
- **Reinforcement Learning:** Reinforcement learning is especially relevant for adaptive control systems in microgrids. In this approach, the system learns optimal strategies through trial and error, constantly improving its decisions based on real-time feedback. For example, reinforcement learning can be used to optimize the dispatch of energy storage systems, ensuring that energy is stored during periods of low demand and released during peak times to balance the grid.
- **Challenges and Benefits:** While ML offers significant advantages, such as improved accuracy in predictions and automation of complex processes, there are challenges in its implementation. These include the need for large and high-quality datasets, the computational resources required for training complex models, and the potential risks of overfitting, where models perform well on training data but fail to generalize to new data. Additionally, the interpretability of ML models is crucial, particularly in critical applications like energy management, where decisions must be transparent and justifiable.

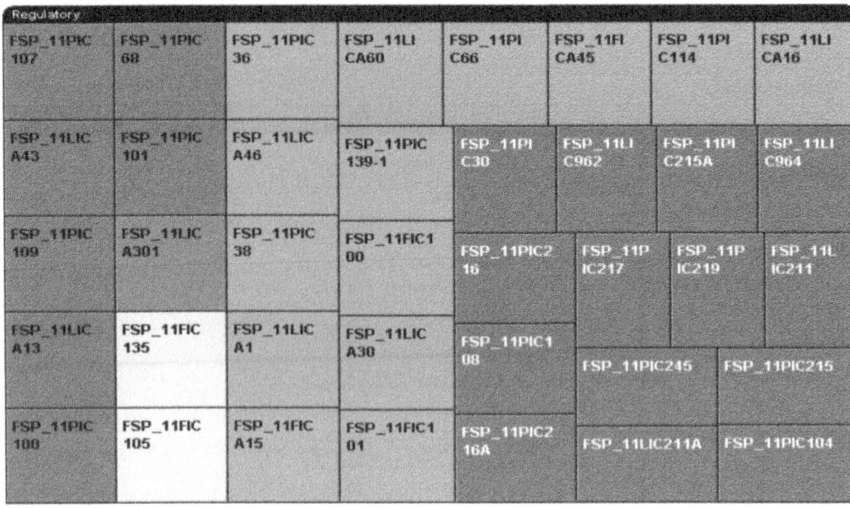

Figure 14.9 Equipment performance enhancement models – AI-based.

In summary, ML is a powerful tool that enhances the efficiency, reliability, and resilience of microgrid operations. By integrating ML into energy systems, microgrids can become more adaptable and capable of handling the complexities of modern energy distribution and consumption.

Exercise 14.6 Describe challenges and benefits of the AI system.

14.9.3 Data Analytics in Microgrid Control

Data analytics plays a crucial role in microgrid control by enabling the analysis of vast amounts of data generated by energy production, consumption, and grid performance. This data-driven approach allows grid operators to make informed decisions that optimize energy distribution, enhance grid stability, and improve overall efficiency.

- **Energy Production Data:** Data from renewable energy sources, such as solar panels and wind turbines, is often highly variable. Analyzing this data allows operators to predict production levels based on factors like weather forecasts and historical trends. This predictive capability is essential for planning energy distribution and ensuring that supply meets demand.
- **Energy Consumption Data:** Detailed consumption data provides insights into how energy is used by different sectors within the microgrid, such as residential, commercial, and industrial users. By analyzing this data, operators

can identify patterns and trends, optimize load distribution, and implement demand response strategies to balance the grid during peak usage times.

- **Grid Performance Data:** Monitoring grid performance is critical for maintaining stability and preventing outages. Data on voltage levels, frequency, and the health of grid components can be continuously analyzed to detect potential issues before they escalate. Real-time analytics platforms can process this data as it is generated, enabling immediate responses to any deviations from normal operation.
- **Predictive Analytics:** Predictive analytics uses techniques, such as time series analysis and regression models, allowing operators to forecast future energy needs and adjust supply strategies accordingly. For example, ML models trained on historical data can predict spikes in demand, enabling the grid to prepare for increased energy consumption and avoid overloading.
- **Real-Time Data Analysis:** Real-time analytics is essential for maintaining grid stability, especially in microgrids with high levels of renewable energy integration. By analyzing data as it is generated, operators can quickly adjust energy flows, activate backup systems, and respond to unexpected changes in production or consumption. This capability is crucial for preventing disruptions and ensuring a reliable energy supply.
- **Big Data Technologies:** The vast amounts of data generated by microgrids require robust storage, processing, and visualization tools. Big data technologies like Hadoop and Spark enable the efficient handling of large datasets, while advanced visualization tools help operators make sense of complex data. By leveraging these technologies, microgrids can optimize their operations and improve decision-making. Data analytics transforms the way microgrids are managed, providing the insights needed to optimize performance, enhance reliability, and reduce costs. However, implementing these technologies requires overcoming challenges related to data quality, integration, and privacy. As microgrids continue to evolve, the role of data analytics will become increasingly important in ensuring their success.

14.9.4 Machine Learning Applications in Microgrid Control

Machine learning (ML) offers practical solutions to many challenges faced in microgrid operations. By applying ML techniques, microgrids can improve their efficiency, reliability, and adaptability, leading to better overall performance.

- **Predictive Maintenance of Microgrid Components:** Predictive maintenance uses ML models to analyze data from sensors monitoring the condition of grid components such as transformers, inverters, and batteries. By identifying patterns that precede equipment failures, these models can predict when maintenance is needed, allowing operators to address issues before they lead to costly outages. For example, neural networks can be trained to recognize early warning signs of component wear and tear, enabling pre-emptive maintenance that reduces downtime and extends the lifespan of critical infrastructure.

- **Load Forecasting Using Machine Learning Models:** Load forecasting is crucial for balancing supply and demand within a microgrid. ML models, such as decision trees, gradient boosting, and artificial neural networks (ANNs), can predict future energy demand based on historical data, weather forecasts, and other relevant variables. Accurate load forecasting enables grid operators to plan energy distribution more effectively, reducing the need for expensive peak power plants and minimizing energy waste. In practice, these models can be deployed in real-time systems that continuously update predictions as new data becomes available, ensuring that the grid can respond quickly to changes in demand.
- **Optimization of Energy Resources with Machine Learning:** Optimizing the use of energy resources is essential for maximizing the efficiency of a microgrid. ML algorithms, such as genetic algorithms, particle swarm optimization, and reinforcement learning, can develop strategies for balancing the use of renewable energy sources, traditional generation, and energy storage systems. For example, reinforcement learning can be used to determine the optimal times to charge and discharge energy storage systems, ensuring that energy is stored when it is abundant and released when it is needed most. These optimization techniques help microgrids reduce costs, increase the use of renewable energy, and improve overall system performance.
- **Anomaly Detection in Microgrids:** Anomaly detection is critical for identifying irregularities that could indicate potential issues within the grid, such as equipment failures or cyber threats. ML techniques like clustering algorithms, principal component analysis (PCA), and autoencoders can detect anomalies in energy consumption patterns, grid frequency, and voltage levels. By identifying these anomalies early, operators can take corrective action before they lead to significant problems. For example, an anomaly detection system might alert operators to unusual voltage fluctuations, allowing them to investigate and resolve the issue before it causes an outage. These applications demonstrate the practical impact of ML on microgrid operations, showing how these technologies can be used to enhance efficiency, reduce costs, and improve the reliability of energy systems. By integrating ML into their operations, microgrids can become more resilient and better equipped to handle the challenges of modern energy distribution.

14.9.5 Case Studies and Practical Implementations

To illustrate the real-world impact of ML and data analytics on microgrid operations, this section presents several case studies and practical implementations.

These examples highlight how these technologies are being used to solve complex challenges and improve the performance of microgrids.

Case Study 1 **Predictive Maintenance in a Renewable Energy Microgrid:** In this case study, a microgrid that integrates solar and wind power implements predictive maintenance to monitor the health of critical components. Sensors installed on transformers, inverters, and batteries continuously collect data on temperature, vibration, and electrical currents. ML models, such as neural networks, analyze this data to predict when equipment is likely to fail. By addressing potential issues before they lead to failures, the microgrid reduces downtime, lowers maintenance costs, and extends the lifespan of its infrastructure. The case study provides quantitative data showing the reduction in unplanned outages and the cost savings achieved through predictive maintenance.

Case Study 2 **Load Forecasting in an Urban Microgrid:** This case study examines an urban microgrid serving a diverse population of residential, commercial, and industrial customers. The microgrid uses ML-based load forecasting models to predict energy demand on an hourly and daily basis. Historical consumption data, weather forecasts, and other relevant factors are used to train models that provide highly accurate predictions. By forecasting demand more accurately, the microgrid can optimize its energy generation and distribution, reducing the need for expensive peak power plants and minimizing energy waste. The case study highlights the improvements in grid efficiency and cost savings achieved through advanced load forecasting.

Case Study 3 **Optimization of Energy Resources in a Hybrid Microgrid:** In this case study, a hybrid microgrid combines renewable energy sources with traditional generation and energy storage systems. The microgrid uses ML optimization algorithms, such as genetic algorithms and reinforcement learning, to develop strategies for balancing energy supply and demand. The optimization process involves determining the best times to charge and discharge energy storage systems, as well as deciding how much energy to draw from renewable versus nonrenewable sources. The results show that the microgrid achieves significant energy savings, reduces its reliance on fossil fuels, and improves overall system efficiency.

- **Practical Implementation: Anomaly Detection System in a Smart Microgrid:** This section provides an example of a smart microgrid that has implemented an anomaly detection system using ML techniques. The system monitors various parameters, including voltage levels, frequency, and energy consumption patterns, to detect anomalies that could indicate potential issues such as equipment failures or security breaches. When an anomaly is detected, the system alerts operators, allowing them to investigate and take corrective action before the issue escalates. The implementation has been shown to improve grid reliability and security, with fewer outages and a quicker response to potential threats. These case studies and practical implementations demonstrate the tangible benefits of integrating ML and data analytics into microgrid operations. By applying these technologies, microgrids can achieve greater efficiency, reliability, and resilience, ultimately contributing to the overall stability and sustainability of energy systems.

14.10 Challenges and Future Directions

While ML and data analytics offer significant advantages for microgrid operations, their implementation is not without challenges. This section explores the obstacles that must be overcome and the future directions for research and development in this field.

Challenges:

- **Data Quality and Availability:** The effectiveness of ML models depends heavily on the quality and availability of data. In microgrids, data is often collected from a variety of sources, including sensors, smart meters, and external databases. Ensuring that this data is accurate, complete, and up-to-date is critical for training reliable ML models. However, issues, such as data gaps, noise, and inconsistencies can pose significant challenges. Additionally, data from different sources may need to be integrated and standardized, which can be a complex and time-consuming process.
- **Computational Resources:** Implementing advanced ML models, particularly for real-time applications, requires significant computational resources. The processing power needed to analyze large datasets and run complex algorithms can be substantial, particularly in systems with limited infrastructure. Additionally, the deployment of ML models in real-time environments necessitates fast and efficient data processing capabilities, which may require investments in high-performance computing resources.
- **Algorithmic Bias and Interpretability:** ML models can sometimes exhibit bias, which can lead to unfair or suboptimal outcomes. This is particularly concerning in critical applications like energy management, where biased models could result in unequal distribution of energy resources or incorrect predictions. Additionally, many ML models, particularly deep learning models, can be difficult to interpret. Ensuring that these models are transparent and explainable is essential for building trust and ensuring that decisions made by the models are understood and justified.
- **Integration with Existing Systems:** Integrating ML and data analytics into existing microgrid infrastructure can be challenging, especially in systems that were not designed with these technologies in mind. Retrofitting older systems to

support ML applications may require significant modifications in hardware and software, as well as the development of new protocols for data collection and analysis. Additionally, there may be resistance to adopting new technologies, particularly if they require significant changes to established workflows.

Future Directions:

- **Advanced ML Algorithms:** Research and development efforts are likely to focus on creating more sophisticated ML algorithms that can better handle the complexities of microgrid operations. This includes the development of algorithms that are more resilient to variations in data quality, capable of real-time processing, and adaptable to changing grid conditions. Additionally, there is a need for algorithms that can integrate with other emerging technologies, such as IoT and blockchain, to create more interconnected and intelligent energy systems.
- **Edge Computing and Distributed Analytics:** The future of ML in microgrids may involve more distributed computing approaches, where data analysis and decision-making occur nearer to the data sources (i.e., at the edge of the network). Edge computing can reduce latency and improve the responsiveness of ML models, making them more effective in real-time applications. Distributed analytics can also help to balance the computational load, allowing for more efficient use of resources and reducing the need for centralized processing.
- **Integration with Emerging Technologies:** The integration of ML with IoT, blockchain for energy trading, and advanced cybersecurity technologies is another area ripe for exploration. These integrations will enable the development of more secure, efficient, and flexible energy systems, where microgrids can operate seamlessly within larger, decentralized networks. For example, IoT devices can provide real-time data on energy consumption and production, which can be used to optimize grid operations. Blockchain can facilitate secure and transparent energy trading, while cybersecurity technologies can protect against threats to grid integrity.
- **Policy and Regulatory Considerations:** The adoption of ML and data analytics in microgrid operations also necessitates attention to policy and regulatory frameworks. Ensuring data privacy, addressing ethical considerations related to AI, and developing standards for ML applications in critical infrastructure are all important areas for future work. Policymakers and regulators will need to work closely with industry stakeholders to create frameworks that support innovation while ensuring the safety and reliability of energy systems.
- **Collaboration and Innovation:** The advancement of ML and data analytics in microgrids will require ongoing collaboration between academia, industry, and government. Cross-disciplinary partnerships will be essential for developing the next generation of energy management technologies and for addressing

the technical, social, and regulatory challenges that accompany their implementation. By fostering collaboration and innovation, stakeholders can ensure that these technologies are developed and deployed in ways to maximize their benefits while minimizing their risks.

14.11 Conclusion

The integration of ML and data analytics into microgrid operations represents a significant advancement in the management of modern energy systems. These technologies offer powerful tools for optimizing grid performance, improving reliability, and enhancing resilience in the face of dynamic and complex environments.

- **Recap of Key Points:** Throughout this chapter, we have explored the application of ML and data analytics in various aspects of microgrid operations, including predictive maintenance, load forecasting, energy resource optimization, and anomaly detection. Each of these applications demonstrates the potential of these technologies to address critical challenges in energy management, from the variability of renewable energy sources to the need for efficient and reliable grid operation.
- **The Evolving Role of Machine Learning in Microgrid Operation and Control:** As ML continues to evolve, its role in microgrid operation and control is likely to become increasingly central. ML models can continuously learn and adapt to changing conditions, allowing microgrids to respond more effectively to fluctuations in supply and demand. This adaptability is particularly important in the context of renewable energy, where production can be unpredictable. By processing and analyzing vast amounts of data in real-time, ML can enable autonomous grid management systems that optimize energy distribution without the need for constant human intervention.

  Furthermore, the integration of ML with other emerging technologies, such as IoT, blockchain, and advanced cybersecurity measures, promises to create a more interconnected and intelligent energy network. In this network, microgrids can operate not only as isolated entities but as integral parts of a decentralized energy system that is both resilient and efficient.

  Future Outlook for Research and Development

  Looking ahead, there are several key areas where further research and development are needed to fully realize the potential of ML and data analytics in microgrids:

- **Advanced ML Algorithms:** There is a need for the development of more sophisticated ML algorithms that can better handle the complexities of microgrid operations. These algorithms should be capable of real-time processing, adaptable to changing conditions, and resilient to variations in data quality.

- **Edge Computing and Distributed Analytics:** The future of ML in microgrids may involve more distributed computing approaches, where data analysis and decision-making occur closer to the data sources. This approach can reduce latency, improve responsiveness, and make ML models more effective in real-time applications.
- **Integration with Emerging Technologies:** The integration of ML with other emerging technologies, such as IoT, blockchain, and advanced cybersecurity measures, offers exciting possibilities for the future of energy systems. These integrations can create more secure, efficient, and flexible microgrids that operate seamlessly within larger, decentralized networks.
- **Policy and Regulatory Considerations:** As ML and data analytics become more prevalent in microgrid operations, there will be a need for new policy and regulatory frameworks to ensure that these technologies are used safely and responsibly. Policymakers and regulators will need to work closely with industry stakeholders to create standards that support innovation while ensuring the safety and reliability of energy systems.
- **Collaboration and Innovation:** The advancement of ML and data analytics in microgrids will require ongoing collaboration between academia, industry, and government. By fostering cross-disciplinary partnerships, stakeholders can ensure that these technologies are developed and deployed in ways to maximize their benefits while minimizing their risks.
- **Implications for the Protection of Power Distribution Systems:** The advancements in ML and data analytics have significant implications for the protection and resilience of power distribution systems. These technologies enhance the ability to predict and prevent faults, optimize the flow of energy, and detect and respond to security threats in real-time. As microgrids become more prevalent and interconnected, their effective and secure operation will be critical to the overall stability of the power grid. By integrating ML and data analytics into microgrid operations, we can create energy systems that are not only more efficient and reliable but also more secure and resilient in the face of future challenges. These technologies represent a key component of the ongoing evolution of microgrids and power distribution systems, and their continued development and implementation will be essential for ensuring a sustainable and reliable energy future.
- **Call to Action:** This chapter concludes with a call to action for further innovation and research in the field of ML and data analytics for microgrid operations. As the energy sector undergoes significant transformations, there is a growing need for new technologies and approaches that can address the challenges of

modern energy systems. Researchers, industry professionals, and policymakers are encouraged to explore the possibilities that ML and data analytics offer, to push the boundaries of what is possible, and to contribute to the ongoing evolution of microgrids and power distribution systems.

The potential of ML and data analytics to revolutionize microgrid operation and control is vast. By continuing to develop and implement these technologies, we can create energy systems that are not only more efficient and reliable but also more secure and resilient in the face of future challenges.

14.12 Probability AI Models Configuration

We can easily compute the Joint probability from a Bayes Net! For any Bayes Net:

$$p(V_1 \ldots ..V_2) = \prod_{i=1}^{n} p\left(V_i | Parents(V_i)\right)$$

In other words, Bayes Nets is really an encoding of the conditional dependencies of a set of random variables. All variables are independent of other variables given their parents. For our example:

$$p(H, A, U, S, B) = p(H)p(A)p(U|H)p(S|H, A)p(B|A)$$

So, for the specific setting above:

$$P(H = T, A = F, U = T, S = T, B = F) = P(H = T)P(A = F)$$
$$P(U = T|H = T)P(S = T|H = T, A = F)P(B = F|A = F)$$
$$= 0.05 \times (1 - 0.025) \times 0.5 \times 0.3 \times (1 - 0.15) = 0.006215625$$

We can compute any arbitrary probabilities from joint probabilities by the method of "marginalization" = summing out variables that we do not want.

$$P(S) = \sum_{H,A,U,B} P(H, A, U, S, B)$$
$$P(S) = \sum_{H,A,U,B} P(H)P(A)P(U|H)P(S|H, A)P(B|A)$$
$$P(S) = \sum_{H}\sum_{A}\sum_{U}\sum_{B} P(H)P(A)P(U|H)P(S|H, A)P(B|A)$$

Next move the sums so that a sum is placed only before all the terms that depend on it. For example, P(S|H,A) depends on sum A and sum H, so those sums occur left of it:

$$P(S) = \sum_{A} P(A) \sum_{H} P(H)P(S|H, A) \sum_{U} P(U|H) \sum_{B} P(B|A)$$

Notice that: $\sum_B P(B|\,A) = 1$ and same for $\sum_U P(U|\,H) = 1$
So, dropping the B and U terms, we get the final summation:

$$P(S) = \sum_{A} P(A) \sum_{H} P(H)P(S|H,A)$$

which works out to:

$$P(S) = P(H)P(A)P(S|H,A) + P(H)P(A)P(S|H,A) + P(H)P(\overline{A})P(S|H,\overline{A}) \\ + P(\overline{H})P(\overline{A})P(S|\overline{H},\overline{A}) = 0.0347$$

14.13 Self-Assessments Questions

Self-Assessment Question 14.1

A) Describe the evolving role of machine learning in microgrid operation and control and implications for the protection of power distribution systems.

Self-Assessment Question 14.2

1) Describe the advanced ML algorithms, edge computing, and distributed analytics.
2) Integraton with emerging technologies, policy and regulatory considerations, collaboration, and innovation.

Self-Assessmentd Question 14.3

A) Describe the predictive maintenance of microgrid components and load forecasting using machine learning models.
B) Optimization of energy resources with machine learning and anomaly detection in microgrids.

Bibliography

Gergely Hollósi, Dániel Ficzere, Attila Frankó, Máté Bancsics "Enabling Efficient Knowledge Sharing for Industrial AI" published in IEEE Network Operations and Management Symposium 2024.

M. B. Sergeeva, V. V. Voskobovich, A. M. Kukharenko "Data Processing in Industrial Internet of Things (IIoT) Applications : Industrial Agility" published in Wave Electronics and its Application in Information and Telecommunication Systems (WECONF) 2022.

Parikshit Vasisht, Sudhakar Ranjan, Rahul J. Jadhav "Artificial Intelligence as a Driver of Industrial Competitiveness" published in International Conference on Smart and Sustainable Technologies in Energy and Power Sectors (SSTEPS) 2022.

Martin Hollender, Chaojun Xu, Ruomu Tan "Engineering Challenges in Industrial AI" published in IEEE/ACM 3rd International Conference on AI Engineering – Software Engineering for AI (CAIN) 2024.

15

Cybersecurity and Information Program's Power System

15.1 Objectives

At the end of this section, you will have had an introduction to the general principles of control and mitigation plan of the cybersecurity components used for industrial applications, particularly in the energy sector:

- Cybersecurity Introduction
- Cybersecurity Operations
- Cybersecurity Impact and Solution
- Validate Cybersecurity Process Mapping
- Data Collection and Mapping
- **Project Scope for IT Penetration and Assessment**
- **Project Scope for OT Penetration and Assessment**

The scoping exercise, a critical determinant of the effectiveness of cybersecurity measures for a selected company's critical infrastructure, is not just a task. It is a significant and impactful effort to ensure the exactness and completeness of operation and control data. With the rapidly growing demand for electric power relative to the available supply and, in many countries, the aging of power system infrastructure, there is an increased concern about suppliers' ability to meet the electric needs, and the risks of greenhouse gas emissions. These concerns, combined with the ongoing limited capital investment related to new transmission systems, make it necessary to develop new cost-effective and intelligent power systems. Additionally, the development of renewable energy sources (RES), such as photovoltaic (PV) energy harvesting technologies, has led to a rapid increase in the integration of PV systems into residential and commercial premises.

Protection of Electrical Power Distribution Systems: Smart grid, Microgrid, AI, and Cyber Security,
First Edition. Tariq Masood, Jamil Abdo, and Atif Iqbal.

It's crucial to meticulously validate that all Qatar Energy (QE)[1] process steps are accurately mapped to the reference model. Similarly, all industrial control systems must be rigorously mapped to QE, leaving no room for error.

15.2 Introduction

The consequences of security breaches in the smart grid can be severe and may affect the entire infrastructure, such as a country's economy and people's lives. These risks are further elevated at times when there are growing concerns of malicious intentions for electronic intrusions on the grid's nodes or consumer devices. Such attacks have already been recently reported in several countries, including **Finland, Turkey, and Ukraine**. Therefore, a robust and comprehensive security solution is highly demanded as part of the demand side management (DSM) and demand response (DR) strategies.

To address the above-mentioned challenges and their potential impacts on the power distribution infrastructure, a convex optimization system is proposed to derive secured DSM and DR strategies that minimize the cost of energy for consumers and power utilities while protecting against overloads in distribution transformers, meeting the demands of fluctuating power from power sources, such as electric vehicles (EVs), plug-in EVs, and augmenting the existing smart grid evolution with smart meters. Distribution transformers are power grid assets often subjected to overloading conditions, which may result in their failure sooner than expected. Currently, grid operators do not have access to the actual loading conditions of these transformers and hence their aging conditions. Often, during periods of intense power consumption, excessively aged and thermally overloaded transformers can fail, causing interruption to the electric power supply. Moreover, distribution transformers may be overloaded due to the shifting of the customers' loading behavior following the market incentives.

Network modeling is based on smart meter data, allowing visibility into the distribution network. Smart meters are part of the advanced metering infrastructure (AMI) that are deployed as gateways to the smart grid. AMI consists of energy consumption meters and two-way communication facilities capable of analyzing received measurements from customers by a utility's local servers. Thus, AMI has been utilized mainly to provide full access to both utilities and customers to visualize, process, and manage their energy needs, which will make use of smart meter readings to estimate actual loading of the distribution transformers, and lead to a change in the operation and planning of distribution systems.

---

1 Having been an Operations Advisor at Qatar Energy for over 25 years, I can attest to the successful implementation of this original work at Qatar Energy.

Exercise 15.1 Describe the DSM and DR strategies to address the cybersecurity issues.

15.3 Cybersecurity Operations

The proliferation of cyber infrastructures, such as power electronic transformers, phasor measurement units (PMUs), and remote terminal units (RTUs) into electric distribution networks exposes them to greater cyber vulnerability and increases the risk of cyberattacks. The anticipated contribution of smart grid power distribution systems can be divided into two main parts. The first part aims to push the state-of-the-art cybersecurity of distribution grids by contributing a novel, applicable, and scalable moving target defense (MTD) framework against false data injection (FDI) attacks. The second aim is to enhance the distribution system unbalance using a three-to-single phase power electronic transformer for single-phase laterals tapped on three-phase main feeders.

Most common vulnerabilities are as follows:

1) Active interception to drop or alter messages sent by the smart meter to the utility, such as reporting false data. If this attack is performed at a large scale, the smart grid might very well go down as it will be acting on demands that do not exist and will surely take wrong shedding decisions and actions. A similar attack can target data sent from the utility to the smart meters.
2) Passive interception of messages that consequently gain private information of the user. For example, the room in which the electric device is being used. This information can be used to stage more severe attacks. To tackle this issue, a two-layer security mechanism for the smart grid is proposed. The first layer being a secure-by-construction prevention mechanism uses blockchain as the underlying technology to achieve security and privacy. Whereas the second layer is a detection mechanism that analyzes the data received by the utility servers and the smart meters, and looks for any behavioral abnormality.

15.4 Cybersecurity Impact and Solution

This section will address the following significant challenges and address the issue by implementing different approaches:

1) A hidden moving target defense (HMTD) strategy is proposed to remain stealthy against attackers. Existing moving target defense (MTD) strategies often overlook the fact that alert attackers can detect MTD activation before launching a false data injection (FDI) attack. The HMTD strategy addresses

this by remaining undetectable even in the presence of measurement noise, enhancing its applicability in real-world scenarios.

2) Analytical feasibility conditions must be developed to rapidly assess the stealthiness of MTD strategies by leveraging the structural characteristics of attackers' bad data detection problems. These conditions will form the theoretical groundwork for a systematic understanding of the stealth capabilities of MTD strategies, enabling more precise detection and countermeasure implementations.
3) Modeling the cyber interaction between the system operator and attackers as a repeated Bayesian Stackelberg Game, where an approximate dynamic programming approach will be applied to solve it. This framework will help predict attackers' strategies and optimize the system's defense mechanisms by considering both historical behavior and dynamic responses.

Exercise 15.2 Describe the HMTD, MTD, and FDI strategies to determine impact and solutions.

4) Developing a robust HMTD-based reconfiguration model for unbalanced radial distribution networks. To solve this, an alternating direction method of multipliers (ADMM)-based semidefinite programming approach will be employed, ensuring efficient and scalable optimization for enhanced network security and resilience.
5) Developing a high-performance computing (HPC) and power-hardware-in-the-loop (PHIL)-enabled testbed for proof-of-concept validation and demonstration. These methods will utilize a test system built at Kansas State University, enabling the simulation and validation of the designed models in realistic power grid environments.
6) Developing a new topology for single-phase power distribution that ensures balance across the three-phase power system. This topology will aim to maintain system efficiency and stability while addressing the challenges of unbalanced loads in modern distribution networks.
7) Develop a three-to-single-phase power electronic transformer topology that allows remote connection control, voltage regulation, power factor correction, metering, and protection of any single-phase lateral tapped on a three-phase power distribution feeder. Study the impact of unbalanced enhancement in conjunction with the proposed MTD strategy on an example of the selected electrical power distribution system.

Innovation: This section discusses the applicable and scalable MTD framework against the FDI attacks and enhancement of phase balancing in the three-phase electric distribution network using three-to-single-phase power electronic transformers. It will include theoretical analysis, model development, algorithmic design, numerical validation, testbed development, and real-time simulation. It will extend current understanding of MTD to the field of power systems. It will provide distribution system operators with valuable insights on how to securely operate the distribution network.

Expected Impacts: This section will demonstrate in outreach and educational activities to increase public awareness of cyberattacks and to facilitate the understanding of cybersecurity in the smart grid among stakeholders, ratepayers, regulators, utility executives, and market participants. It would also promote societal embracing of transformative new technologies, and potentially herald a new age of grid cybersecurity that will empower the next generation of power grids.

Exercise 15.3 Describe the innovation and expected impact to vulnerability of the power operations network.

15.4.1 Validate Process Mapping

In this section, we discuss how the DSM and the DR technologies are integrated to deliver required power and to meet the commercial and domestic consumer requirements in a large and selected electrical power system operation. Some important aspect of recent improvements in the distribution networks is the involvement of customers in the DSM and the DR programs. Most existing DSM and DR programs are based on reducing the power consumption of customers according to predetermined policies of load priorities during peak times. The smart grid is considered a communication network with smart meters, control devices, computers, automation, and emerging technologies interacting together. All entities in the smart grid must be able to communicate efficiently with each other, making the smart grid a big, complex, heterogeneous network inheriting most of the security vulnerabilities of cyber systems. As of January 2017, the US Department of Energy released an alarming report, confirming that the US electric grid is still at risk of cyberattacks.

Exercise 15.4 Describe the mapping framework to integrate the DSM and DR program.

My previous organization provided information about 423 systems as compared to the initial 371 systems. New systems are discovered, and others are no longer considered relevant. Figure 15.1 indicates the Data collection framework.

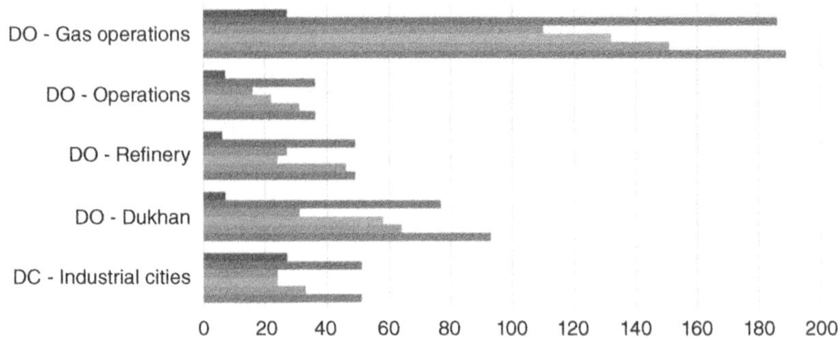

	DC - Industrial cities	DO - Dukhan	DO - Refinery	DO - Operations	DO - Gas operations
Number of new systems	27	7	6	7	27
Number of systems assigned to unit	51	77	49	36	186
Number of systems with network diagram	24	31	27	16	110
Number of systems with process overview	24	58	24	22	132
Number of systems with questionnaire	33	64	46	31	151
Total number of systems	51	93	49	36	189

Figure 15.1 Data collection framework.

15.4.2 Data Collection Frame

It effectively emphasizes the vital role of prioritizing risk assessments to safeguard operational integrity and ensure the efficient functioning of our services, as shown in Figure 15.2 and mapping Figure 15.3.

Before moving to the next phase of the program, the CIS Cyber & Information Security Department wants to understand:

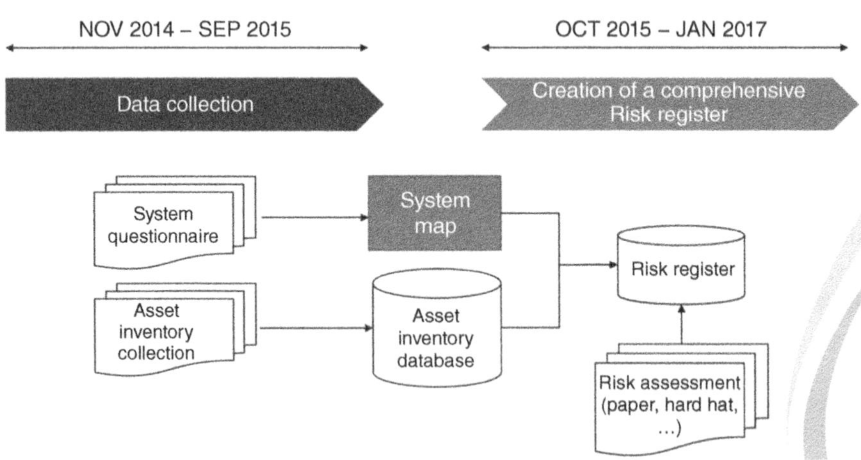

Figure 15.2 Data collection and mapping framework.

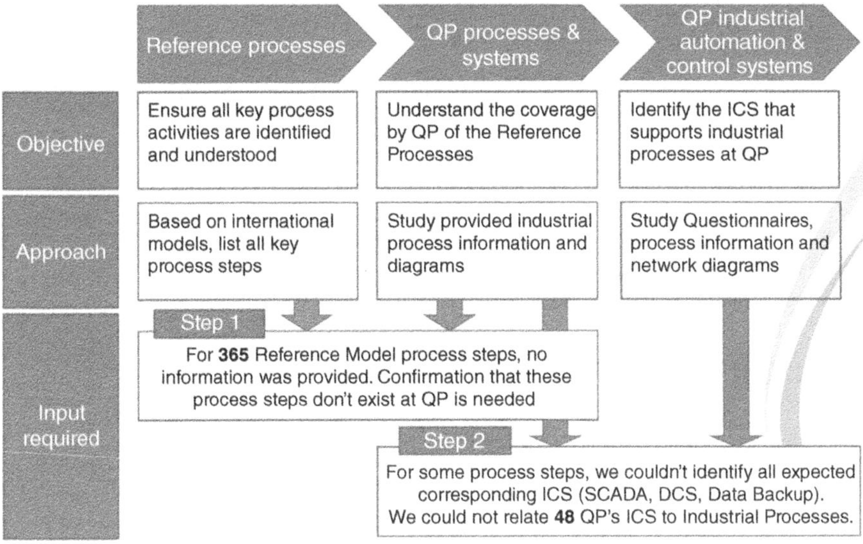

Figure 15.3 Data collection and map structure framework.

1) Which part of my company's critical operations are covered by the data provided?
2) What is the "inherent" mistake made by focusing on the provided systems?

15.4.3 Data Collection – Introducing System Mapping

A system map is a model of my energy company's ICS, providing the spectrum – a link from industrial automation and control systems to business processes. The mapping system indicates the quality of available information on ICS and shows the importance of prioritizing risk assessment.

System Map Structure

Exercise 15.5 Describe the system mapping structure and its operational decision tree. Also describe the validation process mapping framework stepwise with its elements.

15.5 How Can You Help?

The quality of the scoping exercise will have a direct impact on the effectiveness of cybersecurity measures to secure the selected company's critical infrastructure. Therefore, we would highly appreciate your review of the prepared information

Table 15.1 Validating process map structure framework.

Location	Site	Domain	Operation	System	QP System mapping
DKN	DYB	Onshore Fields	Primary process	Gathering system manifolds	Diyab manifold - Gathering system manifolds
DKN	DYB	Onshore Fields	Primary process	Lift gas manifold	Diyab manifold - Lift gas manifold
DKN	DYB	Onshore Fields	Primary process	Chemical injection system (corrosion inhibitor/demulsifier)	Diyab manifold - Chemical injection system (corrosion inhibitor/demulsifier)
DKN	DYB	Onshore Fields	Utility	Drains system	Diyab manifold - Drains system
DKN	DYB	Onshore Fields	Primary process	Launchers for export pipelines	Diyab manifold - Launchers for export pipelines
DKN	DYB	Onshore Fields	Primary process	Receiver for liftgas pipeline	Diyab manifold - Receiver for liftgas pipeline
DKN	DYB	Onshore Fields	Utility	Instrument air packages for LG manifolds	Diyab manifold - Instrument air packages for LG manifolds
DKN	JDGS	Degassing	Primary process	Oil gathering and separation	Jaleha, degassing station - Oil gathering and separation
DKN	JDGS	Degassing	Primary process	Oil storage and pumping	Jaleha, degassing station - Oil storage and pumping

to validate its exactness and completeness. Table No. 15.1 presents the Production facilities and process Mapping, Table 15.2 outlines the process mapping structure framework, and Table 15.3 displays the risk level determined during the mapping and assessment process.

More precisely:

- Validate that all QE process steps are correctly mapped to the reference model
- Validate that all industrial control systems are correctly mapped to QP's processes
- Provide information about the "**Orphan System**"

Step 1 – Validate Process Mapping

1) Compare columns E to J that describe the Reference Model, with column K, which is the equivalent system within Qatar Petroleum.
2) **Write any relevant comments in column V.**
3) If the information is aligned, state **"Reference and QP Systems Aligned" in column W.**
4) If the information is not aligned, please **state this in column W and write an explanation in column V.**

Step 2 – Validate mapping ICS/SCADA/DCS to processes

1) Compare the Qatar Petroleum systems **column K** with **columns R** to **T** that describe the corresponding **SCADA/ICS** systems.

Table 15.2 Map process structure framework.

Site	Unit	System label	System name	System full name
Dukhan	Gas Recycling Plant	RGD-DCS	RGD - Distributed Control System (DCS)	RGD - Distributed Control System (DCS)
Dukhan	Gas Recycling Plant	RGD-FGS	RGD - Fire & Gas Monitoring System (FGS)	RGD - Fire & Gas Monitoring System (FGS)
Dukhan	Gas Recycling Plant	RGD-ESD	RGD - Emergency Shutdown System (ESD)	RGD - Emergency Shutdown System (ESD)
Dukhan	Gas Recycling Plant	RGD-AUX	Aux. Ctrl Sys	Aux. Ctrl Sys - Gas Recycling Plant
Dukhan	Gas Recycling Plant	RGD-TBN	Turbine Ctrl Sys.	Turbine Ctrl Sys. * Gas Recycling Plant
Dukhan	Gas Recycling Plant	RGD-ATS	Antisurge Ctrl. Sys.	Antisurge Ctrl. Sys. - Gas Recycling Plant
Dukhan	Gas Recycling Plant	RG-DCS-SRV	DCS-E-server	DCS-E-server - Gas Recycling Plant
Dukhan	Gas Recycling Plant	RGD-PLC-MIAC	RGD - Main Instrument Air Compressor PLC Package	RGD - Main Instrument Air Compressor Package
Dukhan	Gas Recycling Plant	RGD-PLC-HOH	RGD - Hot Oil Heater Package	RGD - Hot Oil Heater Package
Dukhan	Gas Recycling Plant	RGD-PLC-HWC	RGD - Hydraulic wellheads control package	RGD - Hydraulic wellheads control Package
Dukhan	Gas Recycling Plant	RGD-PLC-RNCP	RGD - Reinjection Compressor Packages	RGD - Reinjection Compressor Packages
Dukhan	Gas Recycling Plant	RGD-PLC-RECP	RGD - Recycle Compressor Packages	RGD - Recycle Compressor Packages

Table 15.3 Map process structure framework.

QP system mapping	Operational ICS	Safety ICS	Auxiliary ICS	Overall consequences rating
Diyab Manifold – Lift gas manifold	Emerson DeltaV ICSS	Emerson DeltaV ICSS	STM network (for DIC) MPLS network DIC Data Collection Node	Very Low
Diyab Manifold – Chemical Injection System (corrosion inhibitor/demulsifier)	Emerson DeltaV ICSS	Emerson DeltaV ICSS	STM network (for DIC) MPLS network DIC Data Collection Node	Very Low
Diyab Manifold – Drains system	Emerson DeltaV ICSS	Emerson DeltaV ICSS	STM network (for DIC) MPLS network DIC Data Collection Node	#N/A
Diyab Manifold – Launchers for export pipelines	Emerson DeltaV ICSS	Emerson DeltaV ICSS	STM network (for DIC) MPLS network DIC Data Collection Node	#N/A
Diyab Manifold – Receiver for lift gas pipeline	Emerson DeltaV ICSS	Emerson DeltaV ICSS	STM network (for DIC) MPLS network DIC Data Collection Node	#N/A

2) Please advise if the information is incomplete in **column V**.
3) If the information is aligned, state "QP Systems and **SCADA/ICS** systems Aligned" in **column W**. If the information is not aligned, please state this in column W and write an explanation in **column V**.

Observation:

1) The "ICS LIST" Tab in the spreadsheet contains a Systems List table.
2) Most systems are identified using a "System Questionnaire ID." This internal reference is used to organize and identify the documentation received from QP.
3) The format of the reference is in the form "D0XXX" and can be seen in the first three columns of the QP Systems List table.

15.6 Cybersecurity Tailored Models

15.6.1 Project Plan Work For (IT Penetration Testing, Vulnerability, and Risk Assessment Framework)

Table 15.4 shows the project timeline for IT Penetration Testing, Vulnerability, and Risk Assessment Framework.

15.6.1.1 Operational Summary

This project plan outlines the approach for conducting a comprehensive IT penetration testing, vulnerability, and risk assessment for a Power Plant of CCGT (combined cycle gas Turbine). The plant features multiple energy blocks, various facilities, and critical IT infrastructure components, including distributed control systems (DCS) and communication systems. The aim is to assess the security posture of the IT infrastructure to identify vulnerabilities, evaluate risks, and provide actionable recommendations.

Table 15.4 IT Penetration Assessment and Framework.

Phase	Duration	Start date	End date
Preparation and planning	2 weeks	TBD	TBD
Reconnaissance	1 week	TBD	TBD
Vulnerability scanning	2 weeks	TBD	TBD
Penetration testing	3 weeks	TBD	TBD
Risk assessment	1 week	TBD	TBD
Reporting	2 weeks	TBD	TBD
Review and finalization	1 week	TBD	TBD

15.6.1.2 Objectives

- **Identify Vulnerabilities:** Discover potential security weaknesses in the IT infrastructure.
- **Assess Risks:** Evaluate the risks associated with identified vulnerabilities.
- **Evaluate Security Posture:** Determine the effectiveness of existing security controls.
- **Recommend Improvements:** Provide actionable recommendations to enhance security.

15.6.1.3 Scope of Work

IT infrastructure to be assessed for the following:

- Firewalls
- Switches and routers
- Servers (physical and virtual)
- Workstations
- Web applications
- Wireless network
- Remote access systems
- Surveillance network and tools

DCS systems to be considered for the following:

- **Siemens SPPA-T3000**
  GE/Alstom DCS

15.6.1.4 Methodology

Preparation and Planning:

Define Scope: Confirm scope and objectives with stakeholders.
Resource Allocation: Assemble the assessment team and allocate resources.
Initial Assessment: Gather information on the IT infrastructure and DCS systems.

Reconnaissance:

Network Discovery: Map out the network topology and identify active devices.
Information Gathering: Collect details on system configurations, software versions, and security controls.

Vulnerability Scanning:

Automated Scanning: Use vulnerability scanners to identify known vulnerabilities.
Manual Testing: Conduct manual testing for complex vulnerabilities not covered by automated tools.

Penetration Testing

Exploit Testing: Attempt to exploit identified vulnerabilities to assess their impact.
Access Control Testing: Evaluate the effectiveness of access controls and permissions.

Risk Assessment

Impact Analysis: Assess the potential impact of each vulnerability on business operations.
Likelihood Assessment: Evaluate the likelihood of exploitation for each identified vulnerability.

Reporting

Executive Summary: Provide a high-level overview of findings and recommendations.
Detailed Findings: Present detailed findings, including severity levels and risk assessments.
Recommendations: Offer specific recommendations for mitigating identified risks.

15.6.1.5 Tools Used

- Vulnerability Scanners: Nessus, OpenVAS
- Penetration Testing Tools: Metasploit, Burp Suite, Nmap
- Network Analysis Tools: Wireshark, TCPDump
- Web Application Testing Tools: OWASP ZAP, Burp Suite
- Risk Assessment Tools: RiskWatch, FAIR

15.6.1.6 Assessment and Testing Environment

Environment Setup:

Test Environment: Ensure a controlled environment for testing that mimics the production environment.
Data Protection: Implement measures to protect sensitive data during testing.

Testing Phases

Pre-assessment: Confirm readiness and scope of the test environment.
Active Testing: Conduct testing as per the defined methodology.
Post-assessment: Validate results and ensure no disruption to operations.

15.6.1.7 Finding Overview

Severity Levels:

- **Critical:** Immediate action required; severe impact on operations.
- **High:** Significant vulnerabilities with potential for substantial impact.

- **Medium:** Moderate impact; requires remediation but not urgent.
- **Low:** Minor vulnerabilities with minimal impact.
  Risk Assessment Matrix:
- **Severity Versus Impact:** Categorize risks based on their severity and potential impact on business operations.

15.6.1.8 Security Controls and Levels

Current Controls: Review existing security controls and their effectiveness.

Control Recommendations: Propose enhancements or additional controls to address identified vulnerabilities.

15.6.1.9 Training Needs

Staff Training: Identify training needs for IT staff based on assessment findings.

Security Awareness: Develop and deliver security awareness training for all employees.

Technical Training: Provide technical training for system administrators on new security measures and best practices.

15.6.1.10 Deliverables

Comprehensive Report: A detailed report including all findings, risk assessments, and recommendations.

Executive Summary: A concise summary for senior management.

Presentation: Presentation of findings and recommendations to key stakeholders.

15.6.1.11 Time Line

15.6.2 Scope of Work for (OT Penetration Testing, Vulnerability, and Risk Assessment Framework)

Table 15.5 shows the project timeline for OT Penetration Testing, Vulnerability, and Risk Assessment Framework.

Step 1 – Validate Process Mapping as shown in (Table 15.1)
Step 2 – Validate the Mapping of SCADA/DCS to Processes as shown (Table 12.2)

Observation: as shown in (Table 12.3)

Project Timeline: Tables 15.4 and 15.5 IT/OT Penetration Assessment and Framework – A table outlines a project timeline with phases, durations, and unspecified start and end dates. The phases include Preparation and Planning (2 weeks), Reconnaissance (1 week), Vulnerability Scanning (2 weeks), Penetration Testing (3 weeks), Risk Assessment (1 week), Reporting (2 weeks), and Review and Finalization (1 week). All start and end dates are marked as TBD (to be determined).

Table 15.5 OT Penetration Assessment and Framework.

Phase	Duration	Start date	End date
Preparation and planning	2 weeks	TBD	TBD
Reconnaissance	1 week	TBD	TBD
Vulnerability scanning	2 weeks	TBD	TBD
Penetration testing	3 weeks	TBD	TBD
Risk assessment	1 week	TBD	TBD
Reporting	2 weeks	TBD	TBD
Review and finalization	1 week	TBD	TBD

15.6.2.1 Operational Summary

The goal of this project is to evaluate and strengthen the cybersecurity posture of the Operational Technology (OT) environment within a Combined Cycle Gas Turbine (CCGT) power plant. The plant's infrastructure includes various critical systems like gas and steam turbines, balance of plant facilities, and DCS. This project aims to identify vulnerabilities, assess risks, and implement necessary controls to mitigate cyber threats.

Detailed Process Overview

15.6.2.2 Planning and Preparation

Kickoff Meeting: Meet with key stakeholders to understand the operational environment, goals, and security concerns. Define the scope, objectives, and timelines of the assessment. Identify key personnel, including those responsible for plant operations and IT/OT systems.

Documentation Review: Collect and review existing documentation, including network diagrams, DCS configurations, security policies, and previous assessments. Identify potential risks and areas of focus based on initial findings.

15.6.2.3 Assessment Phase

Gap Analysis: Compare current OT systems and controls against industry best practices and relevant standards (e.g. NEPRA, ISA/IEC 62443). Identify gaps between existing security measures and those required to protect against cyber threats. Deliver a gap analysis report with specific recommendations.

Risk Assessment: Conduct high-level and detailed risk assessments to identify and prioritize risks to plant operations. Evaluate risks across different zones and equipment categories, considering factors like impact and likelihood. Use a risk assessment matrix to visually represent risks.

Vulnerability Assessment: Use automated and manual tools to scan OT systems for vulnerabilities (e.g. outdated software, misconfigurations, weak passwords). Perform in-depth analysis to confirm findings and assess their potential impact. Evaluate vulnerabilities against NEPRA and ISA/IEC 62443 standards.

15.6.2.4 Testing Phase

Penetration Testing: Simulate cyberattacks on the OT environment to test the effectiveness of existing security controls. Focus on critical components such as DCS, firewalls, network interfaces, and remote access systems. Document the methods used and the outcomes of each test scenario.

Testing Environment: Ensure testing is conducted in a controlled environment to avoid disrupting plant operations. If needed, create a mirrored test environment using backup systems to conduct tests without affecting live operations.

15.6.2.5 Reporting Phase

Findings Documentation: Categorize findings by severity (Critical, High, Medium, Low) as per CVSS (Common Vulnerability Scoring System). Provide detailed descriptions of vulnerabilities, including their potential impact and recommended remediation steps. Include a risk assessment matrix correlating severity with potential impact.

Corrective Action Plan: Develop a comprehensive plan to address identified vulnerabilities and risks. Outline specific actions required, including hardware/software upgrades, zone segregation, and policy updates. Prioritize actions based on the criticality of risks and operational impact.

Report Delivery: Compile findings into a detailed report covering all aspects of the assessment. Include sections such as Executive Summary, Methodology, Tools Used, Environment Description, and Recommendations. Present the report to key stakeholders and discuss findings and next steps.

15.6.2.6 Remediation and Improvement

Implementation Support: Assist in implementing recommended controls and improvements. Provide technical guidance on hardware and software upgrades, network segmentation, and other security measures. Coordinate with plant operations and IT/OT teams to ensure smooth implementation.

Training: Provide hands-on training to the Operations and Maintenance (O and M) team on identified risks and mitigation strategies. Offer workshops on best practices for OT security, focusing on incident response and system hardening. Develop a training plan tailored to the specific needs of the plant's technical staff.

15.6.2.7 Continuous Monitoring and Follow-Up

Monitoring Setup: Implement monitoring tools and systems to continuously track the security posture of the OT environment. Set up alerts and reporting mechanisms to quickly identify and respond to potential threats.

Follow-Up Assessment: Conduct follow-up assessments to evaluate the effectiveness of implemented controls. Update risk assessments and vulnerability reports based on new findings and changes in the OT environment.

Ongoing Support: Provide ongoing support and consultation to address new vulnerabilities and emerging threats. Review and update security policies and procedures as needed to stay aligned with industry standards and regulatory requirements.

Deliverables

- **Executive Summary:** A concise overview of the entire assessment, including key findings and recommendations.
- **Detailed Assessment Reports:** Separate comprehensive reports for gap analysis, risk assessment, vulnerability assessment, and penetration testing.
- **Corrective Action Plan:** A prioritized plan of action for addressing identified issues, including timelines and responsibilities.
- **Training Documentation:** Materials and resources provided during the training sessions, tailored to the plant's OT environment.
- **Continuous Monitoring Plan:** A strategy and implementation guide for ongoing security monitoring of the OT environment.

Tools and Methodologies

- **Tools Used:** Nessus, Nmap, Wireshark, Metasploit, SCADA-specific tools (such as CyberX or Nozomi Networks).
- **Methodologies:** OWASP testing methodology for penetration testing, NIST SP 800–82 for ICS security, ISA/IEC 62443 standards for industrial control system security.

This detailed plan outlines the comprehensive approach to be taken for securing the OT environment of the power plant, ensuring that all potential vulnerabilities are identified and mitigated, and that plant personnel are well-prepared to manage the security risks of their environment.

15.6.2.8 Scope of Assessment

The assessment includes the following tasks that will be conducted:

- **Gap Analysis:** Conduct a thorough examination to identify gaps between the current system and administrative controls and those necessary for comprehensive protection against cyberattacks.

- **Risk Criteria Establishment:** Develop risk criteria tailored to different plant equipment, zones, and networks, ensuring adherence to NEPRA and ISA-62443 standards.
- **Risk Assessment:** Execute both high-level and detailed risk assessments, following NEPRA/ISA-62443 standards.
- **Vulnerability Assessment:** Perform a vulnerability assessment in accordance with NEPRA and ISA/IEC 62443 standards to identify weaknesses and propose remedies.
- **Domain-wise Gap Analysis:** Provide a detailed gap analysis report based on NEPRA domains.
- **Training Needs Evaluation:** Assess the awareness levels of the existing Operations and Maintenance (O and M) team and identify areas requiring further training.
- **Cybersecurity Improvement Plan:** Develop a plan to enhance cybersecurity measures, including zone segregation and hardware/software upgrades.
- **Methodology and Tools Specification:** Specify the methodologies and tools to be used for gap analysis, vulnerability, and risk assessment.
- **Sensitive Information Handling:** Ensure that all sensitive information collected during the assessment is handled securely and in compliance with applicable privacy regulations.
- **Comprehensive Assessment Report:** Deliver a detailed report containing findings from the gap analysis, vulnerability, and risk assessments, along with recommended actions.

15.6.2.9 OT Environment Equipment List

The assessment will cover the following OT systems:

1) Application Servers: xxx
2) Terminal Servers: xxxx
3) Security Server: xxxx
4) HMI Server: xxxx
5) HMI Historian Server: xxxx
6) HMI Engineering Station: xxx
7) Automation Servers: xxxx
8) Workstations: xxxx
9) Firewall Switches: xxx
10) SCALANCE Switches: xxxx
11) Time Servers: xxxx
12) LAN Interfaces: xxxx
13) NAS (Network Attached Storage): xxx

15.7 Self-Assessment Questions

Self-Assessment Question 15.1

Describe the system map and its operational model of selected energy company's ICS, providing the spectrum.

Self-Assessment Question 15.2

To evaluate and strengthen the cybersecurity posture of the operational technology (OT) environment within combined cycle gas turbine (CCGT) power plant.

Self-Assessment Question 15.3

Develop resilience approach for conducting a comprehensive IT penetration testing, vulnerability, and risk assessment for a power plant with a capacity for combined cycle power generation.

Bibliography

Yaofu Cao, Xiaomeng Li, Junwen Liu, "Research on Interacted Response Technology of Cyber Security Protection Devices Based on Deep Reinforcement Learning Oriented to New Generation of Power System" published in 2nd International Conference on Electrical Engineering and Control Science (IC2ECS) 2022.

Dabin Sun, Bowei Wang "Research on the Design of the Implementation Plan of Network Security Level Protection of Information Security" published in 7th International Symposium on Mechatronics and Industrial Informatics (ISMII) 2021.

Dong Liu, Yang Liu, Zhicheng Liu "Analysis and Reflection on the Situation of Industrial Information Security Ransomware Attacks" published in 8th International Conference on Data Science in Cyberspace (DSC) 2023.

Xinli Xiong, Qian Yao, Qiankun Ren "Mission-Oriented Security Framework: An Approach to Embrace Cyber Resilience in Design and Action" published in 7th International Conference on Cryptography, Security and Privacy (CSP) 2023.

Self-Assessment Question's Answers Section-wise

1. INTRODUCTION

Protection of Electrical Power Distribution Systems: Smart grid, Microgrid, AI, and Cyber Security,
First Edition. Tariq Masood, Jamil Abdo, and Atif Iqbal.

2. THE PROTECTION RELAY

Self-Assessment Question 2.1

Considering the protection scheme for the transformer and feeder circuit shown in Figure 2.11, describe how the scheme would respond to a fault on the transformer and a fault on the feeder, given that:

2.1a → CTA and CTB have the same ratio.
2.1b → CTA has a ratio which is slightly higher than that of CTB a small mismatch of, say, 10%.
2.1c → CTB is omitted.

2.1 SAQ 2.1a

With the same CT ratio, the relay *R* would respond to a fault on the transformer but would not see any fault current for a fault on the feeder. This is a classical unit protection scheme.

2.2 SAQ 2.1b

With a small mismatch between the turns ratios of the current transformers CTA and CTB, the protection scheme is described as the "rough balance scheme" and the overcurrent relay is chosen with IDMT characteristics. For a fault on the transformer, the fault current through the relay is the sum of both I'_{CTA} and I'_{CTA}; thus the relay has a high driving current, and in the case of an IDMT relay, it would be $I'_{CTA} - I'_{CTA}$, and a low current will pass through the relay, and it will respond *more slowly* the effective settings for a feeder fault are chosen to grade with any downstream protection.

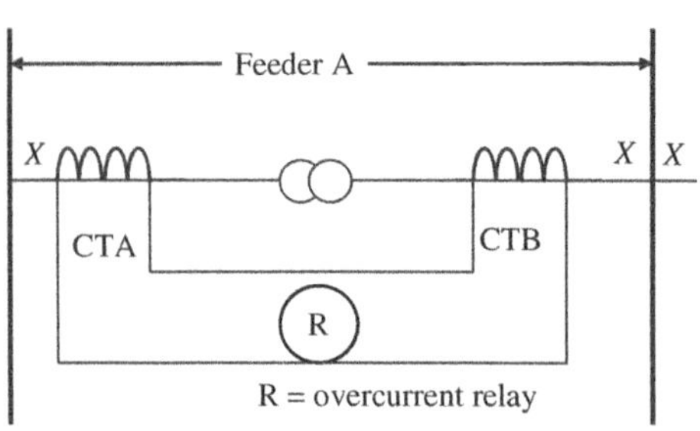

Figure 2.11 Protection scheme associated with a transformer and feeder circuit.

The scheme is a composite and offers a "high" speed protection for the transformer which has many of the features of unit protection, whereas it offers nonunit type protection for the feeder circuit.

2.3 SAQ 2.1 c

If CTB were omitted, the protection scheme would be a classical nonunit protection for the transformer and feeder circuit.

3. NONUNIT PROTECTION OF DISTRIBUTION FEEDERS

Self-Assessment Question 3.1

Determine the optimum settings for the very inverse characteristic IDMT relays positioned at A, B, C, and D used to protect the feeder network shown in Figure 3.10, given the power system data shown in the diagram and that the relays have very inverse characteristic, are nominally rated at 5 A, and have PSM settings of 50, 75, 100, 125, 150, 175, and 200% and TMS settings from 0.1. Use a grading margin of 0.4 s.

3.1 SAQ 3.1

Source impedance "Source capacity varies between 500 and 100 MVA."

33 kV system

$$Z_s = \frac{V^2}{Z}$$

$$Z_{s\ 500} = \frac{33 \times 33}{500} = 2.178\ \Omega$$

$$Z_{s\ 1000} = \frac{33 \times 33}{1000} = 1.089\ \Omega$$

Relay location	In-zone fault impedance minimum	In-zone fault impedance maximum	In-zone fault current maximum	In-zone fault current minimum	Maximum load current	Ratio
A	1.089	2.928	17,495	6507	1050 A	1200:5
B	1.839	4.078	10,360	4627	750 A	1000:5
C	2.989	6.578	6374	2896	350 A	500:5
D	5.489	8.078	3471	2359	150 A	250:5

	Relay location	[a] Maximum relay setting	Maximum relay settings	Chosen relay setting	Chosen PSM%
Note 1	A	2336	2100	2100	2100
Note 2	B	1448	1500	1500	150
	C	1180	700	700	150
	D	Note 3	300	375	150

a) Chosen to provide backup protection for the following feeder.

Note 1: A setting of 200% would have given a primary equivalent settings of 2400 A. the settings of 175% is in range.

Note 2: A compromise is required, but when the available settings are considered, 150% is the only choice.

Note 3: The minimum fault current expected in the load circuit is unknown. A sensitive setting is therefore preferred if it is greater than $2(I_{LOAD})_{MAX}$.

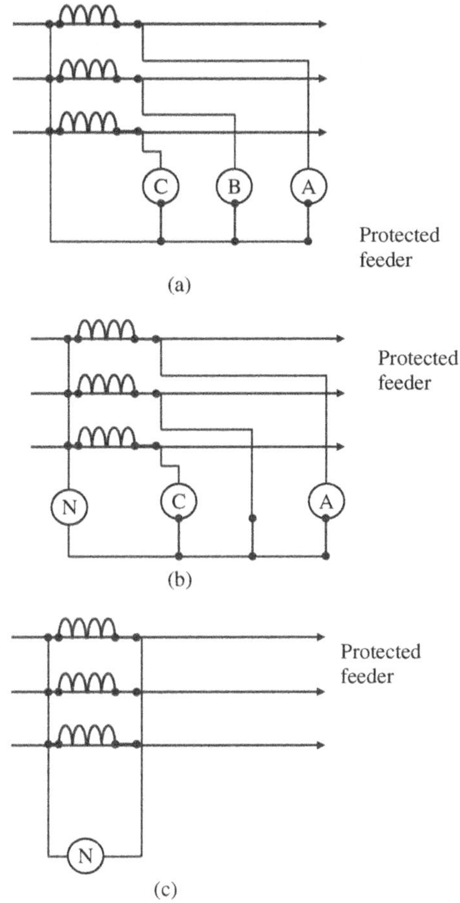

Figure 3.18 Feeder network model for SAQ 3.1. (a) Full scheme configuration, (b) phase fault relay configuration, and (c) earth fault relay only configuration.

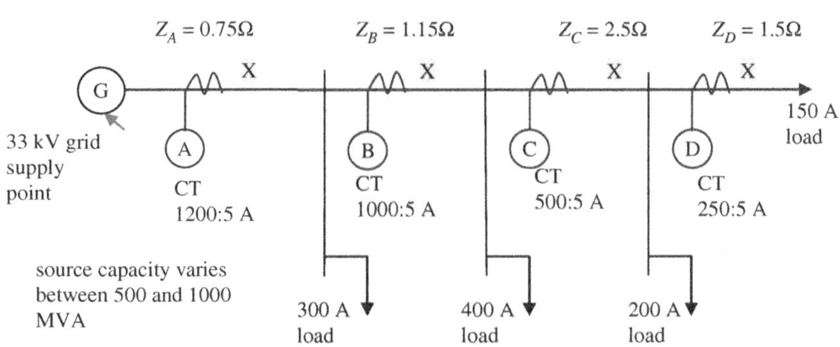

Figure 3.19 Earth fault protection using overcurrent relays.

TMS SETTINGS FOR RELAY C

Grading between relays C and D is required for the maximum fault current experienced by relay D. For this current

$$t_D = \frac{13.5\ T}{\left[\frac{I_f}{PI_N}\right] - 1}$$

$$t_D = \frac{13.5 \times 0.1}{\left[\frac{3471}{375}\right] - 1} = 164\ \text{ms}$$

For the grading margin of 400 ms, t_c operating time is therefore

$$t_c = 164 + 400 = 564\ \text{ms}$$

The TMS settings for relay C t_c is therefore derived from

$$t_C = \frac{13.5\ T_c}{\left[\frac{I_f}{PI_N}\right] - 1}$$

that is

$$0.564 = \frac{13.5\ T_c}{\left[\frac{3471}{375}\right] - 1}$$

$$T_c = 0.152$$

This setting I_s is not available. The alternatives are 0.1 and 0.2. A setting of 0.1 would dramatically erode the grading margin, whereas a setting of 0.2 would provide an enhanced grading margin. The choice of T_c is

$$\therefore T_c = 0.2.$$

TMS SETTINGS FOR RELAY B

Similarly for relay B, its setting is chosen from the grading between relays B and C at the maximum fault current experienced by relay C.

$$\therefore t_c = \frac{13.5\ T_c}{\left[\frac{I_f}{PI_N}\right] - 1}$$

$$t_c = \frac{13.5 \times 0.2}{\left[\frac{6374}{750}\right] - 1} = 360\ \text{ms}$$

$$t_c = 360 + 400 = 760\ \text{ms}$$

T_B is obtained from

$$t_B = \frac{13.5\ T_B}{\left[\dfrac{I_f}{PI_N}\right] - 1}$$

$$0.760 = \frac{13.5\ T_B}{\left[\dfrac{6374}{750}\right] - 1}$$

$$T_B = 0.183$$

As before, the available settings are 0.1 or 0.2. The chosen setting is

$$T_B = 0.2$$

TMS SETTINGS FOR RELAY A

Similarly for relay A, its setting is chosen form the grading between relays A and B at the maximum current experienced by relay B.

$$t_B = \frac{13.5\ T_B}{\left[\dfrac{I_f}{PI_N}\right] - 1}$$

$$t_B = \frac{13.5 \times 0.2}{\left[\dfrac{10{,}360}{1500}\right] - 1} = 457$$

$$t_A = 457 + 400 = 857 \text{ ms}$$

T_A is obtained from

$$t_A = \frac{13.5\ T_A}{\left[\dfrac{I_f}{PI_N}\right] - 1}$$

$$0.857 = \frac{13.5\ T_A}{\left[\dfrac{10{,}360}{2100}\right] - 1}$$

$$\therefore T_A = 0.250$$

The available settings are 0.2 or 0.3. In order to maintain the grading margin, the chosen setting is

4. DISTANCE PROTECTION

Self-Assessment Question 4.1a

Outline the fundamental difference between an ordinary CT and a CT used in protection. In the case of the latter, what measures are taken in the design procedures to ensure that a high accuracy is maintained under a wide range of fault conditions?

4.1 SAQ 4.1a

An ordinary CT operates over a range of currents up to the normal current rating and has specific errors defined only at that value. However, a protection CT is required to operate over a range of currents many times the current rating and therefore requires special design considerations to meet the accuracy requirements for protection purposes.

Since the CT exciting current (responsible for producing the magnetic flux) is primarily responsible for both ratio and phase errors, the protection CT core is of very high-quality material to give a very high magnetizing impedance, thus keeping the exciting current as low as possible. Furthermore, to cater for large fault currents, it is important to design a CT so that the maximum core flux developed stays below the knee-point value for the worst-case fault current; otherwise, the CT will go into saturation and no longer transform the primary current satisfactorily.

Self-Assessment Question 4.2b

A 100/1 A CT is to be used in conjunction with an earth fault relay which has a normal burden of 5 VA at 100% setting and is set to 50% for detecting earth faults. If the worst case primary fault current is 1 kA, what class of CT would be suitable for this application, assuming that a CT error of <5% is required and that the relay does not saturate?

4.2 SAQ 4.1b

A 1 A relay set to 50% equates to a 0.5 A relay setting.

The voltage across coil $= \frac{5\text{ VA}}{0.5\text{ A}} = 10\text{ V}$

CT secondary current $= \frac{1000\text{ A}}{100} = 10\text{ A}$, *i.e.*10 *times* normal current

Therefore, the maximum voltage across burden $= (10 \times 10)\text{ V} = 100\text{ V}$

A CT suitable for this application would be 10 VA *Class* 5*P*10.

5. UNIT PROTECTION OF DISTRIBUTION FEEDERS

Self-Assessment Question 5.1a

Explain why it is necessary to condense information into single phase quantities at each end of a multiphase feeder employing a unit protection scheme of the current differential type. Outline the principles of the two main methods, i.e. the summation transformer and the phase sequence segregating network, used in practice and briefly discuss the reasons why the former can sometimes lead to "blind spots" under certain fault conditions.

5.1 SAQ 5.1a

In unit protection scheme of the current differential type, the use of three separate pairs of pilot wire channels for transmission of the three-phase quantities from one end of the feeder to the other would be very expensive. Hence we condense the three-phase currents into a single quantity for transmission overt just the one pair of pilot wires. SAQ 5 figure shows a typical technique, involving a summation transformer, used in practice for condensing the information.

Use of summation transformer shown below **SAQ 5.1 Figure 5.1a.**

SAQ 5.1

The winding ratio of the transformer is **1:1:*n*** so that the summated output is given as

$$I_s \propto (n+2)I_a + (n+1)I_b + nI_c$$

Self-Assessment Question 5.1b

Give reasons for the need to condense information into a single phase quantity at each end of a multiples feeder employing a unit protection scheme of the current differential type. Outline the principle of a summation transformer used in practice.

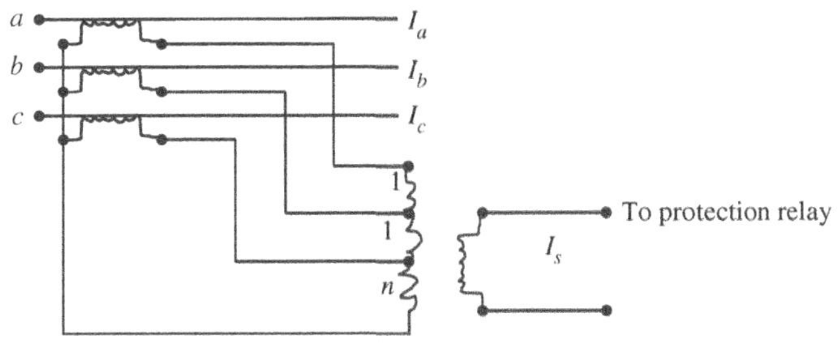

Figure 5.1a Use of summation transformer.

5.2 SAQ 5.1b

When $n = 3$ the summated current is proportional to as stated below.

$$I_s \propto 5I_a + 4I_b + 3I_c$$

Now

$$I_{C1} = I_{c2}\ and\ \frac{I_{C1}}{I_{c0}} = 4,$$

and taking "c" phase as reference, we obtain

$$I_s \propto 5(I_{c0} + a^2 I_{c1} + aI_{c2})I_a + 4(I_{c0} + aI_{c1} + a^2 I_{c2}) + 3(I_{c0} + I_{c1} + I_{c2})$$

where

$$a^2 = \left(-\frac{1}{2} - j\frac{\sqrt{3}}{2}\right), a = \left(-\frac{1}{2} + j\frac{\sqrt{3}}{2}\right)$$

This gives

$$I_s \propto 12I_{c0} - 3I_{c1} = I_{c0}\left[12 - 3\frac{I_{C1}}{I_{c0}}\right] = 0$$

Hence for this fault condition, the transformer ratio with ($n = 3$) is unsatisfactory.

Self-Assessment Question 5.2a

With the aid of circuit diagram, explain the principle of the *Merz-price* circulating current system used to protect a three-phase feeder. Give reasons why biasing is necessary in such system.

5.3 SAQ 5.2a

A diagram of a Merz-Price circulating current protection system for feeders is as shown below in SAQ 5.2 Figure 5.2b.

SAQ 5.2

Where R = relay operate element and B = relay bias element,

The relay differential current $|I_{\text{diff}}|$ and relay bias current $|I_{\text{bias}}|$ are given as

$$|I_{\text{diff}}| = |I_A - I_B|$$

$$|I_{\text{diff}}| = K|I_A - I_B|$$

where K = relay setting and normally varies from about 5 to 50%.

It is apparent that the relay operates when:

$$|I_{\text{diff}}| \geq |I_{\text{bais}}|$$

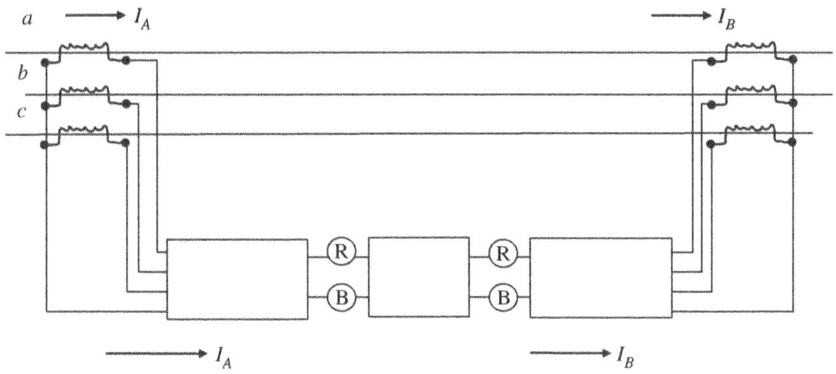

Figure 5.2b

This is the basic principle of current differential protection as applied to feeders. There are two main reasons why bias is required.

- Small difference is the CTs of the two ends can cause a "spill" current to flow through the relay under healthy conditions thus causing unwanted operation
- The impedance characteristic of the pilot circuit can cause current imbalance between two ends, again under healthy condition, with possibility of relay operation.

Self-Assessment Question 5.2b

A 33 kV transformer feeder arrangement is protected with a Merz price circulating current protection scheme. The transformer on the high-voltage side is star connected and is delta connected to the low-voltage side. If the CT on the low-voltage side has a ratio 600/5 A, find the CT ratio on the high voltage.

5.4 SAQ 5.2b

Since the main transformer is star/delta connected, there is a phase shift between the high- and low-voltage line current. This may be compensated for in the protection circuitry by making the CTs delta star connected. Thus, for a base current of 600 A on the low-voltage side, the pilot wire current is 5 A. this is so because the CT ratio is 600/5 A and the CT here are star connected. On high-voltage side, the CTs are delta connected. Hence the CT secondary current is given as:

$$I_{\text{hs}} = \frac{5}{\sqrt{3}} = 2.89 \text{ A}$$

$$I_{\text{hs}} = 600 \times \frac{11}{33} = 200 \text{ A}$$

Hence high-voltage CT ratio is: 200/2.89 A.

6. TRANSFORMER PROTECTION

Self-Assessment Question 6.1

Studying the unbiased differential scheme shown below in Figure 6.14, calculate the suitable value for the relay sensitivity, i.e. minimum operating current. Explain your reasoning and comment on the answer.

6.1 SAQ 6.1

First, we must calculate the CT currents on both the hv and lv sides of the transformer, assuming full load current.

$$I_{\text{lv}} = \frac{M}{\sqrt{3} \times V_L} = \frac{50 \times 10^6}{\sqrt{3} \times 33 \times 10^3} = 874.8 \text{ A}$$

At 10% tap change position,

$$I_{\text{hv}} = \frac{M}{\sqrt{3} \times 1.1 \times V_H} = \frac{50 \times 10^6}{\sqrt{3} \times 132 \times 10^3} = 198.8 \text{ A}$$

At −15% tap change position:

$$I_{\text{hv}} = \frac{M}{\sqrt{3} \times 0.85 \times V_H} = \frac{50 \times 10^6}{\sqrt{3} \times 0.85 \times 132 \times 10^3} = 257.3 \text{ A}$$

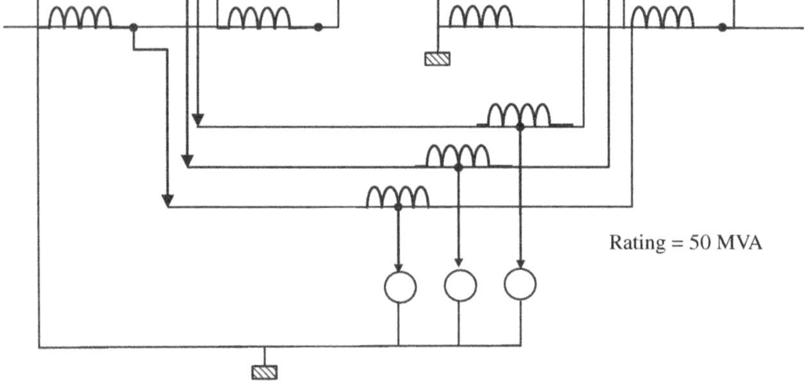

Figure 6.1 Calculate the suitable value for the relay sensitivity.

We now calculate the spill currents:

+10% tap

$$I_{\text{spill}} = \frac{874.8}{900} - \frac{198.8}{\sqrt{3} \times 120} = 0.0153\ \text{A}$$

+15% tap

$$I_{\text{spill}} = \frac{874.8}{900} - \frac{257.3}{\sqrt{3} \times 120} = -0.266\ \text{A}$$

Clearly, the −15% tap position gives the largest spill current. We will set the relays assuming that the scheme has to withstand 10 times full load current under external fault conditions. Hence the minimum relay operating current will be 2.66 A. This is a relatively insensitive setting and will not respond to all faults, particularly faults close to the earthing point on the star windings. A biased scheme would be better in this situation.

7. BUSBAR PROTECTION

Self-Assessment Question 7.1

Figure 7.11 shows a high impedance differential busbar protection scheme. Using the data available, calculate a suitable value for the stabilizing resistance R_S.

7.1 SAQ 7.1

First, we calculate the relay setting based on the minimum fault current, which is stated as 1500 A.

$$I_R = I_S + nI_{eS}$$

$$\frac{1500}{2000} = I_S + 3 \times 0.015$$

Hence

$$I_S = 0.705 \text{ A}$$

Next, we calculate the relay voltage, V_s, using

$$V_s = \frac{I_f}{n}(R_w + R_{CT})$$

It will be apparent that the highest value of V_s will occur if CT3 is saturated, since the CT has the largest lead resistance. Furthermore, we will use the maximum breaker rupturing current, which is given as 30 kA. Hence

$$V_s = \frac{30{,}000}{2000}(2.1 + 2.2) = 64.5 \text{ V}$$

We can calculate the stabilizing resistance using

$$R_s \geq \frac{V_s}{I_s} - R_R$$

Table 7.1 Current versus coil resistance.

Setting current (A)	Coil resistance (Ω)
0.4	2.5
0.6	1.67
0.8	1.25

Min. fault current = 1500 A.

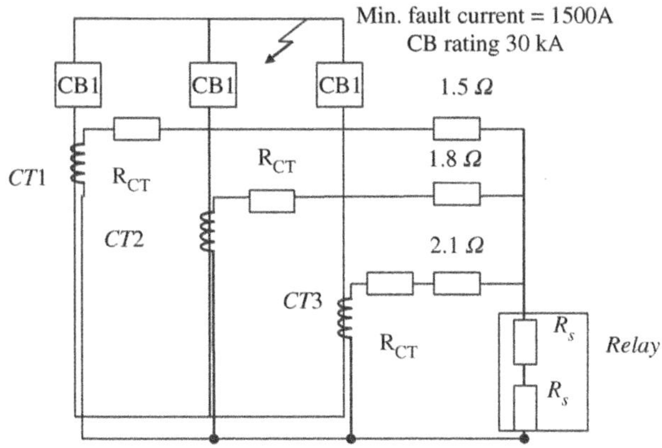

Figure 7.1 Equivalent circuit of high impedance differential scheme.

However, we need a value for R_R, the relay coil resistance. From the figure, we can estimate R_R as being 1.4 Ω. Hence

$$R_s \geq \frac{64.5}{0.705} - 1.4\ \Omega$$

which gives $R_S \geq 90\ \Omega$.

8. MOTOR PROTECTION

Self-Assessment Question 8.1

8.1a → Why is it important to know the motor operating characteristics in motor protection design work?

8.1b → What is the significance of nps current in the rotor of a motor?

8.1c → The rotor heating characteristics of 415 V, 75 kW motor due to nps current are given approximately as

$$I_2^2\, t = 2 \times 10^4 \tag{8.1}$$

where t = time (s).

If nps current of 30% full load current flows in the motor due to phase imbalance, estimate the maximum time before which the imbalance must be detected in order to prevent rotor damage

8.1 SAQ 8.1a

In motor protection design, it is important to know the motor starting and thermal characteristic, first, because the starting current may contain quite large peaks which can cause CT saturation and therefore and unbalanced condition. Thus, if the relay for protecting winding earth fault, for example, is set too low, then the residual current flowing in the unbalanced system can cause unnecessary relay operation. And, second, because of the significant disparity between the motor's and overload relay's thermal characteristic the latter can cause a trip in the time-period between motor start initiation and the actual starting of the motor. This being more likely if the relay operating time is set too low.

8.2 SAQ 8.1b

The nps current is caused by phase unbalance in the system resulting in a phase rotation, opposite to the normal phase rotation. This produces a magnetic field which induces currents in the rotor at twice the system frequency and the latter can cause considerable heating of the motor's rotor if allowed to persist.

8.3 SAQ 8.1c

Motor full load current $= \frac{75{,}000}{\sqrt{3} \times 425} = 104$ A

Thus, nps current $I_2 = 0.3 \times 104 = 31$ A

Therefore t, the maximum time allowed $= \frac{20{,}000}{(31)^2} - 21$ s

9. EMBEDDED GENERATION PROTECTION SYSTEM

Self-Assessment Question 9.1

Briefly explain why phase by phase current differential protection is unable to provide 100% protection for an embedded generator's stator winding and describe what action is required should a stator winding fault be detected. Some research (recommended) may be required to answer this question.

9.1 SAQ 9.1

Considering one phase of an embedded generator's stator winding as shown in SAQ 9.1 Figure 9.1a, the differential current resulting from a phase-to-earth fault is the current generated by the voltage xV_{ple} across the earthing impedance, the winding impedance between the fault and the earth point, and the fault impedance.

SAQ 9

The fault current I_F resulting from the above is given by

$$I_F = \frac{xV_{\text{ple}}}{(Z_E + xZ_w + Z_F)}$$

The current passing through the relay is directly proportional to this, where $I_{\text{RF}} = NI_F$ and N is the current transformer ratio

The relay will have a minimum setting I_s and therefore will not operate $I_{\text{RF}} < I_s$. Therefore, it is a proportion of the winding for which, if a fault occurs within that area of the winding, the protection cannot operate. This proportion is defined by x, which is obtained from

$$I_s = \frac{NxV_{\text{ple}}}{(Z_E + xZ_w + Z_F)}$$

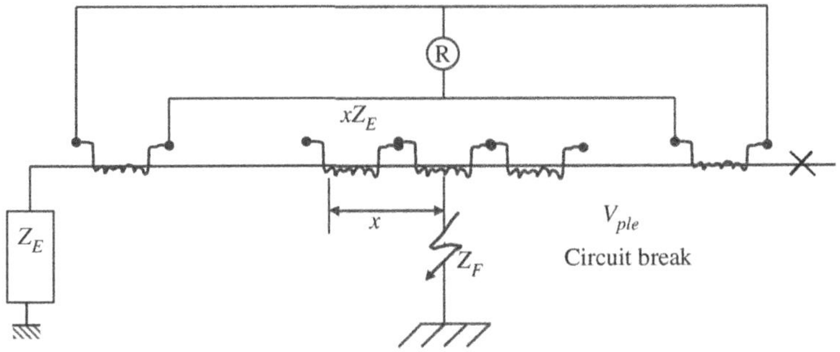

Figure 9.1a One phase of an embedded generator stator winding.

$$X = \frac{I_s(Z_E + xZ_w + Z_F)}{NV_{\text{ple}}}$$

Should a fault occur on the stator winding and the protection trip, it is not sufficient to open the generator's circuit breaker. This action will not remove the fault and the source of energy driving it. The prime mover must also be tripped and stopped to prevent further damage.

Self-Assessment Question 9.2

Briefly explain why loss of grid protection presents a major challenge to protection relay designers.

9.2 SAQ 9.2

Loss of grid protection presents a major challenge to protection relay designers because it is not a fault condition, nor an abnormal operating condition, yet it is an undesired operating condition that could lead to future problems. These future problems are related to possible out-of-step reclosure between the embedded generator and the utility grid supply, and the possible loss of earth connection for the islanded utility section, including the embedded generator.

An embedded generator control system may well be able to maintain the supply within the required voltage and frequency limits, and there are no fundamental reasons why the system needs to be earthed. Provided that the load is being met and the system is stable, few parameters can be used to differentiate between a grid-connected and an isolated system.

The response of the system to a disturbance can provide an insight into the source of power and hence provides a method of detecting loss of grid. However, can the system be designed such that it can differentiate between loss of grid and fault conditions, major system disturbances, or system instability.

10. AUTORECLOSE RELAYING

Self-Assessment Question 10.1

Briefly describe the characteristics of the following types of faults and comment on the possible success of autoreclosing a circuit breaker following the protection trip:

a) Transient
b) Permanent
c) Semipermanent

10.1 SAQ 10.1a

A transient fault A transient fault is characterized by an arc that collapses once the source of power is removed and that does not restrike when the source of power is reconnected. It is commonly produced by a lightning strike close to an overhead distribution feeder, which causes temporary insulation failure.

An autoreclose of the circuit breaker has a high probability of restoring the pre-fault conditions to the system.

10.2 SAQ 10.1b

A permanent fault A permanent fault is characterized by permanent damage to the power system plant and a loss of insulation. It can be caused by ageing, mechanical damage, or pollution. Any reclosure of the circuit breaker will restore the short-circuit fault and could lead to extending the damage.

A semipermanent fault A semipermanent fault is characterized by foreign objects fouling the insulation and causing a short-circuit fault. There is a possibility that these foreign objects can be burned away by the short-circuit fault current and hence cleared from the system.

The classical semipermanent faults are associated with vegetation, typically creepers and tree-tops. Autoreclosure of the circuit breaker and restoration of power to the system can be successful. Although multishot autoreclose may be required to burn the fault material away and restore the insulation.

Self-Assessment Question 10.2

Describe the duties of the lockout relay in an autoreclose scheme commenting on its operation in the single-shot and multishot autoreclose schemes shown in Figures 10.2 and 10.3 respectively.

10.3 SAQ 10.2

When an autoreclose sequence is unsuccessful, the lockout relay is tripped, and the autoreclose scheme is immobilized until it is manually reset. Failure of the autoreclose sequence implies that the power system has been permanently damaged, and repairs are required before it can be returned to service.

In a single-shot autoreclose (Figure 10.2) a second protection trip during the reclaim time will cause the lockout relay to be operated, opening contacts LO-1 and blocking further action. As with the single-shot scheme, contacts LO-1 requires resetting manually before the scheme can be reset.

11. COORDINATED PROTECTION AND CONTROL

Self-Assessment Question 11.1

What are the functions found in a substation? Why are conventional substation systems difficult to maintain, test, and expand? How are these problems overcome in a modern coordinated substation?

11.1 SAQ 11.1

The main substation functions are as follows.

- Protection of power system equipment
- Control and interlocking of switchgear
- Generation of alarms for system/local control
- Event recording
- Disturbance recording
- Automatic reclosing/restoration
- Synchronizing checks
- Load shedding

These can be broadly categorized into protection, control, or monitoring functions.

Conventional substation protection systems are implemented from individual relays, each having its own interface with the plant e.g. connections to voltage transformers, current transformers, and breaker operating gear. Control systems are generally made from hard-wired logic, i.e. auxiliary relays suitably interconnected. Both these systems use large amounts of wiring and give rise to the following problems.

- When commission mistakes are frequently made by human technicians on systems using long runs of wiring that are duplicated many times, since humans are not good at dealing with monotonous tasks.
- If a fault develops in the system, it is very difficult to trace.
- Alteration to the system usually means redesigning the whole system due to its hard-wired nature.

In a modern coordinated substation, relay interfaces to the plant can be provided through serial digital communications systems, thus reducing the amount of wiring needed in the substation. Although redundancy of the plant interfaces will be needed for operational dependability, several relays can share access to substation information. The logic of control functions in a substation is no longer hard-wired since the task is performed by a computer. Thus, the control logic will be implemented in software. Since software can be easily modified compared to hardware, an addition or change to the control system is easily accommodated.

Self-Assessment Question 11.2

Describe how data is communicated

11.2a → Within a bay
11.2b → Between bays

11.2 SAQ 11.2a

All terminals within a bay are connected to an object bus which is specific to a single bay. Due to the proximity of the bay to the power system plant, the object bus is implemented with optical fiber links to prevent electromagnetic interference (emi) from switchgear operation affecting data transmission. Generally, the throughput of data along the object bus is not great since its prime purposes include allowing all terminals access to plant interfaces and to allow communication with the rest of the substation through the bay control terminal. A typical data rate along the object bust is 9600 bits per second (bit/s).

11.3 SAQ 11.2b

Communication between the bays occurs on the station bus. The station bus has less risk from (emi) than the object bus since is does not need to pass close to plant. Thus the station bus is implemented from coaxial cable and typical runs ethernet. The station bus networks all bays in the substation together with the substation control system and a gateway onto a wide area network which typically allows system control centers to operate the substation, and consequently a large amount of monitoring to the performed, the station bus need a relatively high bandwidth. The specification for ethernet is a bandwidth of 10 Mbit.

12. SMART GRID OPERATIONS AND CONTROL

Self-Assessment Question 12.1

12.1 SAQ 12.1a

Describe the **Smart grid system for PV and Wind operations, and its key elements.**

12.2 SAQ 12.1b

Describe the Fault Detection, Isolation, and Restoration (FDIR) and FLISR[1]

Refer to Sections 12.6 and 12.12.

Self-Assessment Question 12.2

12.3 SAQ 12.2a

Describe the following:

- SCADA system and its operational functions, and also discuss one or two cases studies related to Power Distribution System.
- Challenges in Regulatory and Standardization Implementation.

For Answer: Refer to Sections 12.8 and 12.12.

---

1 Fault Location, Isolation, and Service Restoration.

13. MICROGRID OPERATIONS AND CONTROL

Self-Assessment Question 13.1

13.1 SAQ 13.1

A) Describe the classification of microgrid and AC microgrid operations and control.
B) Describe the grid following and forming inverters.

For Answer: Refer Sections 13.3, 13.5.6, and 13.5.7.

Self-Assessment Question 13.2

13.2 SAQ 13.2

Describe the following.:

- Advantages of short-circuit ratio study
- Strong and weak microgrid
- Traditional short-circuit ratio study

For Answer: Refer Sections 13.6 and 13.7.

Self-Assessment Question 13.3

13.3 SAQ 13.3

Describe the following:

- Types of circuit breakers in the AC and DC Microgrid
- Discuss the Time-Current curve
- Terminology of LSI (Long time, short time, and Instantaneous)

For Answer: Refer Section 13.8.

14. ARTIFICIAL INTELLIGENCE AND MACHINE LEARNING POWER SYSTEM

Self-Assessment Question 14.1

14.1 SAQ 14.1

Describe the Evolving Role of Machine Learning in Microgrid Operation and Control and Implications for the Protection of Power Distribution Systems.

Refer Sections 14.7 and 14.9.4.

Self-Assessment Question 14.2

14.2 SAQ 14.2

Describe the Advanced ML Algorithms, Edge Computing, and Distributed Analytics.

Also describe Integration with Emerging Technologies, Policy and Regulatory Considerations, and Collaboration and Innovation.

For Answer: Refer Sections 14.7 and 14.9.4.

Self-Assessment Question 14.3

14.3 SAQ 14.3

1) Describe the predictive maintenance of microgrid components, load forecasting using machine learning models.
2) Describe the Optimization of energy resources with machine learning, anomaly detection in microgrids

For Answer: Refer Section 14.12.

15. CYBERSECURITY AND INFORMATION PROGRAM'S POWER SYSTEM

Self-Assessment Question 15.1

15.1 SAQ 15.1

Describe the system map and its operational model of the selected Energy company's ICS, providing the spectrum.

For Answer: Refer Sections 15.3 and 15.4.

Self-Assessment Question 15.2

15.2 SAQ 15.2

To evaluate and strengthen the cybersecurity posture of the Operational Technology (OT) environment within the Combined Cycle Gas Turbine (CCGT) power plant.

For Answer: Refer Section 15.8.2.

Self-Assessment Question 15.3

15.3 SAQ 15.3

Develop a resilience approach for conducting a comprehensive IT penetration testing, vulnerability, and risk assessment for a Power Plant with a capacity for combined cycle power generation.

For Answer: Refer Section 15.8.

1

Tutor Marked Assignments

1.1 Assignment A

TMA A.1

Plot the standard current/time characteristic curve used to describe the operation of a fuse and identify the key points on the diagram? **[6 marks]**

TMA B.1

Explain why the standard current/time characteristic curve for a fuse is limited in predicting the actual fault clearance times of a fuse protecting a power circuit. **[4 marks]**

TMA C.1

For the 11 kV/415 V supply system shown in Figure 1, suggest the most appropriate choice of fuse rating for the fuses at points A, B, C, D, E, F, G, H, J, and K given that fuses are available with the following ratings. **[10 marks]**
415 V fuses: 31.5 A, 63 A, 160 A, 315 A, and 630 A
Low voltage fuses: 1 A, 3 A, 5 A, 30 A, 50 A, and 100 A CT ratio = 10.1

TMA A.2

Describe the characteristics of the wound current transformer used for:

i) Protection function
ii) Metering applications

[8 marks]

Protection of Electrical Power Distribution Systems: Smart grid, Microgrid, AI, and Cyber Security, First Edition. Tariq Masood, Jamil Abdo, and Atif Iqbal.

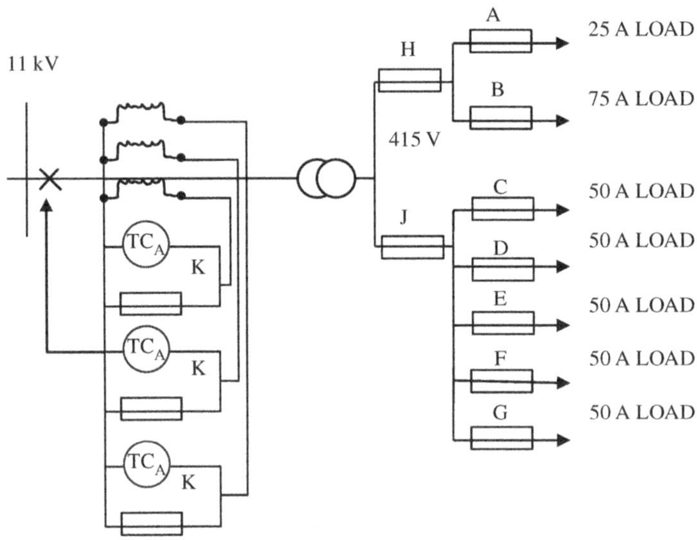

Figure 1 11 kV/415 V Distribution system.

TMA B.2

Define the term accuracy limit a applied to a standard distribution system protection current transformer. **[4 marks]**

TMA C.2

Explain the importance of class X current transformer and the concept of knee point voltage. Comment on the need to operate a current transformer below its knee point voltage. **[8 marks]**

TMA A.3

The 50 MVA delta star voltage transformer shown in Figure 2 is protected by an unbiased current differential relay. Shown how the current transformer are connected and determine suitable CT ratio and the relay's current setting. Assume that CTs are available in ratio which are a multiple of 25 A:1 A. Stat all other assumptions. **[20 marks]**

TMA A.4

Define the term "high speed auto reclose," "delayed autoreclose," "dead time", "reclaim time," and lockout as used in autoreclose scheme. **[8 marks]**

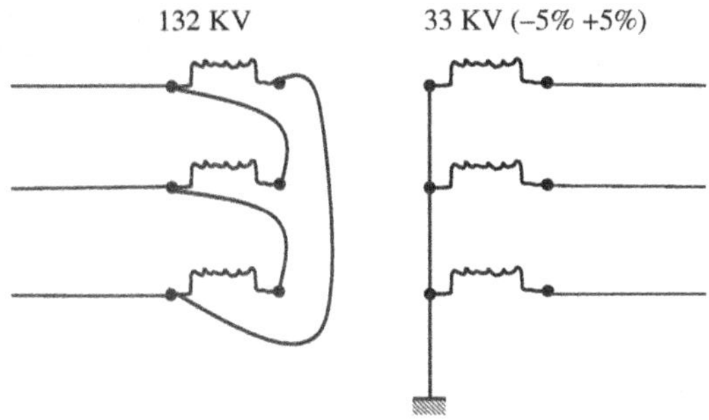

Figure 2 50 MVA delta/star power transformer.

TMA B.4

With the aid of timing diagram, explain how a four-shot autoreclose scheme would respond to a permanent fault, a semi-permanent fault, and transient fault. **[12 marks]**

TMA A.5

Comment on the impact of digital electronic systems and microprocessors on modern protection and control scheme highlighting those areas where "new" features have been introduced. **[10 marks]**

TMA B.5

Describe the features of a coordinated protection and control scheme in a modern substation. **[10 marks]**

1.2 Assignment B

TMA B.1

Traditional organization have used a pyramid structure to ensure that there is no confusion about lines of command and communications. With the growth of multi-disciplinary projects in companies, there has been a general move to using matrix structure. Describe the organization of a matrix-structured company and explain the potential advantages and disadvantages of using this compared to a pyramid structure. **[20 marks]**

TMA B.2

A detailed analysis of a project at the outset can minimize the number of problems that could be encountered as the project proceeds. Described the most common problems that arise during the execution of a project that can be attributed to inadequate project analysis in the early stages and the processes that can be used to minimize them. **[20 marks]**

TMA B.3

Table A describes the activities associated with a project: **[6 marks]**

a) Use a precedence matrix to reorganize these activities into a more logical structure. **[6 marks]**
b) Calculate the best guess and the expected duration of the project, noting any points of concern. **[8 marks]**
c) Calculate the probable time required for the completion of the project with a 90% level of confidence. **[6 marks]**

Activity	Dependency	duration	duration	duration
A	F,G	1	2	3
B	D	3	4	7
C	A,B,F	2	3	4
D	H	4	6	7
E	—	1	2	3
F	G,A	1	2	3
G	H	3	4	4
H	E	2	2	3

TMA B.4

It is often argued that the true secret of good project management is ensuring that there "no surprise" for those involved with the project including both the sponsors and those doing the work. Traditionally however, the argument has been that the secret of success is being both "on time and on budget." Discuss these arguments and identify the key points in these arguments. **[20 marks]**

TMA B.5

A project is generally considered to have been finished when the final project is delivered to the client and invoices are settled. The need for an audit of the project's execution is often overlooked. Describe the reasons why this is the case and the benefits that the company and the project team would gain by undertaking a proper project audit. **[20 marks]**

About the Authors

Tariq Masood

Dr. Tariq Masood is a distinguished professional with a wealth of academic qualifications in electrical power system engineering. His credentials **include an MEng, an MPhil**, and a **PhD (Smartgrid/Microgrid Renewable Energy Distributed Power Generation)** from the esteemed University of Bath, Bath, UK. He also obtained **Scientific Teaching Fellowship in 2022**, accredited and sponsored by the National Science Foundation **(NSF)** and Howard Hughes Medical Institute (**HHMI**) in the **United States**. These qualifications, coupled with his status as a chartered engineer registered with the Royal Engineering Council in the UK, a senior member of IEEE-USA, a senior member of AIChE-USA, and a member of IET-UK, reassure his extensive expertise. His professional affiliations also include being a certified CET International USA member and a registered ASCET (American Society of Certified Engineers and Technicians) USA member.

Dr. Masood is an Assistant professor **(Tenured-Track)** at Frostburg State University in Frostburg, Maryland. His leadership roles at the university, including serving as a Senator **of the Faculty Senate**, a Member of the Academic Affairs Committee, and a Member of the Academic Standard Committee, demonstrate his commitment to fostering academic excellence.

Dr. Masood is a versatile, collaborative, accomplished, and knowledgeable professional. His successful track record as a **Professional Engineer/Academician. His collaborative approach makes everyone feels included and part of a team. He has also c**ontributed as a Visiting and Adjunct Professor and **Invited Speaker at OIT, UTT, Texas A&M Qatar, UTA, TTU, UND, U.B., and UDC** in the United States of America. He is also certified by the Lean Six Sigma Green Belt. In the past, he also served as **Vice President of Technical Operations** at Energy and Utilities Service in New Jersey, USA.

Protection of Electrical Power Distribution Systems: Smart grid, Microgrid, AI, and Cyber Security,
First Edition. Tariq Masood, Jamil Abdo, and Atif Iqbal.

He joined Qatar Petroleum in 1997 as a Technical Coordinator Operations/ Advisor. He is a member of the Best Practices Technical Committee (BPTC) and other production integrity committees under the patronage of **H.E. Dr. Mohd Saleh Al-Sada**, Minister of Energy and Industry, Qatar. His contributions to the field are evident in his more than 68 technical research papers published in IEEE conferences, journals, papers, and books (**ISBN-978-3-659-40995-0)**. He has received several awards for his outstanding performance and dedication to improving Qatar Petroleum's production operations and control, including two Mubarak Awards from H.E., the Minister of Energy and Industry, and three Al-Hasba Awards from the Vice President of Operations. He was the secretary for the GCC oil-producing companies (**QP – Qatar, PDO – Oman ARAMCO – Saudi Arabia, KOC – Kuwait, TATWEER – Bahrain, and ADNOC – United Arab Emirates**) Production and Maintenance Technical Committee in 2008 and 2011. **He is also serving as an Associate Editor of IEEE Access (Impact Factor 3.44)**.

You may reach him by email: t.masood.dr@ieee.org

Jamil Abdo

Dr. Jamil Abdo is a distinguished Mechanical Engineering professor and the current Chair of the Department of Engineering at Frostburg State University, which comprises Electrical Engineering, Mechanical Engineering (in collaboration with the Department of Mechanical Engineering at the University of Maryland – College Park), and Material Engineering programs. He earned his MS and PhD degrees from the Department of Mechanical Engineering and Energy Processes, Southern Illinois University, Carbondale, . With over 22 years of experience, Dr. Abdo combines academic and industry expertise, having served as the principal investigator in numerous industrial projects in collaboration with prestigious entities including MIPS, ARC, Shell, and the Qatar Foundation. Throughout his career, he has led and contributed to many significant research projects funded by federal and state agencies as well as industrial partners. As Chair, Dr. Abdo also serves as the Director at the Center of Product Design and Advanced Manufacturing at Frostburg State University. Dr. Abdo's primary research interests include product design and development, machine design, and mechatronics. His recent projects cover a broad spectrum of research areas and applications, such as renewable energy, autonomous vehicles, and smart grids. Previously, Dr. Abdo served as a professor of mechanical engineering at Sultan Qaboos University and chaired the Mechatronics Engineering Department at Hashemite University in Jordan. He also served as the Director of Companies Innovation at the Industrial Innovation Center in Oman.

Dr. Abdo has published an extensive number of journal articles and conference proceedings and edited two books. He has been a guest editor for several journals, including the *Journal of Material and Manufacturing Processes* and the *Journal of Symmetry*. Dr. Abdo has frequently been a keynote speaker and conference chair at various academic events. Dr. Abdo has received many distinguished awards and recognitions

Atif Iqbal

Atif Iqbal (Fellow IEEE, USA; Fellow IET, UK; Fellow IE, India; Fellow AIAA, Asia Pacific) received his BSc and MSc degrees in Electrical Engineering from Aligarh Muslim University (AMU), Aligarh, India, in 1991 and 1996, respectively, and PhD from Liverpool John Moores University, Liverpool, UK, in 2006. He received his DSc degree (Habilitation) in control, informatics, and electrical engineering from the Gdansk University of Technology, Gdansk, Poland, in 2019. He is a Full Professor of the Department of Electrical Engineering, Qatar University, Doha, Qatar, and a former Full Professor of the Department of Electrical Engineering, AMU, Aligarh, India. Dr. Iqbal has been listed in top 2% highly cited scientists of the world (Stanford University, USA) since 2019. He has published widely in international journals and conferences on his research findings related to power electronics, variable speed drives, e-mobility, smart grid, complex energy transition, microgrids and nanogrids, and renewable energy sources. He has authored or coauthored more than 580 research articles, 8 patents, and 4 books and several chapters in edited books. He has supervised several large research and development projects worth several million USD. He has supervised 21 PhD students. He was also a recipient of the Outstanding Faculty Merit Award for 2014–2015 and the Research Excellence Awards at Qatar University, in 2015 and 2022. He has received Research Excellence Award from the College of Engineering, Qatar University in 2019. He has received several best research papers awards in top International Conferences. He is serving as the Vice-Chair of the IEEE Qatar Section. He is an Associate Editor of the *IEEE Transaction on Industrial Electronics* and Senior Editor of *IEEE Access*.

Index

Protection of Electrical Power Distribution Systems: Smart grid, Microgrid, AI, and Cyber Security, First Edition. Tariq Masood, Jamil Abdo, and Atif Iqbal.